Topics in Applied Physics

Volume 148

Topics in Applied Physics is a well-established series of review books, each of which presents a comprehensive survey of a selected topic within the domain of applied physics. Since 1973 it has served a broad readership across academia and industry, providing both newcomers and seasoned scholars easy but comprehensive access to the state of the art of a number of diverse research topics.

Edited and written by leading international scientists, each volume contains high-quality review contributions, extending from an introduction to the subject right up to the frontiers of contemporary research.

Topics in Applied Physics strives to provide its readership with a diverse and interdisciplinary collection of some of the most current topics across the full spectrum of applied physics research, including but not limited to:

- Quantum computation and information
- Photonics, optoelectronics and device physics
- Nanoscale science and technology
- Ultrafast physics
- Microscopy and advanced imaging
- Biomaterials and biophysics
- Liquids and soft matter
- Materials for energy
- Geophysics
- Computational physics and numerical methods
- Interdisciplinary physics and engineering

We welcome any suggestions for topics coming from the community of applied physicists, no matter what the field, and encourage prospective book editors to approach us with ideas. Potential authors who wish to submit a book proposal should contact Zach Evenson, Publishing Editor:

zachary.evenson@springer.com

Topics in Applied Physics is included in Web of Science and indexed by Scopus.

Eun Ha Choi
Editor

Plasma Biosciences and Medicine

Springer

Editor
Eun Ha Choi
Plasma Bioscience Research Center
and Applied Plasma Medicine Center
Kwangwoon University
Seoul, Korea (Republic of)

ISSN 0303-4216 ISSN 1437-0859 (electronic)
Topics in Applied Physics
ISBN 978-981-19-7934-7 ISBN 978-981-19-7935-4 (eBook)
https://doi.org/10.1007/978-981-19-7935-4

This Springer imprint is published by the registered company Springer Nature Singapore Pte Ltd.
The registered company address is: 152 Beach Road, #21-01/04 Gateway East, Singapore 189721, Singapore

Contents

Contributors

Henrike Brust Leibniz Institute for Plasma Science and Technology, Greifswald, Germany

Eun Ha Choi Plasma Bioscience Research Center (PBRC), Department of Electrical and Biological Physics, Kwangwoon University, Seoul, Korea

Jinsung Choi Plasma Bioscience Research Center, Department of Electrical and Biological Physics, Kwangwoon University, Seoul, Korea

Philine Henriette Doberschütz Department of Orthodontics, Greifswald University Medicine, Greifswald, Germany

Alexander A. Fridman Nyheim Plasma Institute, Drexel University, Philadelphia, USA

Ihn Han Plasma Bioscience Research Center, Department of Electrical and Biological Physics, Kwangwoon University, Seoul, Korea

Young June Hong Plasma Bioscience Research Center, Department of Electrical and Biological Physics, Kwangwoon University, Seoul, Korea

Nagendra Kumar Kaushik Plasma Bioscience Research Center (PBRC), Department of Electrical and Biological Physics, Kwangwoon University, Seoul, Korea

Neha Kaushik Department of Biotechnology, College of Engineering, The University of Suwon, Hwaseong-Si, Korea

Michael Keidar Mechanical and Aerospace Engineering Department, George Washington University, Washington, DC, USA

Sehoon Ki Plasma Bioscience Research Center, Department of Electrical and Biological Physics, Kwangwoon University, Seoul, Korea

Jae-Sung Kwon Department and Research Institute of Dental Biomaterials and Bioengineering, Yonsei University College of Dentistry, Seoul, Korea;
BK21 FOUR Project, Yonsei University College of Dentistry, Seoul, Korea

Junsup Lim Department of Electrical and Biological Physics, Plasma Bioscience Research Center (PBRC), Kwangwoon University, Seoul, Korea

Kai Masur Leibniz-Institute for Plasma Science and Technology, Greifswald, Germany

Hans-Robert Metelmann Department of Oro-Maxillo-Facial and Plastic Surgery, Greifswald University Medicine, Greifswald, Germany

Sohail Mumtaz Department of Electrical and Biological Physics, Plasma Bioscience Research Center (PBRC), Kwangwoon University, Seoul, Korea

Gyungsoon Park Department of Plasma-Bio Display and Department of Electrical and Biological Physics, Kwangwoon University, Seoul, Republic of Korea

Christian Seebauer Department of Oro-Maxillo-Facial and Plastic Surgery, Greifswald University Medicine, Greifswald, Germany

Yungoh Shin Plasma Bioscience Research Center, Department of Electrical and Biological Physics, Kwangwoon University, Seoul, Korea

Nicola Wannicke Leibniz Institute for Plasma Science and Technology, Greifswald, Germany

Chapter 1
Introduction to Nonthermal Atmospheric Pressure Plasma: Physical and Chemical Basis

Alexander A. Fridman, Michael Keidar, and Eun Ha Choi

Abstract Plasma bioscience and medicine today is an interdisciplinary division of modern science and technology. It includes fundamental physics essential to develop new plasma sources and its diagnostics applicable for biological applications, clinical treatments, medicine to apply the technology not only on cells but also on the whole organism and medical testing, and finally bioscience and environmental issues. Nonthermal atmospheric pressure plasma can be efficacious in several biological applications such as blood coagulation treatment, sterilization, tissue bioengineering, modification of biomaterials, wound healing, agriculture, food processing, eradication of environmental issues, and various other applications. It is important to note that this chapter explained the safety and stability of plasma devices used for various applications, which is the main concern in this developing field. In the first part, the history of plasma bioscience and medicine area including clinical trials in cancer and other diseases is included. Specifically, in this chapter detailed diagnostic procedures for various plasma sources and plasma-based substances are incorporated. The last section of this chapter described optical emission spectroscopy (OES) for the reactive species detection, measurement of the plasma electron density and temperatures by interferometry and collisional radiative models, and rotational and vibrational temperatures of molecules in plasma gases by using an Boltzman plot from optical spectroscopy, the plasma radical densities by absorption spectroscopic method, and the plasma parameters for industry biomedical plasma products.

A. A. Fridman
Nyheim Plasma Institute, Drexel University, Philadelphia, USA

M. Keidar
Mechanical and Aerospace Engineering Department, George Washington University, Washington, DC, USA

E. H. Choi (✉)
Plasma Bioscience Research Center (PBRC), Department of Electrical and Biological Physics, Kwangwoon University, Wolgye Dong, Nowon Gu, 01897 Seoul, South Korea
e-mail: ehchoi@kw.ac.kr

E. H. Choi (ed.), *Plasma Biosciences and Medicine*, Topics in Applied Physics 148,
https://doi.org/10.1007/978-981-19-7935-4_1

1.1 Introduction to Plasma Bioscience and Medicine

1.1.1 Plasma Medicine is a New Division of Medical Science and Technology

Innovative concepts in medical technology give new possibilities: plasma bioscience and medicine are certainly one of those hopes. Latest advancements in physics and engineering have stemmed in several crucial clinical developments. Among several clinical strategies that have been extensively reported in the earlier literature are administration of high energy radiations, ultrasound, lasers, electromagnetic waves. Plasma based strategies are comparatively newbie in the medical areas. Latest exponential advances in electrical physics and engineering endorsed subsequent considerable developments in the cold atmospheric pressure plasma area. Space-uniform and controlled nonthermal atmospheric plasma devices turn out to be a reality. All that created a prospect to safe and control application of plasma to organisms. All that stimulated innovation of a new thrilling field of medicine—plasma medicine. Investigations performed at various main academics and hospitals across the world since from a decade shows that nonthermal plasma can deliver innovative solutions of tricky health issues. Plasma is efficient in decontamination of various surfaces as well as living tissues, sterilizes large amount of air and water, neutralizes hazardous microbes together with those in food materials, and capable to halt severe hemorrhage without affecting normal tissue. Plasma can be precisely utilized to induce wound healing and to cure various diseases such as cancer, disorders related to various other human body parts and organs. Nonthermal plasma is also shown to be efficient in blood treatment to regulate properties of blood. It is also proved to be effective in surface exposure for modification of various biological materials and subjects, in diagnostics and even in pharma field by modifying characteristics of existing drug molecules and producing a new one. These plasma discharges, significantly established lately due to the promising developments in electronic engineering areas, is obviously a potential tool to be delivered to health professionals for solving clinical problems earlier unresolved. This is a foundation of great attention at the moment to the plasma medicine, which is the subject of this article.

1.1.2 Controllability and Safety of Plasma

While discussing about the new plasma devices which are likely to be utilized to humans, also for exposure of biological subjects in biomedical investigations, plasma physicist must emphasize "controllability" and "safety" of these plasma sources. For instance, the floating electrode dielectric barrier discharge plasma (FE-DBD) device extensively utilized for clinical utilization, in Nyheim Plasma Institute, applies about 30–40 kV precisely to body (for example a photo of one of the founders of plasma bioscience and medicine with FE-DBD discharge, Fig. 1.1, [1]). Figure 1.2 illustrates interaction of the DBD-based plasma jet with human body.

Definitely, "safety" and "controllability" of the plasma parameters are main concern in this case. The uniform plasmas together with plasma based medical sources established lately can be efficiently monitored and controlled, which is critical for maintaining doses of clinical treatment, and also for mechanism. With physical, chemical, and biological mechanisms understanding, plasma technology has plenty of opportunities for effective therapies in clinics.

Plasma is indeed far from thermodynamic equilibrium. This non-equilibrium ionized gas can be extremely "innovative" in contact with biological factors. As it was originally shown in 1950s by Stanley Miller (Fig. 1.3) and other group members, this discharge is capable to produce amino acids from methane and inorganic compounds.

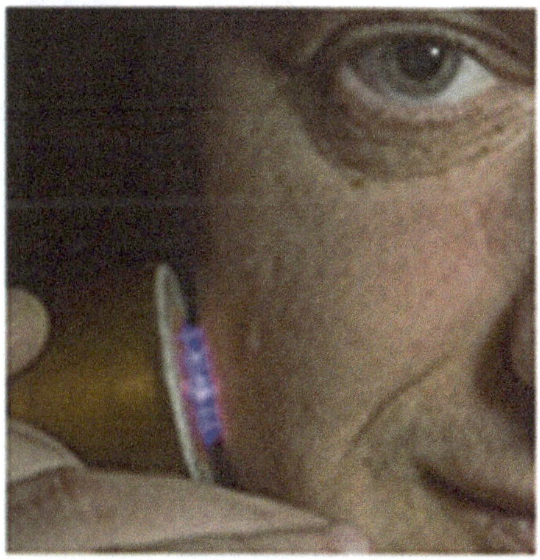

Fig. 1.1 Non-thermal 40 kV FE-DBD plasma was maintained right between an electrode and a person

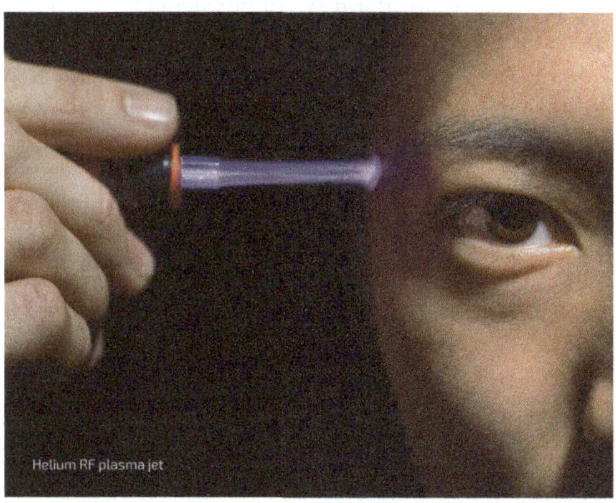

Fig. 1.2 DBD-based atmospheric pressure plasma jet interacting with skin

Fig. 1.3 Stanley Miller from University of Chicago in 1950s produced amino acids using inorganic compounds and methane with plasma

It is quite feasible that nonthermal plasma as multi-parametric and strongly non-equilibrium can be important for origin of life. Latest investigations demonstrate regulated modifications of DNA after the plasma exposure extremely susceptible to physical and chemical parameters of plasma. It describes the exceptional significance of the "controllability" of physical factors and intense knowledge of processes for fruitful development of the plasma medicine area. Accomplishment of plasma technologies needs thorough knowledge of chemical, physician and biological mechanisms of the discharge interaction with biological subjects. Short of basic knowledge of physics, chemistry and biology, the plasma medicine is under the risks of turn into a modernized medieval magic (see Fig. 1.4).

1.1.3 Plasma Application to Biology and Medicine from the Very First Steps Till Today

Earlier, it was observed that discharges in ambience outcome in "strange smells" affecting organism, and humans. These concepts and explanations, which are far older than *plasma bioscience* and *plasma physics* concept itself, can be credited to

Fig. 1.4 Son of Frankenstein, 1939, operating kind of gliding arc discharge plasma

investigation of famous chemist Martinus van Marum in year 1785 with discharges on water. A famous German chemist Christian Friedrich Schönbein detected production of the similar "strange pungent smell" and identified it as ball lighting. He separated in 1839 the gaseous species accountable for this "unusual smell" and known as "ozone". This ionized gas applications in bioscience have been considerably intensify in 1850s when Werner von Siemens initially used dielectric barrier discharge to produce ozone for disinfection of water. Big scale of plasma-based sterilization or disinfection has been realized few decades later, when the first plasma plant for water supply was constructed in France Nice city. Remarkably, ozone was believed as an electric discharge-based constituent: A California city even had as its formal saying "Beaumont: Zone of Ozone" at that time.

First plasma discharge utilization to cure diseases can be credited to investigation of French researcher Jacques-Arsène d'Arsonval in 1900s, as shown in Fig. 1.5. The appropriate sources had been made possible by Nikola Tesla, who performed with exceptionally high-frequency currents at high voltage, producing "remarkable light phenomena which demonstrated safe to human being for direct exposure or treatments".

In early of 1900s in Germany, the plasma sources were more established for the caloric treatment of humans (diathermy). A german researcher Rumpf created a source which differed as of the French ones by applying a capacitively coupled electrode comprising of a Leydener bottle which was exposed to skin of human. This plasma source can be deemed the first dielectiric barrier discharge source in plasma medicine. In the beginning of 1920s, thermal electrosurgical sources and then argon gas based plasma coagulator became an significant move in clinical uses of plasma.

Fig. 1.5 First "pre-historic" steps of plasma medicine

Crucial preliminary progress in this path is linked to investigations of well known American researcher William Bowie in 1920s.

In beginning of 1960s, patent reports and publications in journals began to emerge around the world with a emphasis on nonthermal plasma decontamination, sterilization, regulation of bioativity of microbes, as well as properties of polymer biomaterials. Several relevant literatures started increasing in beginning of 1990s, also several results presented in international conference at that time. On the basis of these preliminary achievements, as well as revolutionary progress of new plasma sources capable to be treat human body in clinics, the novel area of PLASMA MEDICINE was born in early 2000s.

Plasma medicine today is an interdisciplinary division of science and engineering. It includes physics essential to establish new plasma sources applicable for clinical treatments, medicine to apply on human patients and testing, and finally bioscience to acknowledge complicated biological processes and pathways involved in the treatment. Brief review cannot describe major achievements of the plasma medical technology and related fields of plasma agriculture and food processing. The detailed information on plasma medicine fundamentals and applications can be found in the major reviews and books of Fridman et al. [2], Laroussi et al. [3], Fridman and Friedman [1], Metelmann et al. [4], Toyokuni et al. [5], Keidar et al. [6], Kuo [7], Lu et al. [8], Keidar [9].

1.1.4 Plasma as a Helpful Tool for Medicine, Electric Discharges in Plasma Medicine

Nonthermal plasmas are capable to generate high level of reactive species (e. g. electrons, ions, atoms and radicals, excited atoms and molecules, and photons with wide range). Large amount of reactive species are important for plasma uses as plasma-based ignition and combustion, and plasma based ozone formation for water disinfection. In clinical settings, production of the high-level active species can be important, such as, for decontamination, and tissue bioengineering.

Cold plasmas, offering high concentration of the reactive species at room temperature. This characteristic regulates elitism of plasma usage in electronics productions: several components of advance computers, mobile phones, tv panels (including plasma panel), and various other electronic devices are produced applying nonthermal plasma technology. This crucial property also establishes extensive usage of nonthermal plasma in treatment of biomaterials such as polymers: textiles, photographic sheets, packaging materials and many other applications. In clinical settings, plasma generated high concentration of active species can be interesting, for instance, for blood coagulation and modifications of blood components, disinfection of skin and tissues, and ultimately wound healing and various disorders not efficiently cured earlier. these definite plasma characteristics mentioned above permit important alteration of traditional chemical and bio-chemical processes, permit crucial intensification of their effectiveness, and efficacious to stimulate biochemical reactions that cannot be achieved by traditional technologies.

Hot and cold plasmas both are very important for bio-medical applications. High temperatures and energy densities characteristic for hot plasmas reveal their uses for cauterization and tissue removal in the course of surgical procedure. These sources extensively utilized currently in clinical procedure, including tissue disinfection. Hot plasma in air also is very important to generate nitric oxide (NO), which establishes its use in the treatment of wounds and various other diseases. As mentioned above, cold plasma facilitates production of exceptionally high concentration of the reactive species, while keeping bulk temperatures as low as ambient temperature. It regulates particular niche of the cold plasma, which is generally blood coagulation and sterilization of skin and tissues, disinfection of clinical tools and materials and devices, treating of polymers, tissue engineering, and conclusively various illnesses not successfully cured earlier.

Mainly plasma based biomedical applications need operation at ambient pressure, thus necessitate usage of atmospheric pressure plasma devices. The corona discharge is well-known type of plasma. Corona discharge electron temperature is more than 1 eV, however gas temperature stays around room temperature. These plasmas are, specifically utilized in treatment of polymers such as fabrics to support appropriate adhesion before applying dyes.

Atmospheric pressure cold plasmas can be efficacious in blood coagulation and treatment, sterilization, tissue bioengineering, modification of biomaterials, wound healing, and various other diseases treatment. It is crucial to understand the safety of the device, not corona but more complex plasma sources such as DBD plasma or Jet plasma used for various biomedical applications (see Figs. 1.1 and 1.2).

1.1.5 Plasma Chemistry: A Basic Foundation

Nonthermal plasma is a multicomponent system that is extremely active due to a large amount of charged particles, excited atoms and molecules, reactive oxygen and nitrogen species, and UV photons. Every plasma constituent performs its function in plasma-based chemical kinetics. Such as electrons are typically first to get the energy from the electric field and then disseminate it to other constituents. Varying parameters of the electron gas density, temperature, and electron energy distribution function usually allow a way to regulate plasma-based chemical reactions.

Ions are heavy particles, which are capable of give a major influence on plasma-based chemical kinetics due to their capability to put down reaction activation barriers. This feature is responsible for so-called plasma catalysis, which is especially critical in plasma-based combustion, hydrogen production, fuel conversion, exhaust gas cleaning, and even in the treatment of cells and tissues.

Vibrational excitation of molecules usually makes the most important contribution in the plasma-based chemical kinetics due to the high energy (1 eV) of the plasma electrons delivering the majority of the energy in gases such as N_2, CO, CO_2, H_2 e.a. into vibrational excitation. Modulation of plasma-based chemical kinetics via vibrational excitation allows for achieving the ultimate values of energy efficiency. Electrical excitation of gas molecules may also play a substantial role, particularly when the excited molecule lifetime is extremely long. Thus, we can say plasma-produced excited oxygen gas molecules $O_2(^1\Delta_g)$, efficiently contribute to the plasma-based modification of materials and biomedical applications.

The role of atoms and radicals is noticeably important. Thus, we can point out that oxygen atoms and hydroxyl radicals efficiently produced in air plasma discharges can perform crucial functions in several plasma-based oxidation processes. Photons generated by the plasma discharges as well perform significant roles in many utilizations from plasma-based light sources to water sterilization.

Plasma is also a multi-parametric system together with a multicomponent system, although it is far from thermodynamic equilibrium. Cold plasma discharges are usually strong non-equilibrium systems. Concentrations of the reactive species explained previous text can surpass several orders of magnitude than attained in quasi-equilibrium systems at a similar gas temperature. Effective control of discharge allows the direction of the chemical and biological process in the desired way, selectively, and via the best mechanism. Regulation of plasma-based chemical kinetics required a thorough knowledge of fundamental processes and kinetics of the discharges.

As it was described previously, plasma-based chemical kinetics mechanisms are quite complicated. While there is no doubt that, the level of complexity is way higher in the case of plasma-based interactions with cells and tissues. The simplest method to approach this complex is to compare biological results of plasma, which are under the not very old fundamental scientific research, with ionizing radiation (IR). From the first look, there is a basic similarity between the molecular effect of IR and nonthermal plasma. Important to note that both affect biomolecules and organisms via the production of ROS. But there are also key differences in the biological mechanisms induced by IR and cold plasma. Importantly, cold plasma can produce several crucial active species that IR does not. Active species generated by plasma includes reactive oxygen and nitrogen species, high level of ions, and strong electric fields. Therefore, investigating the mechanism of the cold plasma treatment must include characterizing the impacts of other species too. More remarkedly, IR is high-energy penetrating radiation that produces reactive oxygen species within cells and to a certain extent triggers immediate damage to cells. Nonthermal plasma, on the contrary, does not directly produce reactive species intracellularly. It acts by stimulating reaction chains that start in the plasma discharge. Then the effect of plasma progresses via the extracellular medium altering biomolecules and initiating molecular signaling, diffusing via cell membrane to generate noticeable impacts on living organisms (Fig. 1.6).

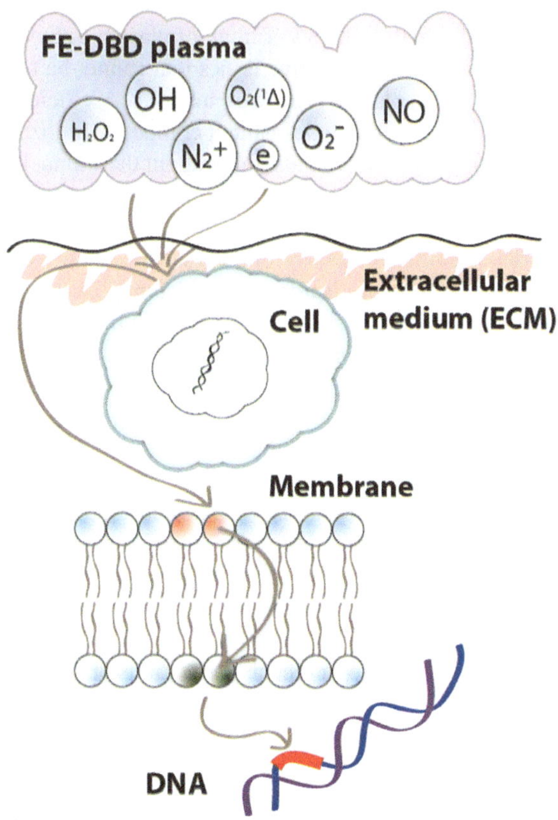

Fig. 1.6 Nonthermal plasma-induced bio-molecular processes: Plasma components diffuse via a membrane from the extracellular media leading to intracellular biological effects

The plasma-medical effect can be analyzed for simplicity as three phases demonstrated in Fig. 1.6 where primary reactive species are produced inside of plasma (first phase). Next, the reactive species are transferred from the gas phase into the liquid phase, finally, the reactive species reach biological cells and activate extremely complex intracellular biochemical processes via various molecular signaling (third phase).

1.1.6 Applied Plasma Medicine

In many applications, nonthermal plasma technologies effectively compete with traditional methods, such as fuel conversion, combustion, coatings, water purification, cleaning of exhaust gases, etc. These plasma applications are fascinating, commercially feasible and usually have made a significant impact on the progress of our society. There is no competition or analogies of the most exciting applications of nonthermal plasma. The plasma abilities in microelectronics processing are exceptional and distinctive. We would not have realized advanced computer systems and mobiles without using plasma. Plasma stays a feasible and effective technology over other traditional strategies. Tuning of plasma's physical properties allows plasma to meet challenges and resolve issues that cannot be resolved in traditional strategies. In some cases, plasma processes are not very efficient, but exceptionally distinctive. Such as other technologies cannot compete with ozone production by plasma (for a century); definitely, we must think about thermonuclear plasma as a distinctive most important energy source.

Similarly, plasma technology attracts considerable attention these days because of opportunities to resolve clinical issues not solved yet, because of the possibilities of treatment of disorders not efficiently treated earlier. The application of cold plasma is capable to shift the paradigm in the medicine area. Those are hopes motivating both fundamental plasma biology and applied to plasma medicine. In ancient times, the application of plasma in the clinical field depends on the thermal effect of discharges for tissue ablation, disinfection etc. Electrocautery is a new method that utilizes regulated heat based on the current passing through the surface of the tissue. However, there is a drawback of this technology treated tissue sometimes adhered to the electrode and causes bleeding again. Plasma can be utilized as an option over electrocautery technology. In case of coagulation, highly conductive plasma is utilized for avoiding the difficulties in tissue adhesion. Thermal plasma is used to cut tissue or skin, though the exact mechanism remains under investigation. Plasma-based thermal effects have also been utilized lately for aesthetic purposes. The main difference in recent applications of plasma over older ones is the utilization of cold discharge effects. Cold plasma is an exciting and encouraging approach to plasma medicine. The key explanation is that cold plasma-based outcomes can be controlled for various biomedical purposes, for example, genetic transfection, cell differentiation or activation, etc. Furthermore, cold plasma can be selective in attaining a required outcome, with little or no side effects. That is why the recent applications such as coagulation and microbial inactivation which does not affect surrounding cells and tissue.

A lot of specific cases show the efficacy of plasma-based wound healing and treatments of various diseases demonstrated. One interesting case of plasma medicine is related to plasma corneal infections treatment, a case when a patient life has been saved as an outcome of plasma exposure. For this situation, a unique micro-plasma source has been built by Dr. Dobrynin and Dr. Gostev for local treatment of skin diseases, and corneal infections. They investigated thoroughly the effect of plasma against bacteria and corneal infection *in vitro* and *in vivo* utilizing micro-plasma treatment. The experiments confirm the effective bactericidal impact of this plasma discharge with negligible changes to surrounding cells or tissue including delicate cornea tissues. In this study on plasma-based treatment of corneal infection on rabbit cornea, two crucial observations made: (1) plasma exposure showed immediate and efficient bacteria inactivation effect, and (2) This plasma exposure also improved wound healing and regeneration of tissues process. This investigation presented a solid ground for an effective application plasma for the treatment of human patient with complex corneal infection and wounds (Fig. 1.7).

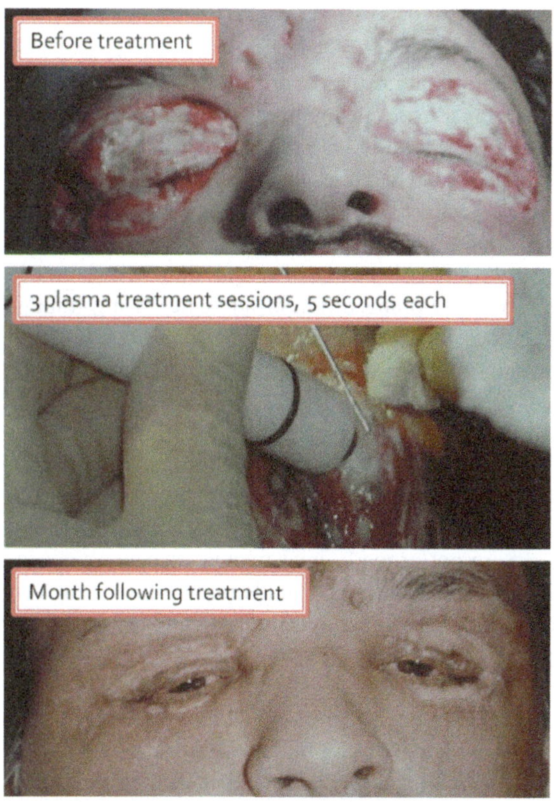

Fig. 1.7 The outcome of treatment (before—on top, after—on bottom) of plasma exposure (shown in the middle) of the complicated ulcerated eyelid wound (in a middle)

Necrotic phlegm on the upper eyelid exposed by air gas plasma for 5s after every few days. After two exposure sessions, the inflammation and edema decreased; and after the third 5s session the plasma exposed area was patient is almost cured and a regenerated tissue appeared. After the 6th session of plasma treatment, the patient was discharged from the hospital (Fig. 1.7). These facts on the first thrilling and inspiring investigation in the applied plasma medicine lead to an opportune moment to take a step from the introduction to the following chapters dedicated to the biomedical area.

As was mentioned, detailed information on plasma medicine fundamentals and applications and especially on the first successful steps in plasma-medical technology can be found in the major reviews and books of Fridman et al. [2], Laroussi et al. [3], Fridman and Friedman [1], Metelmann et al. [4], Toyokuni et al. [5], Keidar et al. [6], Kuo [7], Lu et al. [8], Keidar [9].

1.2 Plasma for Cancer Therapy

1.2.1 Background

Nonthermal atmospheric pressure plasma (NAP) could also called cold plasma is emerging as a promising innovative modality for cancer therapy [6–9]. Cold plasma action is exemplified in several ways, that are mainly related to active species generated in plasma and plasma-based electromagnetic fields by hitting target tissue. The main hypothesis of plasma interaction with biological targets is based on the view constituents of plasma are possibly cytotoxic, like ROS, which may stimulate a "cancer-killing," but RNS could show a "wound healing or immune activation" effect. Developing different strategies using tuning of these modalities can offer huge possibilities for targeting various signaling pathways in human cells. To this end, it has been demonstrated that plasma exposure possesses potent abilities to kill cancers in cell culture and animal experiments. Importantly, the same plasma treatment shows no or negligible toxicity to normal cells [9]. All these biological outcomes are related to plasma-based active species chemistry and other physical properties. Although the role of active species generated by plasma for cancer treatment described several times, however effects of charged particles and an electric field are not properly investigated. Recent evidence suggests that physical factors such as electromagnetic fields might play an important role [10–14]. These recent observations suggest that multiple chemical and physical modalities exist in cold plasma.

The chemical modality of the cold plasma effect on biological objects is related to active species generation and their transport to the liquid phase from the plasma gas phase and other biological barriers. Several active species generated by plasma are important active constituents in cells [10]. It has been debated that the similarity between plasma active species and biologically generated species signifies the main

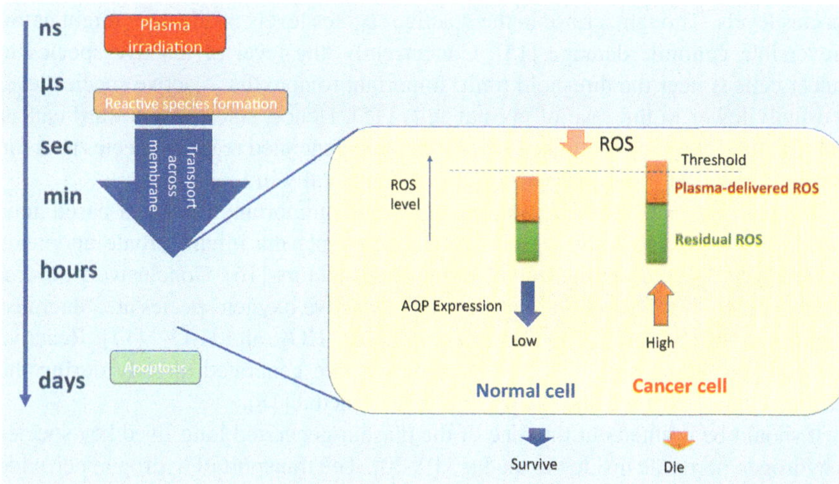

Fig. 1.8 Schematics of plasma interactions at multiscale levels. The magnified insert shows the schematic representation of the plasma interaction with cells explaining selectivity

reason for the application of plasma in cancer treatment and those generated species act as endogenous species in cells [10]. Plasma-generated reactive species show a significant role in 'redox' biology [1].

The outline of plasma interaction with biological objects and schematics explaining selectivity is shown in Fig. 1.8. One can see that plasma action is a multi-scale transfer process traversing from the plasma generation at the nano to micro-second, followed by reactive species generation and transport at the timescale of seconds to minutes, ultimately initiating various signaling process inside cells at hours and days.

Several main biological functions are related to active species. For instance, Watson [11] indicated that reactive species are a "positive force for life" because of their function in cell death processes such as apoptosis. concurrently, reactive species are also well known "for their capability to permanently damage many crucial genetic materials and proteins" As such, inside a healthy normal cell reactive species levels are managed by the anti-oxidant machinery. To this end, Watson noticed that "Mostly all physical or chemical agents such as ionizing radiations, and chemical drugs for direct killing cancers, act through by generating ROS that inhibits important cell cycle steps".

The influence of reactive species on cell processes depends on their levels [12]. Reactive species at less amount can initiate proliferation of cells and support basic cellular functions whereas a high amount of reactive species triggers oxidative stress which may responsible for cell death. Normal cellular function is well-maintained by antioxidant machinery that controls reactive oxygen species levels at a bearable level. A high amount of endogenous reactive species is generated due to the metabolic activity of cancers [12]. To survive, cancer cell alters themselves to control these species levels. Though, a raise in the reactive species levels inside cells might cause irreversible genomic damage [15]. Concurrently, the level of reactive species in cancer cells is near the threshold limit. Important to note that reactive species level is usually lower in the case of normal cells [15]. Hence, selectivity toward cancer cells oxidative therapy is attained when treatment generated reactive species near the threshold level. This hypothesis is shown schematically in Fig. 1.8.

The reactive oxygen species–based approach is important to plasma-based anti-cancer treatments. Reactive species produced by plasma might initiate apoptosis or autophagy by altering the role of intracellular factors [16]. Conclusively several investigations showed that biologically active reactive oxygen species are generated by plasma such as OH, O, O (1D), $O_2(^1\Delta_g)$, O_3, HO_2, and H_2O_2 [17]. Reactive nitrogen species such as NO_2^-, NO^-, and NO^+ are generated directly during the discharge in a gas phase and the plasma-treated media [18].

It should be pointed out that one of the plasma-generated long-lived key species is hydrogen peroxide inside the media [19, 20]. The transport of hydrogen peroxide as well as other ROS is facilitated by cell membrane transport channels such as aquaporins [20, 21]. Multiple studies suggest that tumor cells have more aquaporins than their normal counterpart. Therefore, the transport of hydrogen peroxide is very high in the case of cancer cells through the cell membranes than in normal healthy cells. Such difference in the transport of hydrogen peroxide and rate of consumption might be the potential mechanism of plasma-based cancer treatment. An increase in the reactive species due to the transport of species across cell membranes is related to plasma-generated research species generation outside the cells. An intracellular enzyme such as catalase major factor to control the level of hydrogen peroxide in living cells [22]. It is recently described that the rate of hydrogen peroxide consumption and catalase activity correlate in many cancer and normal cell lines.

1.2.2 In Vivo Applications

Over the last decade, many in vivo investigations have been executed to obtain the anti-cancer impact of cold plasma exposure. In most studies, plasma treatments were carried out by exposing the skin to cancers. Nevertheless, in many investigations, micro-plasma sources were used to direct the microscopic plasma jets to influence the treated cancer below the skin or the cranium [23–25]. All prior investigations described the potential influence of cold plasma on cancer progression.

The initial animal studies were executed by Vandamme et al. [23–25]. They utilized the brain cancer xenograft mice model to assess the anti-cancer outcomes of cold plasma exposure. This groundbreaking study showed a substantial reduction (56%) of tumor volume in the mice exposed to the pulsed floating electrode DBD. The exposed mice survival rate is enhanced by 60% after exposure to the FE-DBD. In this investigation, both tumor volume detection and bioimaging have been utilized to evaluate the cancer inhibition effect of plasma exposure. In this investigation, it is shown that fractioned doses of plasma exposure are much better than a single long exposure. These kinds of strategies are also famous in the case of radiation biology. Plasma treatments in fractions were further used in several investigations for the treatment of cancers (Fig. 1.9a). Subsequently, the investigation of bladder cancer treatment in xenograft mice models using jet plasma treatment (Fig. 1.9b) was executed [25]. Plasma exposure for 2 min reduced the tumor size significantly. Additionally, researchers executed comparable investigations on skin cancer in a mice xenograft model and attained encouraging outcomes that cancer is inhibited fully after 21 days of plasma exposure. Likewise, the subsequent survival rate is enhanced in the murine model after plasma exposure.

These initial investigations reveal the encouraging capability of plasma exposure as a cancer treatment strategy with no or negligible side effects. Additionally, the same kind of cancer inhibitory effect has been noticed in several other investigations on xenograft animal models. Overall, about 27 in vivo investigations have been recognized and concluded the reduction of tumor size and survival rate improvement [27]. It is important to note that all in vivo investigations for plasma therapy against cancer are mainly executed on xenograft mice models.

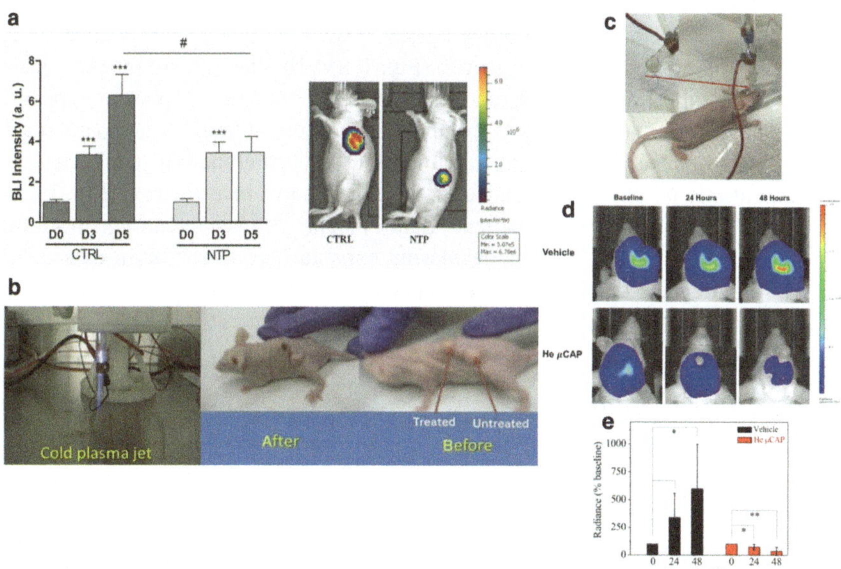

Fig. 1.9 a The initial animal studies for cancer treatment using cold plasma. Once the tumor attained 150 ± 50 mm^3 sizes, plasma exposure is given to a 6 min plasma exposure (frequency 200 Hz) group in fraction for 5 days. **a** Bioluminescence imaging before plasma exposure (Day 0), in between the treatment period (Day 3), and 24 h after the last treatment (Day 5). **b** Characteristic imaging of untreated control and plasma exposed on the 5th day [24]. **b Jet plasma and control and plasma treated mice images with multiple tumors (shown after 24 h)** [25]. **c Treatment of glioblastoma with μ-plasma.** Image displaying plasma transport via an intracranial endoscopic tube; **d** Images of cancer volume; **e** Outline of radiance intensity helium gas treatment as vehicle and plasma exposure [26]

To investigate the anticancer effect of plasma exposure on glioblastoma a microplasma source is developed. Also, the first intracranial mouse model was used in this plasma-based treatment strategy. The plasma source directly delivers active species to brain cancer using the endoscopic tube for the first time in the plasma medicine area, as shown in Fig. 1.1c [26]. In this investigation cancer volume was measured by a real-time bio-imaging method as shown in Fig. 1.1d. The outcome of these investigations indicated that the cancer volume enhanced only by 50% in the case of the treatment group which is much lower as compared to the untreated control which is 600% after 2 days, as shown in Fig. 1.1e.

To further understand the penetration of reactive agents, EM waves, and plasma species through the skin and scalp, the potential anti-tumor properties of CAP jet non-invasive in an intracranial model have been investigated [28]. To this end, the sensitization effect of a combination of CAP + TMZ was also studied. It has been shown that the jet plasma can seep into the bone, together with active species for plasma strategy over chemotherapy. In this investigation, brain cancer cells were implanted intracranially and allowed to proliferate for 1 week. Consequently, the skull is exposed to the jet plasma directly for 1 min at 1 L per minute helium gas, 12.5 kHz frequency, and 10 V with a 1 cm of distance. Further, TMZ is an anticancer drug injected immediately at a 6 mg/kg/day dose for 2 weeks. Anticancer drug TMZ alone did not inhibit cancer growth (Fig. 1.10a, b), due to resistance of cancer against TMZ exposure as earlier reported. However, one dose of plasma exposure alleviated the cancer growth by 40% compared to control, although this did not reach statistical significance (Fig. 1.10a, b). Importantly, a combination of plasma and TMZ potentially inhibited brain cancer progression in this investigation (Fig. 1.10a, b). Altogether, these outcomes indicate that: (1) The jet plasma can penetrate the biological barriers, and (2) A single plasma exposure sensitizes brain cancers to successive anticancer drug TMZ treatment.

1.2.3 Clinical Studies

In 2017, the outline of the first medical investigation was reported [28–30]. The clinical trial registered six patients with advanced (pT4) squamous cell carcinoma of the oropharynx with open infected ulcers. Patients were exposed to jet plasma for 3 times treatments in 3 weeks, each treatment followed by an interval of a week. Plasma exposure triggers a decrease in odor and pain medicine necessity, which enhanced the social activity of patients and initiated an encouraging emotional impact. The partial relapse of cancer in 2 patients within 9 months has been detected. In a biopsy, enough amount of cancer apoptotic cells and desmoplasia were observed in the surrounding tissue. This medical investigation underlines the medical importance and prospective plasma treatment could have in clinical oncotherapy going forward. In general, the medical utilization of potential strategies in other areas of oncology is presently under examination. Plasma-induced cancer inhibition is a potential outcome that will be investigated further In the United States, Keidar et al. George Washington

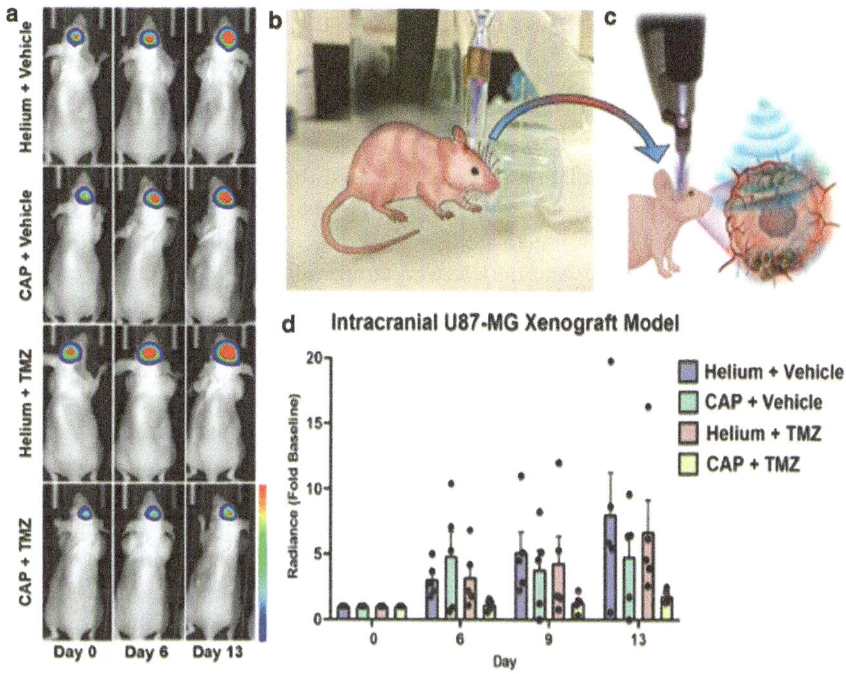

Fig. 1.10 **a** Characteristic BLI images at initial time (day 0) and 6 and 13 days after plasma exposure with or with anticancer drug TMZ injection. **b** Graphic description of the plasma treatment to mice model and **c** penetration of electromagnetic waves across the skin and cranium to kill brain cancer. **d** Outline of emitted radiance throughout the investigation

University together with US Medical Innovation conducted a medical application by treating remaining cancerous tissue without affecting normal cells after surgery of the final stage of colon cancer in a patient in 2015. Next, this group together with the Canady group used Canady Helios plasma source and plasma scalpels in the clinical liver resection to remove and kill remaining cancer cells. Very recently the US Food and Drug Administration has given authorization for 25 patients to undergo plasma treatment as an adjunct therapy to treat tumors [31]. Achieving additional awareness into plasma acceptable side effect summary along with its efficacy as the selective anticancer agent will steer more clinical trials.

1.3 Non-thermal Atmospheric Pressure Plasma Diagnostics

1.3.1 Optical Emission Spectroscopy (OES) for Plasma

In PBRC, the collision radiation models were used for Ar and nitrogen plasmas to obtain plasma temperatures and densities of atmospheric plasmas. In the case of Ar plasma, we referred to simple Ar collisional radiative (CR) model [32, 33]. This can be exploited to investigate the electron temperature of nonthermal plasma by optical emission spectroscopy (OES) with four metastable Ar state intensities. This method is assumed that the plasma follows a Maxwell-Boltzmann distribution and experiment emission intensities are used for determination of an electron temperature and the excited Ar atom densities. The reference [33] considered excited processes only from metastable state denoted by $1s_3$ and $1s_5$ to 2p excited levels denoted by $2p_1 \sim 2p_{10}$ in CR model. But the excited processes from resonance states denoted by $1s_2$ and $1s_4$ to 2p energy levels should not be ignored because these processes must occur in atmospheric pressure plasma [34]. We propose the modified CR model by inclusion of these resonant excited processes for accurate determination of electron temperature for atmospheric pressure plasma jet [35]. For N_2 (or air) plasma, we referred the collisional radiative model of nitrogen gas [36] for obtaining the electron temperature and density. It is included the physical processes from the ground state $X\ ^1\Sigma_g^+$ to the excited states $A\ ^3\Sigma_u^+$, $B\ ^3\Pi_g$ and $C\ ^3\Pi_u$ [37].

The impact excitation of Ar plasma by electron can be described as the populations of the $2p$ excited Ar states are made by electron collisions from Ar atoms in ground to excited states of the metastable and resonance states. Our CR model included the excited processes of the resonance and metastable states and we selected eight $2p$ emission lines for solving the balance equations. The modified equation for a $2p_x$ in Paschen notation excited level can be written by inclusion of resonant states ($i = 2$, 4) along with metastable states ($i = 3, 5$) in left hand side of Eq. (1.1), as,

$$n_e n_g k_{g,2p_x} + \sum_{i=2,3,4,5} n_e n_{1s_i} k_{1s_i,2p_x} = \sum_{i=2,3,4,5} n_{2p_x} A_{2p_x,1s_i} \tag{1.1}$$

where n denotes number density (n_e: electron density, n_g: neutral gas density, n_{1s}: 1s level density, n_{2p}: 2p level density), $k_{1si,\ 2px}$ is the excitation rate coefficient from level $1s_i$ to $2p_x$, which depend on the electron temperature, and $A_{2px,\ 1si}$ is the transition probability from level $2p_x$ to $1s_i$. The subscript g stands for ground level, e for electrons, and $1s_i$ and $2p_x$ are the Paschen notations [33]. The relative number density for specified wavelength could be represented by the optical emission intensities measured from spectrometer. For example, the n_{2px} can be written as in Eq. (1.2),

$$n_{2px} \approx \frac{I_{2p_x-1s_i} \lambda_{2p_x-1s_i}}{A_{2p_x-1s_i}} \tag{1.2}$$

where I is the measured optical emission intensity, and λ the given wavelength [33]. Equation (1.3) can be used in Eq. (1.2) to replace n_{2px} and Eq. (1.1) can be rewritten by following [33]

$$n_e = \frac{\sum_{i=5}^{2} n_{2p_x} A_{2p_x,1s_i}}{n_g k_{g,2p_x} + \sum_{i=2,3,4,5} n_{1s_i} k_{1s_i,2p_x}} \tag{1.3}$$

This equation can be expressed by the measured intensity for any $2p_x$ level of $2p_1$ (750.387 nm), $2p_3$ (706.722 nm), $2p_4$ (794.850 nm), $2p_5$ (751.500 nm), $2p_6$ (763.510 nm), $2p_7$ (810.040 nm), $2p_8$ (842.600 nm) and $2p_9$ (811.531 nm) occurred by the electron impact excitation in Ar plasma. Because the excited atom density n_{2px} is expressed as the relative intensity, the electron density n_e can be the relative values. It is the same to each other for eight 2_{px} levels. We can establish 4 equations depended on electron temperatures with unknown variable of excited atom densities $1s_2$, $1s_3$, $1s_4$ and $1s_5$. The Ar excited 1s atom densities can be solved for all electron temperatures by using mathematical python library. The electron density in Eq. (1.3) can be written by these specific 1s densities depended on the electron temperature. By comparing the two selected electron density equations, it is possible to obtain a section in which the values are the same at a specific electron temperature. This electron temperature at this time becomes the value we are looking for. This method can find the electron temperature and excited 1s atom densities in atmospheric pressure Ar plasma.

Figure 1.11 shows the process of OES diagnostics method by using Ar emission eight lines in nonthermal atmospheric pressure Ar plasma. This method can be used for the determination of electron temperature and excited atom densities ($1s_2$, $1s_3$, $1s_4$, $1s_5$) of Ar plasma. In the atmospheric pressure Ar plasma, we measured the Ar I emission lines and, as applying Ar CR model, the electron temperature was 1.28 eV, the excited Ar $1s_2$, $1s_3$, $1s_4$ and $1s_5$ densities were 1.92×10^{15} cm^{-3}, 4.03×10^{15} cm^{-3}, 1.88×10^{15} cm^{-3}, and 5.79×10^{15} cm^{-3}, respectively.

In an air (or nitrogen) plasma, we basically used nitrogen molecule collisional radiative model of reference [35–39]. The ground state $X\ ^1\Sigma_g^+$ can be transitioned to the excited upper states $A\ ^3\Sigma_u^+$, $B\ ^3\Pi_g$, and $C\ ^3\Pi_u$ by the electron impact effect of plasma [35–39]. And the excited nitrogen molecules in the A and B states could be exchanged by collision with N_2 molecules in the state $X\ ^1\Sigma_g^+$, respectively [35–39]. Nitrogen molecules in lower states $A\ ^3\Sigma_u^+$ and $B\ ^3\Pi_g$ are corresponding to spontaneous emission from higher states $B\ ^3\Pi_g$ and $C\ ^3\Pi_u$, from which the spectra of N_2 FPS (first positive system) and N_2 SPS (second positive system) are emitted [35–39]. And, the excited molecule of $A\ ^3\Sigma_u^+$ state can have the wall deactivation [35–39]. In addition, for practical application to atmospheric plasma, we need to consider the energy pooling reaction, which represents the transition to the $B\ ^3\Pi_g$, $C\ ^3\Pi_u$, and $X\ ^1\Sigma_g^+$ states due to the collision between two molecules in the $A\ ^3\Sigma_u^+$ state [35–39]. Therefore, we can develop the total balance equations for these excited $A\ ^3\Sigma_u^+$, $B\ ^3\Pi_g$, and $C\ ^3\Pi_u$ state [35–39].

In the case of N_2 (A),

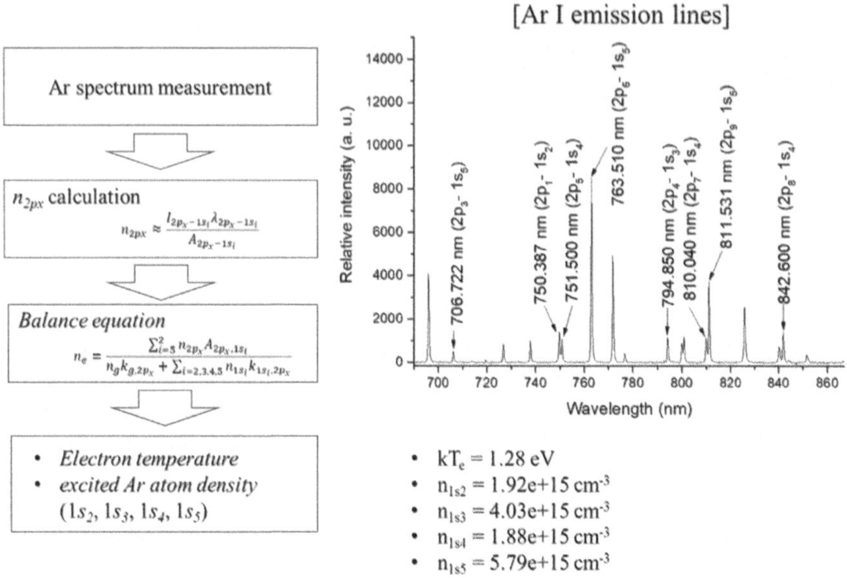

Fig. 1.11 OES diagnostics process of Ar emission lines in atmospheric pressure Ar plasma

$$2n_A^2(k_{\mathrm{AAB}} + k_{\mathrm{AAC}}) + (k_{\mathrm{wall}} + k_{\mathrm{AX}}n_v)n_A - (A_B + k_{\mathrm{BX}}n_g)n_B - n_e n_g Q_A = 0 \tag{1.4}$$

In the case of N_2 (*B*),

$$n_A^2 k_{\mathrm{AAB}} + k_{\mathrm{AX}}n_v n_A - (A_B + k_{\mathrm{BX}}n_g)n_B + A_C n_C + n_e n_g Q_B = 0 \tag{1.5}$$

In the case of N_2 (*C*),

$$n_A^2 k_{\mathrm{AAC}} - A_C n_C + n_e n_g Q_C = 0 \tag{1.6}$$

The n_e, n_A, n_B, and n_C are the density of electron and excited states $A\,^3\Sigma_u^+$, $B\,^3\Pi_g$, and $C\,^3\Pi_u$, respectively. The n_v and n_g are the neutral gas densities for vibrational temperature and ground state depending on the gas temperature. The k_{AX} and k_{BX} are the rate coefficients for transition processes between $A\,^3\Sigma_u^+ - X\,^1\Sigma_g^+$ and $B\,^3\Pi_g - X\,^1\Sigma_g^+$ states, respectively [37]. The k_{AAB} and k_{AAC} are the rate coefficients for the transition process to upper states $B\,^3\Pi_g$ and $C\,^3\Pi_u$ after the collision with excited states $A\,^3\Sigma_u^+$ [37]. The k_{wall} is the wall deactivation rate coefficient that is derived by the diffusion model with the wall reflection [37]. The Q_A, Q_B, and Q_C are the rate coefficients from the ground to excited states $A\,^3\Sigma_u^+$, $B\,^3\Pi_g$, and $C\,^3\Pi_u$, respectively, caused by electron impact excitation [31, 37]. The A_B and A_C are the transition probabilities of excited states $B\,^3\Pi_g$ and $C\,^3\Pi_u$ [37].

We could obtain three 2nd order balance equations for three unknown variables of excited densities n_A, n_B, and $n_C$in Eqs. (1.4)–(1.6). These excited molecule densities for N_2 can be solved simply using mathematical library in python software. Also, we can express the excited molecule densities by assigning an arbitrary electron temperature and density into Eqs. (1.4)–(1.6). The ratio of molecule densities for N_2 SPS and FPS can be written as belows [37],

$$\mathrm{R}(kT_e, n_e) = \frac{A_C n_C}{A_B n_B} \tag{1.7}$$

The magnitudes of Eq. (1.7) could be compared by measured ratio of emission intensities between N_2 SPS and FPS, which depend on the electron temperatures and densities. N_2 SPS emission lines could be selected to be the wavelength of 295.3, 313.6, 315.9, 337.1, 353.7, 357.7, 371.1, 375.5, 380.5, 389.5, 399.8, and 405.9 nm [32]. Also, N_2 FPS intensity could be selected to be the wavelength of 654.5 nm [32]. We can find the 12 lines ration R values, which depended on the arbitrary electron temperature and density, for 12 lines of N_2 SPS and one line of FPS in atmospheric pressure air plasma. The specific electron temperature can be obtained by comparing with measured emission line ratio in OES data of air plasma. These electron temperatures can be expressed according to arbitrary electron densities and obtained for each 12 lines ratio. We can find a specific electron density with the same electron temperature.

Figure 1.12 shows the process of OES diagnostics method by using N_2 emission lines in nonthermal atmospheric pressure air plasma. This method can be used for obtaining the electron temperature and density, rotational and vibrational temperature, and excited N_2 molecule densities (n_A, n_B, n_C) of air plasma. The line ratios of the intensity N_2 plasma were measured and the estimated line ratio between N_2 SPS and FPS can be calculated by using the balance equations of nitrogen collisional radiative model. By comparing these values, the various plasma parameters can be solved in atmospheric pressure air plasma. In the case of plasma temperature in air plasma, the electron, rotation and vibration temperature had values of 1.05 eV, 847.60 k, and 0.76 eV, respectively. Also, in the case of plasma density, the electron, N_2 molecule densities of the excited states A, B, and C had values of 1.12×10^{15} cm^{-3}, 9.78×10^{15} cm^{-3}, 1.31×10^{16} cm^{-3}, and 1.72×10^{15} cm^{-3}, respectively.

Table 1.1 has shown the arrangement for plasma parameters that can be calculated with the collisional radiative model of atmospheric pressure Ar and N_2 plasma.

1.3.2 Measurement of the Electron Density by Using Optical Interferometer

Electron densities which is important parameter of great interest in plasma are responsible for the generation of reactive oxygen or nitrogen species used for industrial, agricultural and medicine processes. But, electron density measurement for weak

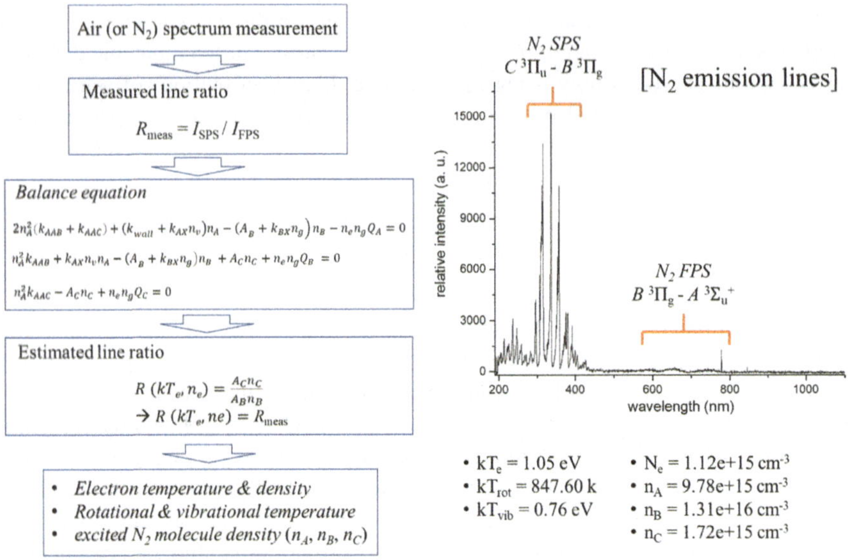

Fig. 1.12 OES diagnostics process of N_2 emission lines in atmospheric pressure Air plasma

Table 1.1 Comparing with Ar and N_2 collisional radiative model

Parameters	Ar CR model	N_2 CR model
Plasma temperature	• Electron temperature	• Electron temperature • Vibrational temperature • Rotational temperature
Plasma density	• Excited Ar atom density($1s_2$, $1s_3$, $1s_4$, $1s_5$)	• Electron density • Excited N_2 molecule density (n_A, n_B, n_C)

ionized plasmas such as atmospheric pressure dielectric barrier discharge (DBD) plasma jet and DBD surface plasma are easily perturbed by small interferences, so diagnosis for electron density is mainly performed by indirect optical methods such as laser interferometry, optical emission spectroscopy, and Thomson scattering [35–41]. Laser interferometers offer the advantages of responsiveness to rapid discharge and the ability to tune sensitivity of electron density by changing the probing laser wavelength. The plasma electron density could be calculated by refractive index obtained from measured laser phase shift [37–41]. In this chapter, we introduce the Michelson interferometry system which perform laser phase shift measurement for estimation of electron density from a plasma without heterodyne system.

Figure 1.13 shows the Michelson interferometry for the plasma electron density measurement. Here M1 to M4 are optical mirrors, BS is a beam splitter, BE is a beam expander, LPF is a line-pass filter (for corresponding to probing laser wavelength of

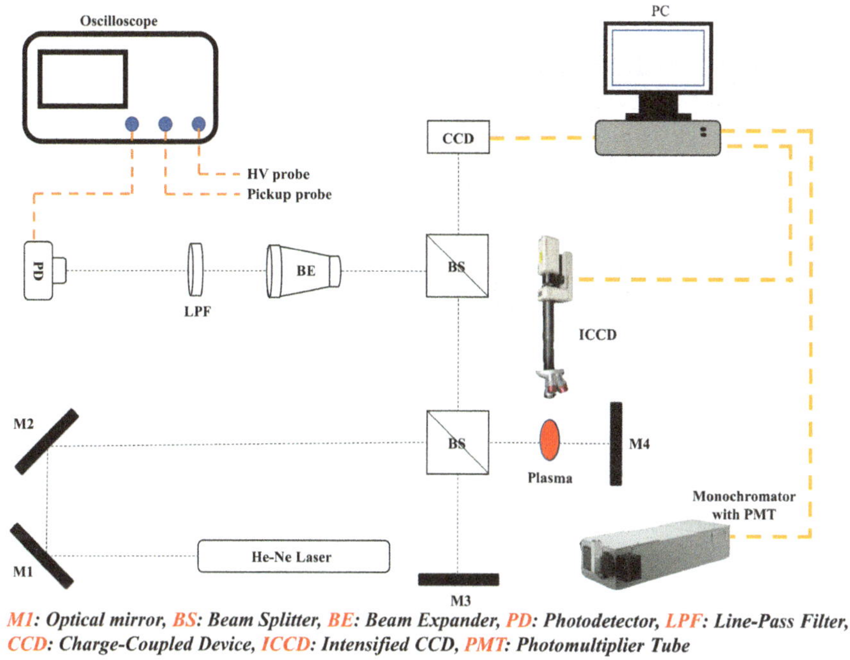

Fig. 1.13 Michelson interferometry for the plasma electron density measurement [43]

632.2 nm), PD is photodetector and CCD is a charge-coupled device. From this interferometry, the phase shift of the laser beam is caused by the optical path difference. The interferogram line intensity I could be denoted by the following equation:

$$I = \frac{I_0}{2}[1 + \cos(k\Delta z + \Delta\varphi)], \tag{1.8}$$

where, I_0 is the initial laser beam intensity prior to division by the beam splitter, k is the wave number, $\Delta\varphi$ is the optical phase difference, and Δz is the optical path difference between the two mirror arms for M3 and M4. The interferent lines consist of constructive and destructive patterns with phases of $2m\pi$ and $(2m - 1)\pi$, ($m = 1$, 2, 3…), respectively, as determined by Eq. (1.8).

Figure 1.14a shows a CCD image for the laser interferent fringes. The relative intensities of the fringes representing constructive and destructive interference are represented by the photodetector signal intensities (in volts) as shown in Fig. 1.14b. The signal of interference pattern would could be shifted from bright (constructive) to dark (destructive) or dark to bright one in this photodetector by change in optical path difference, which has been done by fine adjustment of interferometry mirror (M3 or M4) in Fig. 1.13. From the line shift signals, we can obtain the phase shift of laser interferogram due to the plasma which installed on the fixed mirror arm [41]. This phase shift, $(n - 1)\frac{2\pi}{\lambda}d$, associated with refractive index n, plasma's electron

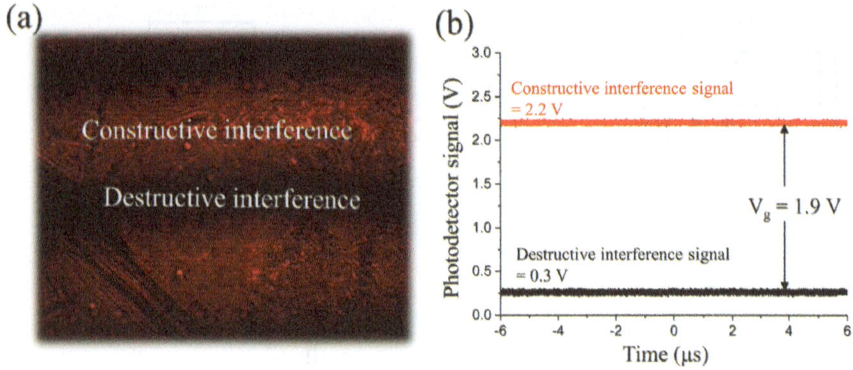

Fig. 1.14 **a** CCD image of interferent fringes with constructive and destructive patterns. **b** Measured photodetector signals in both interferent patterns [43]

density N_e and heavy particle density N_{heavy} is given by the following equation [42]:

$$(n-1)\frac{2\pi}{\lambda}d = -\frac{e^2\lambda d}{4\pi c^2 m_e \varepsilon_0}N_e + \frac{2\pi d}{\lambda}A\left(1+\frac{B}{\lambda^2}\right)\frac{N_{\text{heavy}}}{N_{\text{heavy}_0}}, \quad (1.9)$$

where m_e is the electron mass in kg, λ is the laser wavelength in m, ε_0 is the permittivity in vacuum in F/m, *c* is the speed of light in m/s, and *d* is the laser path in the plasma region inm, which is twice the plasma width because the laser has double transit path from the mirror (M4) to the beam splitter (BS). Further, N_{heavy} and N_{heavy0} are the densities in m^{-3} of heavy particles (ions, molecules, neutral atoms) in plasma state and initial state, respectively, under room temperature and pressure ($T = 290$ K, $p = 1$ atm). *A* and *B* in the second term of right hand side are specific coefficients for heavy particle species [42, 44]. The optical path of the laser beam causes phase shift in the interferogram lines since the refractive index has been changed by plasma.. The slope of phase shift for electron density is very greater than that of heavy particles' one in the atmospheric pressure discharge and heavy mass (72,872 times larger than electron mass in case of argon plasma) [41]. The photodetector signal of phase shift by electron density (ΔV_e), which is caused by fast electrons in the plasma, can be measured by a spatially fixed photodetector. V_g is the gap voltage between constructive and destructive interference. Therefore, the relation of phase shift ($\Delta\varphi$) in Eq. (1.8) between before and after discharge, which can be written as [43]

$$\Delta\varphi = \cos^{-1}\left(\frac{2(V_e - V_d)}{V_g} - 1\right) - \cos^{-1}\left(\frac{2(V_I - V_d)}{V_g} - 1\right), \quad (1.10)$$

where V_I is the initial voltage of photodiode signal before discharge, V_e is the end point voltage of the fast transient signal by plasma electrons after discharge and V_d is the photodiodesignal in destructive interference. For the measured photodiode signal ($\Delta V_e = V_e - V_I$) produced by phase shift ($\Delta\varphi$), the plasma electron density can be measured from $\Delta\varphi = 2\pi d(n-1)/\lambda$ and Eq. (1.9) by elimination of the heavy particles. The electron density N_e could be expressed by [43]

$$N_e = \frac{4m_e\varepsilon_0\pi c^2}{e^2\lambda d}\left\{\cos^{-1}\left(\frac{2(V_e - V_d)}{V_g} - 1\right) - \cos^{-1}\left(\frac{2(V_I - V_d)}{V_g} - 1\right)\right\}[\mathrm{m}^{-3}]. \tag{1.11}$$

The maximum electron density is limited by the V_g value of 1.9 V in this measurement, which corresponds to π radian in phase shift. Therefore, the maximum electron density can be measured to be 5.5 × 10^{18} cm^{-3} in this experiment from Eq. (1.11) under $V_g = 1.9$ V, $d = 320$ μm, and $\Delta V_e = V_g$.

Figure 1.15a, b show the interferogram brightness change by transient spark discharge. After discharge, destructive interference line of interferogram has been changed to constructive interference line by spatial plasma refractive index change. Figure 1.15c shows the total variation of interferogram line intensity, it was increased with discharge from initial level of photodetector signal and restored to initial level after few ~ms. It means that spatial refractive index has been changed by plasma electron density and heavy particle. An initial increase of photodetector signal shows distinguishable slope difference in Fig. 1.15d. Initial increase of photodiode signal, $\Delta V_e + \Delta V_h$, contains both electron and heavy particle effects with slope of 2.94 mV/μs, and late signal ΔV_h contains only heavy particle effect with slope of 0.44 mV/μs, respectively. From these experimental measurement, the electron density could be estimated to be 8.5 × 10^{15} cm^{-3} by Eq. (1.11) by taking into account only the electron slope, 2.50 mV/μs, which can be obtained by subtracting the slope caused by the heavy particle from the initial slope in photodiode signal.

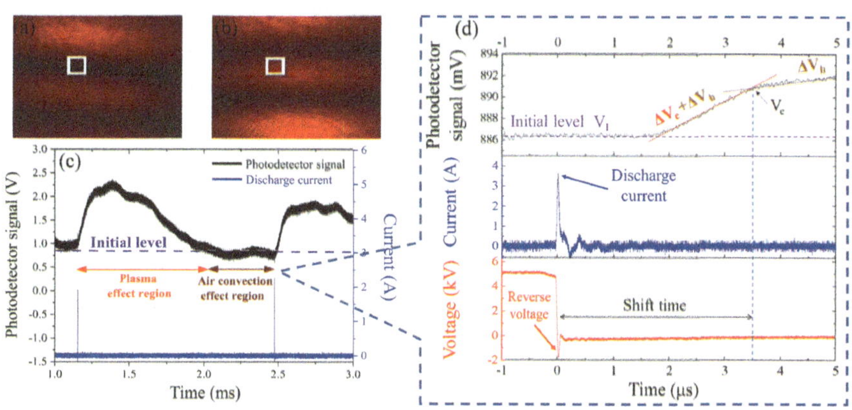

Fig. 1.15 **a** Laser interference pattern before discharge, **b** after discharge (white squares are photodiode sensitive area), **c** Transient photodiode signal (black line) with spark discharge current (blue line). **d** Photodiode signal $\Delta V_e + \Delta V_h$ from plasma electrons and heavy particles with fast slope, and signal ΔV_h from heavy particle with slow slope (top) with voltage (middle) and current (bottom) signal in time [43]

1.3.3 Measurement of the Plasma Radicals by Using Ultraviolet Absorption Spectroscopy

Reactive species in cold atmospheric pressure plasma play an important role in biomedical fields and industry. Because an cold air plasma can generate many kinds of reactive species, the generation of radical species is an important issue for the plasma device. The ultraviolet(UV) diagnostics of the cold plasma is very important to measure the density of reactive species. Also many researchers have widely used ultraviolet absorption pectroscopyin density measurement of reactive species.

The hydroxyl radicals OH of ROS (reactive oxygen species) can be measured by the ultraviolet absorption spectroscopy in CAP plasma. The experimental configuration of this absorption spectroscopy has been represented in references [45, 46]. This setupconsists of the UV lamp, in which a Hg lamp has been used with its power of 0.5 W, centered at 306 nm, and plano-convex lens whose wavelength ranges from ultraviolet to infrared are transmitted [45]. The UV light has been transmitted and focused to the 200 um in diameter in the plasma jet by using the plano-convex lens for UV absorption measurement for the OH species occurred at 307 ~ 309 nm [45]. The Lambert-Beer's law can estimate the density of hydroxyl OH radical species in CAP with UV absorption spectroscopy. The intensity for the incident and transmitted UV light through the plasma region whose thickness is x, are represented to be Io and Iv, respectively. The density of hydroxyl OH radical species produced by the CAP jet is given by [45]

$$N = -\frac{1}{\sigma \cdot x}\ln\left(\frac{I_\upsilon}{I_0}\right) \tag{1.12}$$

where N is the hydroxyl OH density, and σ is the collisional cross-section of about 6 $\times$ 10^{-11} m^2 for OH species [47], and x is 300 μm. The hydroxyl OH radical density could be estimated by Eq. (1.12) from the experimental measurement of I_v/I_o i.e., ratio of the transmitted intensity through the plasma to incident one at wavelengths of 307 ~ 309 nm.

Figure 1.16a The strong UV absorption profile can be seen at wavelenths about 309 nm in the transmitted signal (I_υ) through the plasma (black color line), which is caused by the OH radical species of Ar CAP jet, whose gas flow rates are ranged from 80 to 300 sccm [46]. The incident intensity of UV lamp (I_o) and emission intensity from OH radical species in the CAP jet without UV incidence have been denoted by red and blue lines, respectively, in Fig. 1.16a. The UV emission from the plasma and absorption lines at about 309 nm caused by OH radicals are denoted by the circles in the blue line and the dotted box in the black one, respectively, as shown in Fig. 1.16a [46]. The transmission ratio (I_v/I_o) of UV signals can be converted the OH density at the absorbed wavelength of about 309 nm. Figure 1.16b shows the density of OH radical species at 2 mm over the interfacial water surface versus the Ar gas flow rate from 80 to 240 sccm, where electrical power is 15 W and the driving frequency is 22 kHz [46]. The OH density reaches the maximum to be 2.6 $\times$ 10^{15} cm^{-3} under the

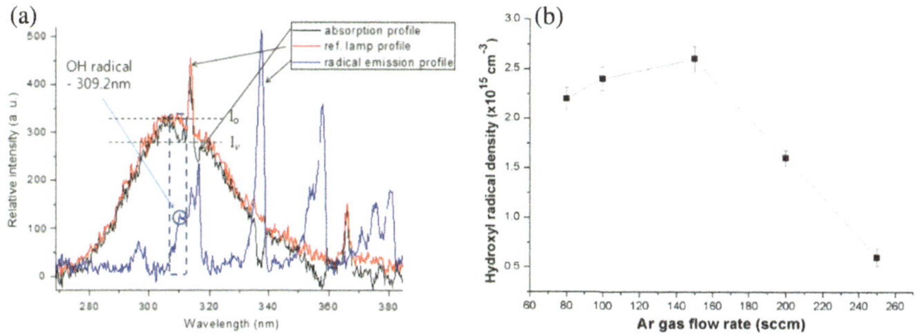

Fig. 1.16 **a** UV transmitted profile (I_v) through the plasma jet (black), in which absorption occurred at wavelengths of 307 ~ 309 nm, caused by the OH radical species. Reference UV lamp profile (I_o) versus the wavelength without nonthermal atmospheric pressure plasma jet (red). Optical emission profiles from the plasma jet versus the wavelength (blue) without UV incidence. **b** Hydroxyl OH radical density at the interfacial water surface region contacted by plasma, which is 2 mm above the water surface versus the Ar gas flow rates from 80 to 240 sccm [45]

gas flow rate of ~150 sccm and it is rapidly decreased to 6.0×10^{14} cm^{-3} for ~250 sccm in this report [46].

For the density measurement of the reactive nitrogen species (NO_x), the cavity-enhanced absorption spectroscopy (CEAS) or Fourier Transform Infrared (FTIR) spectroscopy methods are widely used [48]. For nitrogen dioxide (NO_2) measurement, the visible broad band cavity-enhanced absorption spectroscopy (BBCEAS) method could be used [48]. However, for density measurement of nitric oxide (NO), CEAS should be used for diagnostics of infrared-active molecules by using a mid-infrared laser, whose spectral range is between 3 and 20 μm [49]. The NO_2 absorption profile could be obtained by using UV and visible light sources such as light emitting diode (LED) and Xe or Hg arc lamps. The absorption band of NO_2 molecule includes the electronic transition band in the visible spectra [50]. However, the vibronic absorption band of NO species is located around infrared 5.26 μm and its absorption profile can be detected by a quantum cascade laser (QCL). This kind of NO absorption spectroscopy has been used generally and the QCL can be adjusted to a specific laser wavelength [50]. The BBCEAS and QCL-CEAS techniques could be used well frequently for density measurements of NO_2 and NO respectively, based on these reasons64]. In this chapter, we describe the density measurement of NO_2 and NO generated by air NAP jet by employing the BBCEAS with LED and CEAS with QCL [51]. For the measurement of NO_2 and NO density, a visible LED (660 nm) and a mid-infrared laser diode (LD) (5.2386 μm) could be used, respectively [51]. These radical densities could be measured by using Beer-Lambert law, Eq. (1.12), which is obtained by the absorbed laser intensity passing through the plasma gas region inside an optical cavity of CEAS [51]. We obtain the NO_2 density to be ~2.5×10^{16} cm^{-3} in air plasma et. and Also NO density has a value of ~4×10^{15} cm^{-3} [51] in recent report [51]. To find the NO maximum absorption wavelength the transmission ratio (I_v/I_o) has been measured in the QCL's tunable wavelength ranges [51]. Figure 1.17a

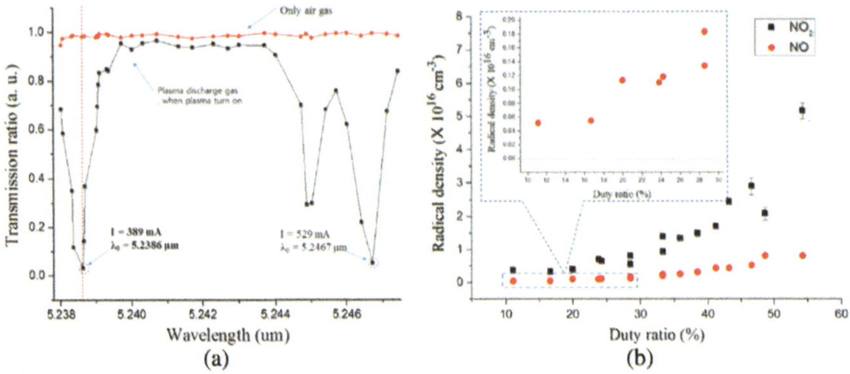

Fig. 1.17 **a** The transmission ratio of mid-infrared LD beam for NO density mesurement versus wavelength of QCL, **b** NO_2 (black) and NO (red) radical densities versus the duty ratio (%)

shows the absorption wavelength 5.2386 μm corresponding to the NO's first peak of absorption [51]. Figure 1.17b shows the densities of NO_2 and NO reactive species in the air NAP jet versus pulse duty ratio [51].

1.3.4 Plasma Parameter Characteristics for Industry and Biomedical Plasma Products

In PBRC (Plasma Bioscience Research Center) of Kwangwoon university, Korea, various plasma biomedical devices have been developed and studied about plasma parameters, plasma density and temperature [46]. For these parameters, we applied to the nitrogen collisional radiative (CR) model with OES data of NAP or CAP [52]. The nitrogen OES can be obtained based on the CR model for the determination of plasma temperatures (electron temperature, rotational and vibrational temperature) and densities (electron density, and excited nitrogen molecule densities) [53]. The OES intensities from the N_2 second positive system (SPS) and first positive system (FPS) are very important in modelling the N_2 CR model, by which the plasma temperatures and plasma densities could be estimated in nonthermal atmospheric pressure plasma [16]. These physical properties are provided by PBRC to a Korean and foreign companies for supplying various plasma products. These physical parameters could be used as evidence for plasma being used for various purposes as well as basic data for performance evaluation.

Figure 1.18 shows that PBRC plasma sources can be largely classified into plasma jet and surface DBD plasma type. Plasma jet is suitable for the treatment of local treatment area in human body (teeth, skin wound) or material sample. On the other hand, surface DBD plasma could be is used for large area plasma processing purposes. In particular, the surface DBD type is divided into three categories, where counter or facing-DBD, coplanar-DBD and floating electrode—DBD. The plasma parameters

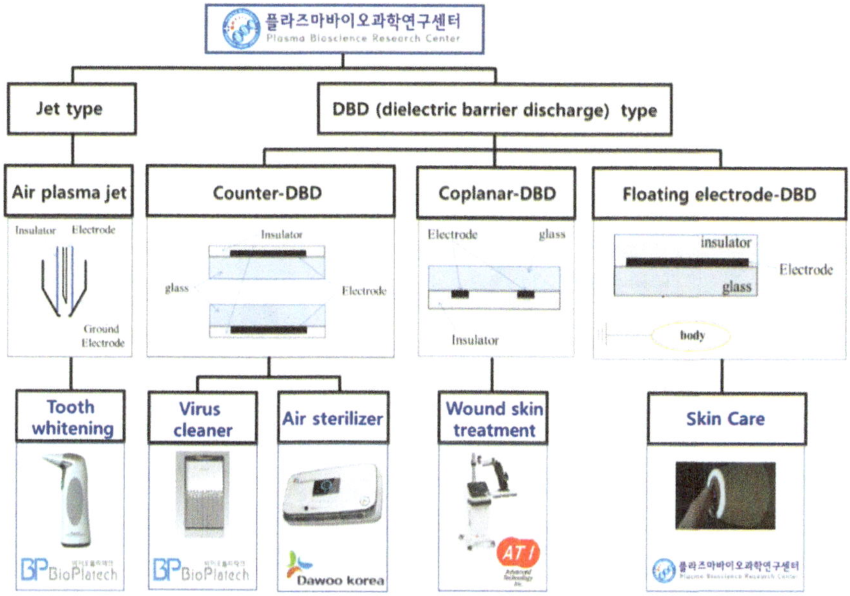

Fig. 1.18 The plasma biomedical devices developed in PBRC of Kwangwoon university

(electron density and temperature, vibrational and rotational temperature, excited nitrogen molecule densities nA, nB, and nC) could be investigated through N_2 CR model for PBRC plasma sources.

The air plasma jet consists of a needle-shaped electrode, a ground electrode covered around the outside, and a glass tube that insulates central power electrode. This technology is currently applied as a plasma tooth whitening device through technical transfer to "Bio-Platech" company, and its development has been completed and is on marketing sale. The counter or facing—DBD is a structure in which two glass substrates printed by electrode material and then covered with dielectric materials, respectively, and then they are installed by facing each other with a finite separation distance, as shown in Fig. 1.18. Plasma discharge occurs in the space between glass substrates by applying opposite polarity to each counter electrode. This plasma source is currently used in the virus cleaner and air sterilizer products. We have completed commercial product development in cooperation with "Bio-Platech" and "Dawoo Korea" company. The coplanar-DBD is a surface discharged structure consisting of two electrodes covered with insulator on a glass substrate. The plasma discharge occurs on insulator surface by applying opposite polarity to each electrode in coplanar surface, as shown in Fig. 1.18. Currently, this technology is developing for the wound skin treatment medical devices in hospital through "ATI" company. The structure of the floating electrode—DBD is similar to that of the counter-DBD except counter substrate with counter electrode. Here, the human body could be grounded as a counter electrode and only homopolar voltage signal has been put onto the floating DBD electrode. Hence the plasma discharge occurs

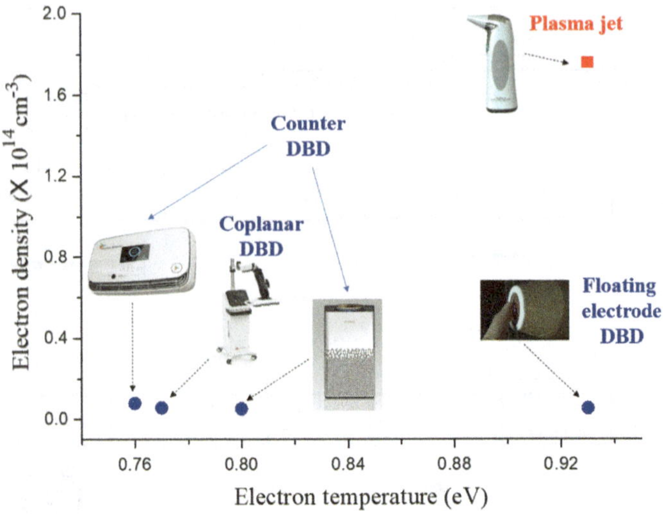

Fig. 1.19 Mapping of electron density and temperature in biomedical plasma devices

in a space between the floating-DBD surface and body. It can be used for skin care medical device since this plasma can directly contact on skin surface.

Figure 1.19 shows the mapping for electron density and electron temperature, which are obtained from the nitrogen CR model for PBRC plasma sources. The electron temperature and density are found to be maximum of 0.93 eV and 1.76×10^{14} cm^{-3} in air plasma jet. The electron temperature of floating electrode-DBD is higher than that of other DBD sources. Also, the electron density shows that they are in the ranges of ~10^{12} cm^{-3} in DBD type.

The rotational and vibrational temperature of the plasma jet and DBD plasma typeare shown in Fig. 1.20. Here these plasma rotational and vibrational temperatures are shown to be more than twice those of the DBD type. The rotational and vibrational temperature of the jet type are approximately maximum of 712 K and 0.76 eV, respectively. In the case of surface DBD plasma, the average value for rotational and vibrational temperature are shown to be 350 K and 0.35 eV, respectively, which are about half of that of jet type.

For the excited nitrogen molecule, N_2, density is shown in Fig. 1.21, where *nA* is higher than *nB*, *nC* for all types of plasma sources. Here nA, nB, and nC denote the N_2 molecular densities for excited energy states of $A\ ^3\Sigma_u^+$, $B\ ^3\Pi_g$, and $C\ ^3\Pi_u$, respectively. At the plasma jet, the order of nA is similar to nB, but in the case of DBD, nA is higher than other species. In plasma jet type, the excited molecule densities have the ranges between 10^{14} and 10^{15} cm^{-3}, also DBD type has the ranges between 10^{12} and 10^{14} cm^{-3}.

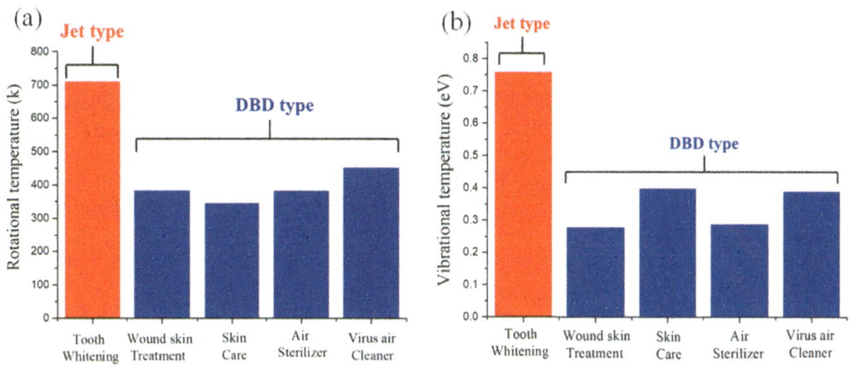

Fig. 1.20 **a** The rotational and **b** vibrational temperature for plasma biomedical devices

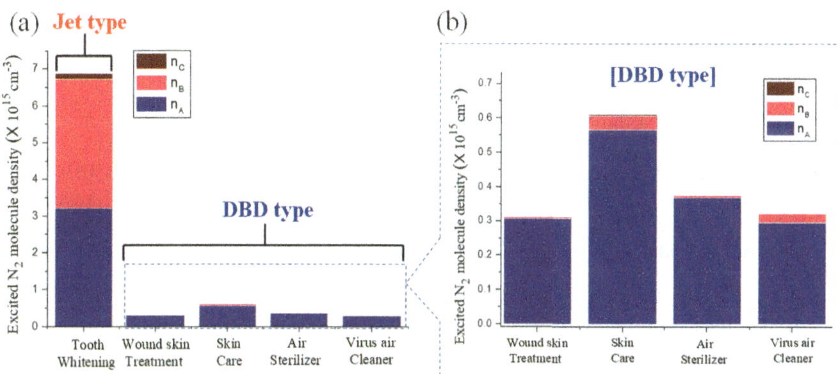

Fig. 1.21 **a** The excited N_2 molecule density of plasma jet and surface DBD plasma type and **b** expanded excited N_2 molecule density of surface DBD plasma type for biomedical devices

Acknowledgements This work was supported by the Korean National Research Foundation (2021R1A6A1A03038785).

References

1. A. Fridman, G. Friedman, *Plasma Medicine* (Wiley, UK, 2013)
2. G. Fridman, G. Friedman, A. Gutsol et al., Applied plasma medicine. Plasma Process. Polym. **5**(6), 503–533
3. M. Laroussi, M. Kong, G. Morfill, W. Stolz (eds.), *Plasma Medicine* (Cambridge University Press, N.Y, 2012)
4. H.-R. Metelmann, T. von Woedke, K.-D. Weltmann (eds.), *Comprehensive Clinical Plasma Medicine* (Springer, Berlin, Germany, 2018)
5. S. Toyokuni, Y. Ikehara, F. Kikkawa (eds.), *Plasma Medical Science* (Elsevier, Acamic Press, UK, USA, 2018)

6. M. Keidar, D. Yan, J. Sherman, *Cold Plasma Cancer Therapy* (IOP e-books, 2019)
7. S. Kuo, *Cold Atmospheric Plasmas, Their Use in Biology and Medicine*, (World Scientific Publications, 2019).
8. X.P. Lu, S. Reuter, M. Laroussi, D.W. Liu, *Non-Equilibrium Atmospheric Pressure Plasma Jets: fundamentals, Diagnostics,and Medical Applications* (CRC Press, Taylor & Francis, 2019)
9. M. Keidar (ed.), *Plasma Cancer Therapy* (Springer, Berlin, Germany, 2020)
10. D. Graves, J. Phys. D: Appl. Phys. **45**, 263001 (2012)
11. J. Watson, Perspective: oxidants, antioxidants and the current incurability of metastatic cancers. Open Biol. **3**, 120144 (2013)
12. R. Cairns, I. Harris, T.W. Mak, Regulation of cancer cell metabolism. Nat. Rev. Cancer **11**, 85–95 (2011)
13. D. Yan., Q. Wang, M. Adhikari, A. Malyavko, L. Lin, D. B. Zolotukhin, X. Yao, M. Kirschner, J. Sherman, M. Keidar, A physically triggered cell death via transbarrier cold atmospheric plasma cancer treatment. ACS appl. Mater. interfaces. **12**(31), 34548–34563 (2020)
14. X. Yao, D. Yan, L. Lin, J. Sherman, K. Peters, S. Keir, M. Keidar, Cold plasma discharge tube enhances anti-tumoral efficacy of temozolomide. ACS Appl. Bio Mater. **5**(4), 1610–1623 (2022)
15. D. Trachootham, J. Alexandre, P. Huang. Nat. Rev. Drug Discov. **8**, 579–591 (2009)
16. E.A. Ratovitski, X. Cheng, D. Yan, J.H. Sherman, J. Canady, B. Trink, M. Keidar, Anti-cancer therapies of 21-st century. Novel approach to treat human cancers using cold atmospheric plasma. Plasma Process. Polym. **11**, 1128–1137 (2014)
17. G. Manda, M.T. Nechifor, T.-M. Neagu, Reactive oxygen species, cancer and anti-cancer therapies. Curr. Chem. Biol. **3**(1), 22–46 (2009)
18. P. Lukes, E. Dolezalova, I. Sisrova, M. Clupek, Plasma Sour. Sci. Technol. **23**, 015019 (2014)
19. N. Kurake, H. Tanaka, K. Ishikawa, T. Kondo, M. Sekine, K. Nakamura, H. Kajiyama, F. Kikkawa, M. Mizuno, M. Hori, Cell survival of glioblastoma grown in medium containing hydrogen peroxide and/or nitrite, or in plasma-activated medium. Arch. Biochem. Biophys. **605**, 102–108 (2016)
20. D. Yan, A. Talbot, N. Nourmohammadi, J. Sherman, X. Cheng, M. Keidar, Toward understanding the selectivanti-cancer capacity of cold atmospheric plasma—A model based on aquaporins. Biointephases **10**, 040801 (2015)
21. D. Yan, H. Xiao, W. Zhu, N. Nourmohammadi, L.G. Zhang, K. Bian, M. Keidar, The role of aquaporins in the anti-glioblastoma capacity of the cold plasma-stimulated medium. J. Phys. D: Appl. Phys. **50**(055401), 2017 (2016)
22. C.M. Doskey, V. Buranasudja, B.A. Wagner, J.G. Wilkes, J. Du, J.J. Cullen, G.R. Buettner, Redox Biol. **10**, 274–284 (2016)
23. M. Vandamme, E. Robert, S. Pesnel, E. Barbosa, S. Dozias, J. Sobilo, S. Lerondel, A. Le Pape, J.M. Pouvesle, Antitumor effect of plasma treatment on U87 glioma xenografts: preliminary results. Plasma Process. Polym. **7**, 264 (2010)
24. M. Vandamme, E. Robert, S. Lerondel, V. Sarron, D. Ries, S. Dozias, J. Sobilo, D. Gosset, C. Kieda, B. Legrain, J.-M. Pouvesle, A. Le Pape, ROS implication in a new antitumor strategy based on non-thermal plasma. Int. J. Cancer **130**, 2185 (2011)
25. M. Keidar, R. Walk, A. Shashurin, P. Srinivasan, A. Sandler, S. Dasgupta, R. Ravi, R. Guerrero-Preston, B. Trink, Cold plasma selectivity and the possibility of a paradigm shift in cancer therapy. Br. J. Cancer **105**, 1295 (2011)
26. Z. Chen, H. Simonyan, X. Cheng, L. Lin, J. Canady, J.H. Sherman, M. Keidar, A novel micro cold atmospheric plasma device for glioblastoma both in vitro and in vivo. Cancers **9**(6), 61 (2017)
27. A. Dubuc, P. Monsarrat, F. Virard, N. Merbahi, J.-P. Sarrette, S. Laurencin-Dalicieux, S. Cousty, Use of cold-atmospheric plasma in oncology: a concise systematic review. Ther. Adv. Med. Oncol. **10**, 1–12 (2018)
28. V. Soni, M. Adhikari, H. Simonyan, L. Lin, J.H. Sherman, C.N. Young, M. Keidar, In vitro and in vivo enhancement of temozolomide effect in human glioblastoma by non-invasive application of cold atmospheric plasma. Cancers **13**(17), 4485 (2021)

29. H.-R. Metelmann, C. Seebauer, V. Miller et al., Clinical experience with cold plasma in the treatment of locally advanced head and neck cancer. Clin Plasma Med **9**, 6–13 (2018)
30. H.-R. Metelmann, T. von Woedtke, K.D. Weltmann, Comprehensive *Clinical Plasma Medicine: cold Physical Plasma for Medical Application* (Springer), pp. 185–195
31. https://www.businesswire.com/news/home/20190731005521/en/USMI-JCRI-ABTS-Receive-FDA-Approval-Conduct-U.S
32. D. Mariotti, Y. Shimizu, T. Sasaki, N. Koshizaki, Method to determine argon metastable number density and plasma electron temperature from spectral emission originating from four 4p argon levels. Appl. Phys. Lett. **89**, 201502 (2006)
33. J.S. Lim, R.H. Kim, Y.J. Hong, P. Lamichhane, B.C. Adhikari, J. Choi, E.H. Choi, Interactions between atmospheric pressure plasma jet and deionized water surface. Results Phys. **19**, 103569 (2020)
34. X. Zhu, Y. Pu, A simple collisional-radiative model for low-temperature argon discharges with pressure ranging from 1 Pa to atmospheric pressure: kinetics of Paschen 1s and 2p levels. J. Phys. D: Appl. Phys. **43**, 015204 (2010)
35. X.-M. Zhu, Y.-K. Pu, Optical emission spectroscopy in low-temperature plasmas containing argon and nitrogen: determination of the electron temperature and density by the line-ratio method. J. Phys. D: Appl. Phys. **43**, 403001
36. X.-M. Zhu, P. Yi-Kang, Determination the electron temperature in inductively coupled nitrogen plasmas by optical emission spectroscopy with molecular kinetic effects. Phys. Plasmas **12**, 103501 (2005)
37. J. T. Gudmundsson, Electron excitation rate coefficients for the nitrogen discharge. Sci. Inst. Univ. Iceland, Reykjavik, Iceland, Tech. Rep. RH-09-2005 (2005)
38. A. Ortiz-Mora, A. Díaz-Soriano, A. Sarsa, M.S. Dimitrijević, C. Yubero, A practical method for plasma diagnosis with Balmer series hydrogen lines. Spectrochim Acta Part B **163**, 105728 (2020)
39. K. Urabe, H. Muneoka, S. Stauss, K. Terashima, Microscopic heterodyne interferometry for determination of electron density in high-pressure microplasma. Plasma Sour. Sci. Technol. **23**(6), 064007 (2014)
40. K. Urabe, H. Muneoka, S. Stauss, K. Terashima, Development of near-infrared laser heterodyne interferometry for diagnostics of electron and gas number densities in microplasmas. Appl. Phys. Exp. **6**, 126101 (2013)
41. J.C. Rost, E.M. Davis, A. Marinoni, M. Porkolab, M.A.V. Zeeland, A combined phase contrast imaging and heterodyne interferometer for multiscale fluctuation measurements in tokamak plasmas. J. Inst. **14**, C12023 (2019)
42. C.W. Allen, A.N. Cox, *Allen's Astrophysical Quantities* (Springer Science & Business Media, 2001)
43. K.H. Becker, U. Kogelschatz, K.H. Schoenbach, R.J. Barker, *Non-Equilibrium Air Plasmas at Atmospheric Pressure* (CRC Press, 2004)
44. Y.J. Hong, C.J. Nam, K.B. Song, G.S. Cho, H.S. Uhm, D.I. Choi, E.H. Choi, Measurement of hydroxyl radical density generated from the atmospheric pressure bioplasma jet. J. Instrum. **7**(03), C03046 (2012)
45. Y.H. Kim, Y.J. Hong, K.Y. Baik, G.C. Kwon, J.J. Choi, G.S. Cho, H.S. Uhm, D.Y. Kim, E.H. Choi, Measurement of reactive hydroxyl radical species inside the biosolutions during non-thermal atmospheric pressure plasma jet bombardment onto the solution. Plasma Chem. Plasma Process. **34**(3), 457–472 (2014)
46. H.P. Dorn, R. Neuroth, A. Hofzumahaus, Investigation of OH absorption cross sections of rotational transitions in the band under atmospheric conditions: implications for tropospheric long-path absorption measurements. J. Geophys. Res.: Atmosp. **100**(D4), 7397–7409 (1995)
47. G. Berden, R. Engeln, *Cavity Ring-Down Spectroscopy Techniques and Applications* (Willy, UK 2009), p. 59
48. S. Iseni, S. Reuter, K.-D. Weltmann, NO_2 dynamics of an Ar/Air plasma jet investigated by in situ quantum cascade laser spectroscopy at atmospheric pressure. J. Phys. D: Appl. Phys **47**, 075203 (2014)

49. J.P. Burrows, A. Dehn, B. Deters, S. Himmelmann, A. Richter, S. Voigt, J. Orphal, Atmospheric remote-sensing reference data from gome: Part I. temperature-dependent absorption cross-sections of NO_2 in the 231–794 nm range. J. Quant. Spectrosc. Radiat. Transf. **60**(6), 1025 (1998)
50. Y.J. Hong, J. Lim, J.S. Choi, K.D. Weltmann, E.H. Choi, Measurement of nitrogen dioxide and nitric oxide densities by using CEAS (cavity-enhanced absorption spectroscopy) in nonthermal atmospheric pressure air plasma. Plasma Process. Polym. **18**(1), 2000168 (2021)
51. M. Keidar, D. Yan, J.H. Sherman, *Cold Plasma Cancer Therapy* (Morgan & Claypool Publishers, 2019), pp. 53–73
52. H. Ahn, K. Kim, N. Hoan, C. Kim, E. Moon, K. Choi, S. Yang, J. Lee, PLoS ONE **9**, e86173 (2014)
53. D. Trachootham, J. Alexandre, P. Huang, Nat. Rev. Drug Discov. **8**, 579–591 (2009)
54. A. Lofthus, P.H. Krupenie, The spectrum of molecular nitrogen. J. Phys. Chem. Ref. Data **6**(1), 113–307 (1977)
55. P.J. Ryan, J.W. Bradley, M.D. Bowden, Comparison of Langmuir probe and laser Thomson scattering for plasma density and electron temperature measurements in HiPIMS plasma. Phys Plasmas **26**, 040702 (2019)
56. T. Akiyama, M.A. Van Zeeland, R.L. Boivin, T.N. Carlstrom, J.A. Chavez, C.M. Muscatello, R.C. O'Neill, J. Vasquez, M. Watkins, W. Martin, A. Colio, D.K. Finkenthal, D.L. Brower, J. Chen, W.X. Ding, M. Perry, Bench testing of a heterodyne CO_2 laser dispersion interferometer for high temporal resolution plasma density measurements. Rev. Sci. Instrum. **27**, 123502 (2016)
57. B. Seo, P.M. Bellan, Spatially translatable optical fiber-coupled heterodyne interferometer. Rev. Sci. Instrum. **88**, 123504 (2017)
58. J.S. Lim, Y.J. Hong, B. Ghimire, J. Choi, S. Mumtaz, E.H. Choi, Measurement of electron density in transient spark discharge by simple interferometry. Results Phys. **20**, 103693 (2021)
59. D. Yan, A. Talbot, N. Nourmohammadi, X. Cheng, J. Canady, J. Sherman, M. Keidar, Principles of using cold atmospheric plasma stimulated media for cancer treatment. Sci. Rep. **5**, 18339 (2015)
60. M. Ishaq, M.D.M. Evans, K.K. Ostrikov, Biochimica et Biophysica Acta (BBA)-molecular cell research **1843**(12), 2827–2837 (2014)
61. F. Leipold, R.H. Stark, A. El-Habachi, K.H. Schoenbach, Electron density measurements in an atmospheric pressure air plasma by means of infrared heterodyne interferometry. J. Phys. D Appl. Phys. **33**, 2268–2273 (2000)

Chapter 2
Cancer Treatment and Immunomodulation by Nonthermal Plasma Technology

Nagendra Kumar Kaushik, Neha Kaushik, and Eun Ha Choi

Abstract Plasma has been broadly developed as an encouraging safe method for cancer treatment and immune modulations. The selectivity of plasma regarding cancer cells in comparison with their normal counterparts has attracted researchers as a novel cancer treatment method. In plasma bioscience and medicines areas, both direct plasma treatment and plasma-treated liquids played a significant role in cancer treatment and immunomodulation strategies. In this chapter, cancer cells' redox imbalance and immune activation (activation of immune cells or immunogenic cell death induction) using plasma devices and plasma-treated liquids has been discussed. It has been also shown that plasma-induced damage-associated molecular patterns or antigens are linked with the secretion of various cytokines/chemokines for the enhancement of immune response against cancers. These plasma-based immunogenic strategies lead to immune cell stimulation which can build advanced future technologies to develop plasma based vaccines preparation as well as immune checkpoint blockade and cell-based therapies.

2.1 Introduction

Plasma is ionized gas and also known as the fourth state of matter. New technologies related to the non-thermal or cold atmospheric plasmas generation have been currently established and implemented in biology and medicine. In general, two main strategies have been broadly used to produce plasma, named indirect and direct plasma discharges. In case of the indirect discharge method, the plasma reactive

N. K. Kaushik (✉) · E. H. Choi (✉)
Plasma Bioscience Research Center, Department of Electrical and Biological Physics, Kwangwoon University, Seoul 01897, Korea
e-mail: kaushik.nagendra@kw.ac.kr

E. H. Choi
e-mail: ehchoi@kw.ac.kr

N. Kaushik
Department of Biotechnology, College of Engineering, The University of Suwon, Hwaseong-Si 18323, Korea

E. H. Choi (ed.), *Plasma Biosciences and Medicine*, Topics in Applied Physics 148,
https://doi.org/10.1007/978-981-19-7935-4_2

species are delivered through gas flow against the main arc discharge whereas, in the case of the direct discharge method, living cells or tissue is another. On these basic principles, mainly two main plasma sources, Jet plasma [33, 49, 98] and other the dielectric barrier discharge (DBD) [97], had been established that were utilized in the field of biology and medicine majorly [21, 22, 32, 34, 39]. Several reactive species such as nitrogen and oxygen-based radicals, along with other constituents are produced during plasma exposure[18, 35, 49]. This complex chemistry results in typical interaction among plasma and biological tissues or cells [24, 48]. Several literature suggests that plasma could be applied clinically in wound repair [4, 8, 27, 30, 47, 82, 100], in blood coagulation [29, 34, 67], as well as cancer cells treatments [21, 40, 42, 81, 88, 89] (Fig. 2.1).

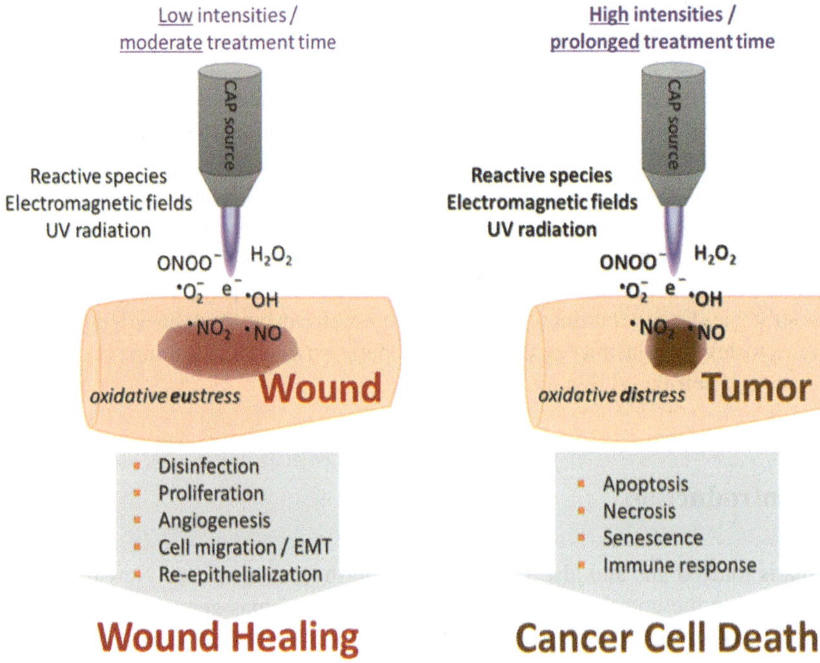

Fig. 2.1 A pictorial description of the application of plasma in wound healing and anticancer approach. Disinfection, wound healing, and tissue repair is triggered via moderate doses of plasma treatment (left panel) whereas prolonged plasma exposure induces oxidative stress for cancer therapy [11]

There was remarkable developments in treatment procedures existing for cancer therapies leading to enhanced patient survival and even remedy for some of the cancers[63, 71]. Radiation therapy is local but not specific with side effects to normal counterparts during therapy. Surgical treatments are also local and specific where only damaged cells are treatment without failure to remove all tumor cells. In the case of chemotherapy, these are systematic and nonspecific having toxicity on non-cancerous cells [63]. Some pieces of evidence suggest when chemotherapy is combined with other locally available treatments such as radiotherapy and surgery, their effectiveness is further enriched [99, 112]. In this case, normal cells were damaged, however, it could repair cellular damage and eventually recover letting the patient stay with the treatment. Besides these conventional treatment approaches, it is discovered that the use of immunotherapy promotes certain anti-cancer effects and produces memory T cells for prolonged safety [65]. Currently, cancer immunotherapy is a widely accepted potential treatment strategy where the human immune system has been employed to aim particularly the cancers without harming normal cells [64]. The immunity plays a key role in carcinogenesis as well as control of cancer where genetically mutated cells were identified and destroyed by inflammatory immune cells. During this process, some cells were escaped due to their low level of immunogenicity, therefore leading to unchecked growth with significant mutations, and forming cancer cells. In this way, they circumvent immune observation and actively provoke factors for immune suppression [65]. The production of immunity in cancer could be self-propagating process. This cyclic process resulting in an aggregation of immune-modulatory factors to improve immune cell functions. The immune system cycle could be defined through suppressive factors which promote immunity regulative mechanisms, that may interrupt the progress or restrict the immunity. This immunity cycle could be split into various steps, starting from the cancer-released antigen

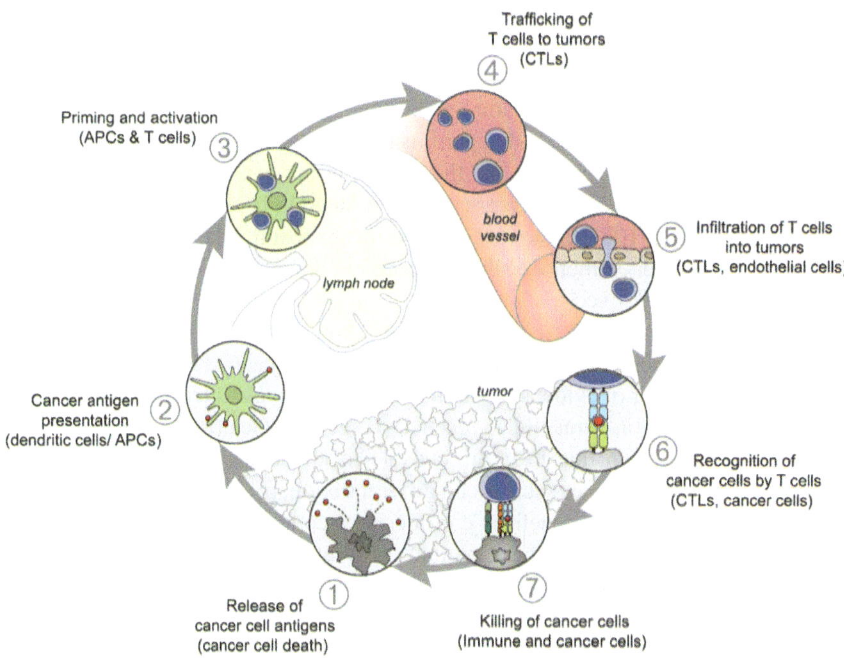

Fig. 2.2 The major steps involved in the cancer based immunity cycle [16]. APCs are antigen-presenting cells whereas CTLs are cytotoxic T lymphocytes

process and closing with the death of tumor cells. Every step has been described in detail in Fig. 2.2.

2.2 Plasma-Induced Anticancer Effects and Signaling Mechanism

As mentioned above, plasma has been broadly developed as an encouraging safe method for anti-cancer treatment. Selectivity of plasma regarding tumor cells in comparison with their normal counterparts has been attracted researchers as a novel cancer treatment. For plasma-based treatments purposes, two fundamental schemes have been established. One is to apply the jet plasma [110] or DBD [21] plasma to expose the cells directly seeded in a culture plate/dishes or the mice xenograft tumors (Fig. 2.3a). The second strategy is to apply the plasma activated liquids (PAL) or solutions (PAS), mostly the plasma-activated medium (PAM) to decrease the cancer cells growth [87, 104] or to impede the development of cancers via applying PAM into the mice cancer mass [94] (Fig. 2.3b). PAS is commonly prepared through treatment of plasma including DBD or jet sources directly to the biocompatible liquids for example PBS, medium, including Ringer's solution [46, 106]. Such type of approach is entirely based upon the reactive species and their stable products made during the reaction amid whole plasma and the unique constituents present in liquids for example amino acids (inside medium) and other chemical entities (in physiological solution) [9, 10, 80, 103]. Importantly PAS or PAL could be stored below certain settings for a long time [2, 101, 102]. This feature makes the plasma appealing as a unique key aspect for the indirect treatment, that permits the use of plasma as a pharmacological means. Most of the studies exploited the direct plasma exposure method, however over the last four years, the indirect approach is steadily becoming an eye-catching topic [1, 12, 68, 101, 103]. In the case of the direct exposure, plasma-related reactive species, and the other elements could influence malignant cells while in the indirect exposure, mostly long-lived reactive species are important to study. Hence, both short or long lived species or other plasma elements should be considered for the robust cancer treatment ability regarding direct plasma exposure.

A growing number of evidence exhibits that plasma could alter the cellular redox status via activation of intracellular reactive oxygen species (ROS) formation [61, 86]. It is widely known that ROS are capable to affect numerous signaling trails controlling several processes such as cell differentiation, proliferation, and cell death [79]. Since plasma has the benefit of considering a controlled precursor of accumulative ROS, therefore it could be exploited for treating several diseases, involving cancer. In cancers, the extent of lipid is frequently reduced in comparison to normal cells making them additional susceptible towards oxidative stress [70, 76, 107]. ROS are barely produced through external tools such as plasma although are also produced by normal by-products during cell metabolism. If the intracellular oxidative stress surpasses the quantity which can be controlled by the cellular antioxidant

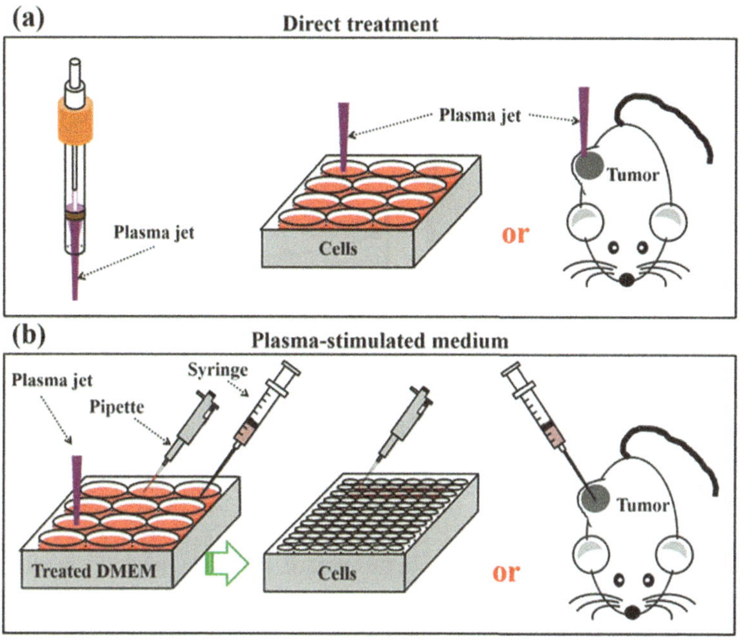

Fig. 2.3 Two prime strategies for plasma application on cancer cells. **a** In vitro or in vivo Direct exposure of plasma on cancer cells or tissue. **b** Indirect exposure of plasma on solutions mostly medium which would be applied to cells seeded in cell culture plate or the cancer tissues [105]

defense system programmed cell death phenomenon will be stimulated by a signaling cascade [92]. Concerning the function of the antioxidant defense system including glutathione, catalases, superoxide dismutases, in counteracting the apoptosis induction this system offers alternative means that could be distinctive among cancer tissue and normal counterpart which might be the consequence of plasma selectivity [6]. Furthermore, researchers also proposed that the alteration in the antioxidant machinery to counteract oxidative stress is accountable for the increase in reactive species and cancer inhibition. It is displayed that leftover ROS and long lived stable species, those are not neutralized through antioxidants in the mitochondria, could damage mitochondria and ultimately trigger cellular damage [38]. It is also claimed that due to the ROS induction in medium after plasma treatment can provoke cell death in leukemic cells when cultured in hypoxic environments, which shows a crucial part in chemotherapy resistance [93]. Understanding the molecular pathways and cellular communications is the furthermost basic study in plasma-based cancer therapy. One important aspect is that plasma indirectly and directly interacts with various factors or molecules present on cell surfaces or inside cells together with signaling pathways (Fig. 2.4).

Regarding the selectivity of plasma for cancerous cells, the effect of cholesterol, aquaporins, and the anti-oxidant machinery on the efficacy of plasma has been widely discussed. Nevertheless, the mechanisms that eventually induce to plasma-activated

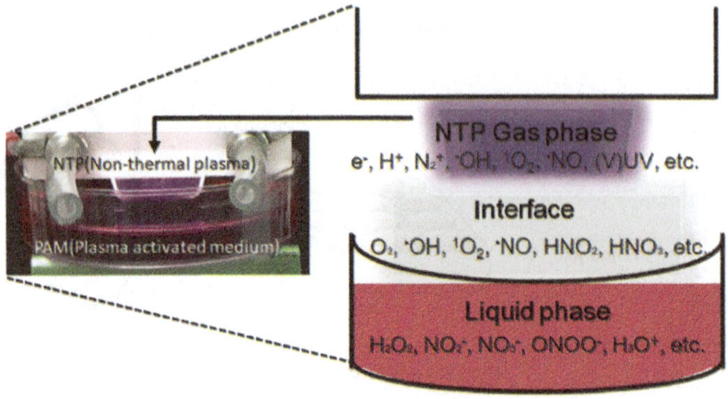

Fig. 2.4 Interactions among plasma and liquids in plasma treated medium. Atomic species combines with nitrogen, oxygen, and water in ambient environment or at the interface of plasma and liquid to generate reactive species. More stable molecules, like nitrates, nitrites, and hydrogen peroxide (H_2O_2), generated in the interface and released into treated medium [90]

cancer inhibition was a big question. In this regard, the exact mechanism after treatments has been investigated. Plasma can alter various intracellular signaling routes which in turn regulate the cellular fate and might elicit cellular death. As a significance of plasma exposure, apoptosis or necrosis could be persuaded however the autophagy, as well as senescence induction, have been detected. These induced processes seem to be plasma-given dose-dependent. For cellular death, it has been proved that plasma could induce detachment of cancer cells [42], cell cycle growth arrest [43, 96], apoptosis by DNA damage[13, 28, 44, 73, 87, 111], and micronucleus [39] in cancer cells. The plasma-induced cellular detachment was initiated through confronting cell-adhesion molecules, for example, integrin [26, 52, 84]. ROS and RNS) are observed in cells [24, 66, 95, 108, 111], and it is assumed that intracellular reactive species cause several physiological outputs. A very instant effect of reactive species is lipid peroxidation on the cellular membrane (Fig. 2.5). This clues to an improved influx of ROS into the cytoplasm. These species further can react with the other factor and molecules which affect a range of activities inside the cell. One of the essential messenger intricated in extra-or intracellular signaling pathways is calcium (Ca^{2+}) which has a important part in the fate of cells. Literature proposed that Ca^{2+} and ROS signaling share a close interaction [23]. Likewise, in these interactions, a rise in reactive species also directly connected with the initiation of DNA damage. These lesions comprise oxidative damage, double-strand DNA breaks [5, 75], in addition to crosslinks amongst proteins and DNA [25]. Moreover, ATM activation intricated in apoptosis signaling has been witnessed in oral carcinoma and skin cancer cells [7, 31]. Notably, senescence, a known irreversible cell growth arrest in respect to oxidative stress [77], could be triggered by short plasma exposure while necrosis and apoptosis are provoked through longer exposure times. Such as skin cancer cells exposed with higher plasma doses by a floating type DBD plasma device killed via

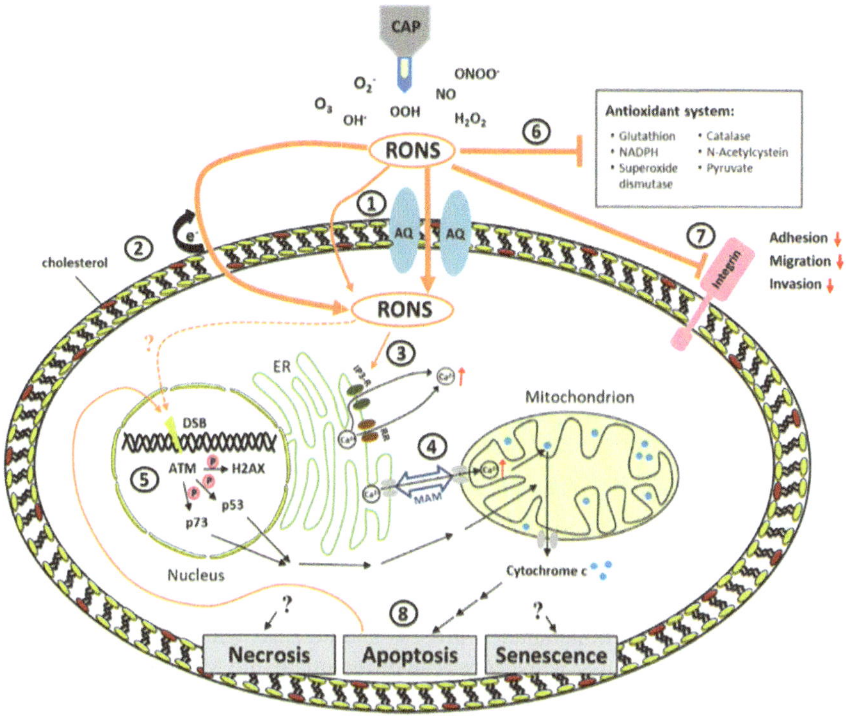

Fig. 2.5 Overview of molecular signaling mechanisms and interactions for the plasma oncology [83]

necrosis and apoptosis phenomenon [21]. Intriguingly, this study shows that even the higher doses did not show any damage to healthy normal tissues [20].

This literature provides the knowledge that the plasma-generated reactive species has the potential to show a variety of biological outcomes with applicability to cancer biology, along with the selective inhibition of cancer cells [45]. For example, regarding tumor ablation, plasma technology could be applied to sterilize the tumor site to promote original normal tissue reformation. The benefit of plasma exposure might be attained by treating cancers that were not removed well during ablation, therefore effectively reducing the tumor relapse or regrowth. Medical professionals assume that plasma exposure directly on the tumor area is exceptionally reasonable towards primary tumor treatment and can be recognized as adjuvant therapy [109].

2.2.1 Plasma-Based Activation of Immune Cells

The cancer microenvironment plays important role in cancer growth and especially in metastatic process [19]. Immunomodulation is a assuring approach that

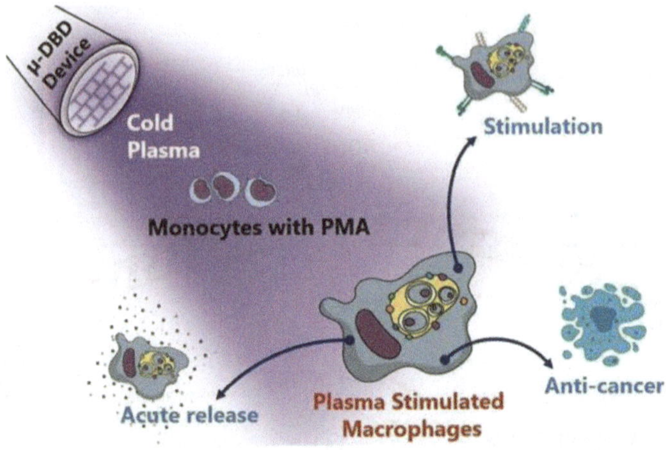

Fig. 2.6 Non-thermal μ-DBD plasma-based activation of the immune cell against cancer [36]

involves the body immunity to affect cancer cells specifically without affecting normal cells or tissue. Immune cells such as dendritic cells or macrophages are very important for the inflammation process and defense against cancers. Though the macrophage-associated with cancer can differentiate into M1 (cancer-killing) or M2 (angiogenic) phenotypes depending on the cancer environment [54]. Targeting immune cells such as dendritic cells, macrophages, lymphocytes, and other immune factors for cancer treatment could be a smart approach to improve current cancer therapy. In recent times, a dielectric barrier discharge and jet plasma devices were applied for studying immuno-stimulatory effects by the plasma medicine community. A recent study showed that plasma treatment promotes the differentiation and activation of M1 macrophages to a significantly levels (Fig. 2.6). Plasma activated macrophages promote cancer inhibitory immune reactions against cancer progression. These modulation also affect metastasis and cancer stemness maintenance in vitro conditions [36].

Also, a microwave plasma system produces nitric oxide in presence of nitrogen and oxygen gas at high temperatures called gaseous nitric oxide (NO). The generated nitric oxide is exposed to water for making NO plasma-activated water (NO-PAW). The role of macrophages or dentritic cells are facilitated by several factors. It is known that nitric oxide (NO) is a key factor for macrophage activation, NO-PAW is proposed as a possible for activation of TAMs (Fig. 2.7). Investigation showed that NO-PAW enhances M1 tumor-killing macrophage as compared to M2 macrophage in cancer co-culture conditions. Characteristic translational and transcriptional level related markers were examined to show molecular changes after exposure to macrophages. In addition, it is shown that NO-PAW upregulated the cancer-killing potential of macrophages. In conclusion, investigation concluded that NO-PAW stimulates macrophages in the cancer microenvironment to stop cancer progression.

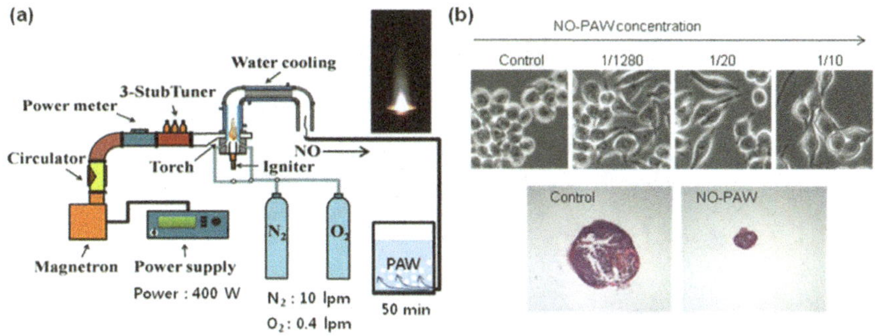

Fig. 2.7 **a**Nitric Oxide Plasma Activated Water (NO-PAW) was produced using a microwave-plasma device with nitrogen and oxygen gases. **b** NO water activated macrophages have a high potential to inhibit skin cancer cell growth [50]

In another investigation, [56] the effects of NO-PAW on cancers and compared them with other plasma devices. The previous study [51] was focused on the effect of plasma based NO water on macrophage differentiation. That study concluded role of NO water to polarize macrophages to M1 or M2 at the molecular level. Further, the cancer inhibitory effect of NO water was also checked in a in vivo mouse model (Fig. 2.8). The outcome indicated that NO water could affect macrophage fate, proposing this strategy as a supportive method for regulation of the task of immune cells in the tumor environment.

Previously plasma researchers also stated that plasma exposure stimulates blood monocyte cells to differentiate into more M1 macrophages. These monocytes differentiated macrophages thus inhibits the progression of cancers in the co-culture system via various immune mediated pathways and signaling (Fig. 2.9). Furthermore, plasma treatment promotes immune cell health as seen in plasma treated monocytes-macrophages study. These important outcomes show that ROS and RNS generated by plasma could stimulate immune cells to kill cancers [37].

Other studies also supported the treatment of macrophages with other plasma devices such as pulsed DBD plasma device to kill specifically cancer cells without affecting normal cells (Fig. 2.10). These outcomes emphasize the medical importance of plasma devices or plasma-based products for tumor immunotherapy. There was a negligible effect of stimulated macrophages on the lung normal cells, indicating that macrophages sustain their selective killing effect towards tumor cells when activated with exposure of plasma. The distinct results on different type of cells may be ascribed in part to released cytokines responses. TNF-α released by macrophages is strongly toxic toward cancer cells however induces normal Beas2B cells to release IL-6 and IL-8 release [85]. These cytokines (IL-6 and IL-8) enhance cell proliferation of normal cell lines [55] for direct reinforcement of immune cells while preserving selectivity function [58].

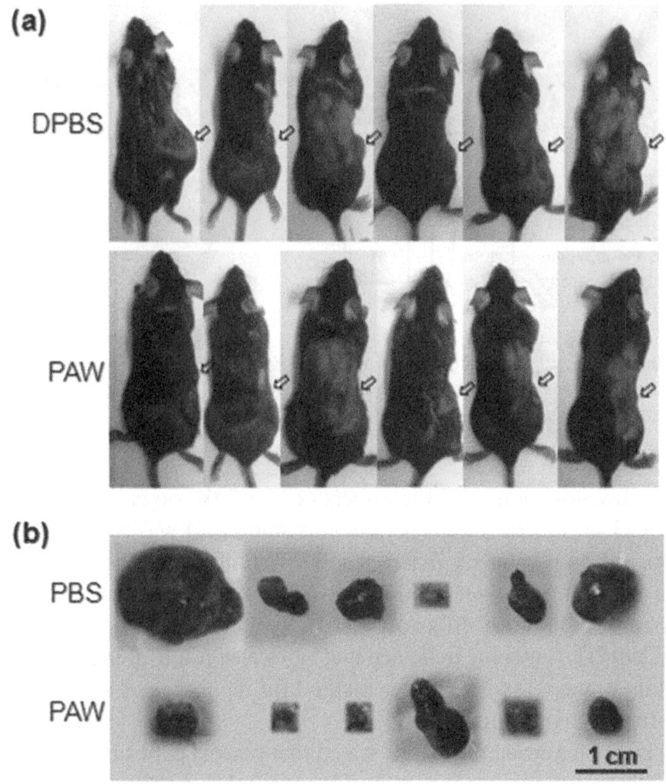

(c)

V (cm^3)	1	2	3	4	5	6	mean±std
PBS	5.76	0.28	0.71	0.01	0.34	1.28	1.39±1.70
PAW	0.27	0.01	0.06	0.69	0.12	0.13	0.37±0.42

Fig. 2.8 Anticancer role of NO water in vivo condition. **a** Syngeneic mouse with skin cancer post 12 days of PAW or PBS administration. **b** Images of tumor tissue mass were taken from each mouse. **c** Details on the volumes of cancer tissue from the of different groups [51]

2.3 Plasma-Based Immunogenic Effect

Previously, most cancer therapeutic strategies were aimed on decreasing cancer load via the application of toxic physical agents or drugs. Though, all these kinds of strategies do not support immune system-based responses to treat cancers. From the last decade, immunity based cancer therapy has accepted the application of stress based strategies such as immunogenic cell death (ICD) and release of damage-associated molecular patterns (DAMPs) in cancers. Recently emerging treatment strategies

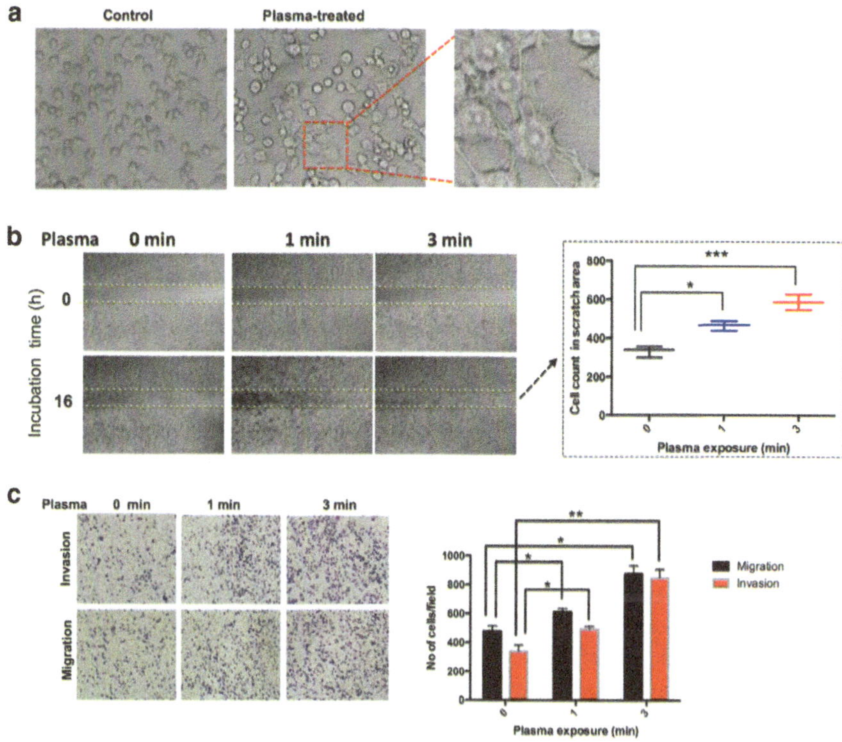

Fig. 2.9 Morphological analysis of plasma-treated RAW macrophages. **a** Microscopic images of macrophages after plasma exposure, **b** a wound healing scratch assay was executed to check movement of macrophage cells after 16 h post-incubation, **c** migration and invasion capacity plasma exposed macrophages detected in a transwell culture system [37].

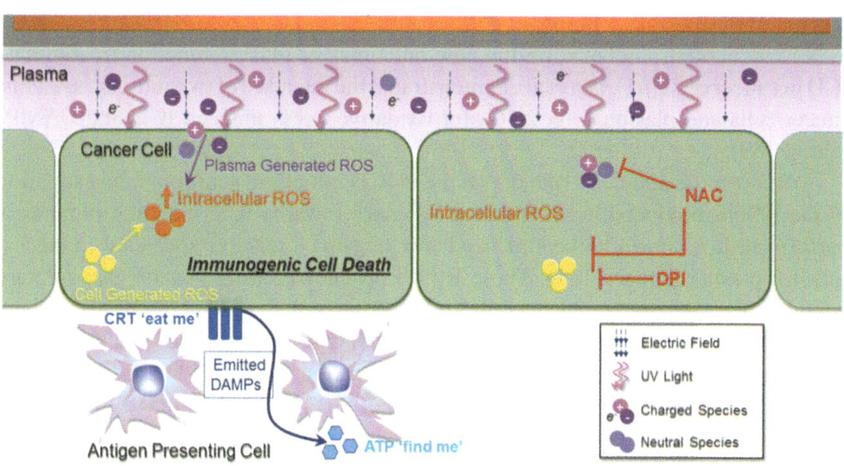

Fig. 2.10 Plasma interaction with cancer cells to induce ICD and DAMPs release [59]

Table 2.1 Important DAMP molecules to enhance immunogenicity of cancers

DAMP	Immunomodulatory role
ATP	Activation of antigen-presenting cells and Act as a 'Find Me' signal
CRT	Increase immunogenicity of cancer cells and act as 'Eat Me' signal
HSP	Heat shock protein attract immune cells and induce activation
HMGB1	Attract immune cells and induce antigen-presenting cells maturation
ILs	Strong pro-inflammatory activity and activation of immune cells
TLRs	Inflammation and recruit and activate cells of immune system

that induce ICD via oxidative stress, suggest the potential to improve outcomes of conventional cancer treatment methods. Therefore cancer-inhibiting activity of nonthermal plasma that is mainly via RONS that are either induced or transported to the cancer cells, is a very practical approach that cold plasma could induce ICD and DAMPs release. The advantage of plasma based treatment is rely on its capability to enhance communication between plasma-treated cancer cells and immune system for its continuous defensive mode against that cancer. Recently, plasma is efficient induced of ICD in melanoma cells [57] in cell culture condition and colon cancers in mice models [60], where hydroxyl radicals and nitric oxide short lived species were found to be effective factors. This investigation offers the approval that nonthermal plasma has the ability for immunomodulation's against cancers which can be employed to medical settings. All these investigations discover that plasma use could be supportive to strategies focusing against resistant cancers through changing the cancer favoring environment. It suggests that plasma can be tuned to induce more ICD and DAMPs, raising their visibility to the human body defense system. Early investigation showed that certain regimes such as reactive oxygen and charged species of plasmas induce release of ATP and enhance presentation of CRT on the cell membrane (Table 2.1) [65].

Researchers also debated on the new outcomes of plasma function to stimulate ICD in cancer cells [41]. Lin et al., demonstrated the interface of two intricate systems, cancer cells and plasma (Fig. 2.10), for oxidative stress induced ICD and DAMPs release [59].

To know more about the role of plasma in ICD initiation, a thorough investigation of the RONS produced by the plasma exposure towards ICD inducing regimes is performed. It is shown both short lived and long-lived reactive species were necessary for plasma induced ICD. These RONS produced during plasma exposure are completely related to the plasma doses. On the basis on these outcomes, researchers can optimize specific parameters of plasma that can offer medical practitioners to make a tool to control RONS for specific application such as ICD-based treatment of cancers at clinical settings, as shown in Fig. 2.11 [57]. Researchers found that plasma-generated oxygen species and charges are the key component for ICD induction and DAMPs release after eliminating various other components using gas or physical barriers (Fig. 2.12). Several important DAMPs, such as ATP release, and

CRT on cell membranes were studied to evaluate role of plasma to enhance immune response against cancers [59].

Also, in previous investigations, various anti-cancer treatment strategies proposed to modulate immune system [17] using various immune factors such as cytokines or chemokines, as well as immune checkpoint blockade (ICB), and cell-based therapies. It is proposed that ICB therapy could be the best tool to improve treatment efficacy and reduce side effects. To investigate this standpoint, Guojun et al. showed strategies using transdermal plasma for the best ICB based treatment strategy. Tumor associated antigen or DAMPs can be generated after release of RONS and molecules for ICB in the targeted tissue via hollow shaped microneedles for the better delivery and implicating improved immunotherapy using plasma. The combined effect of plasma generated RONS and ICB therapy with microneedles suggests a novel system for treatment against various dreadful diseases [15]. To support this proposal recent study

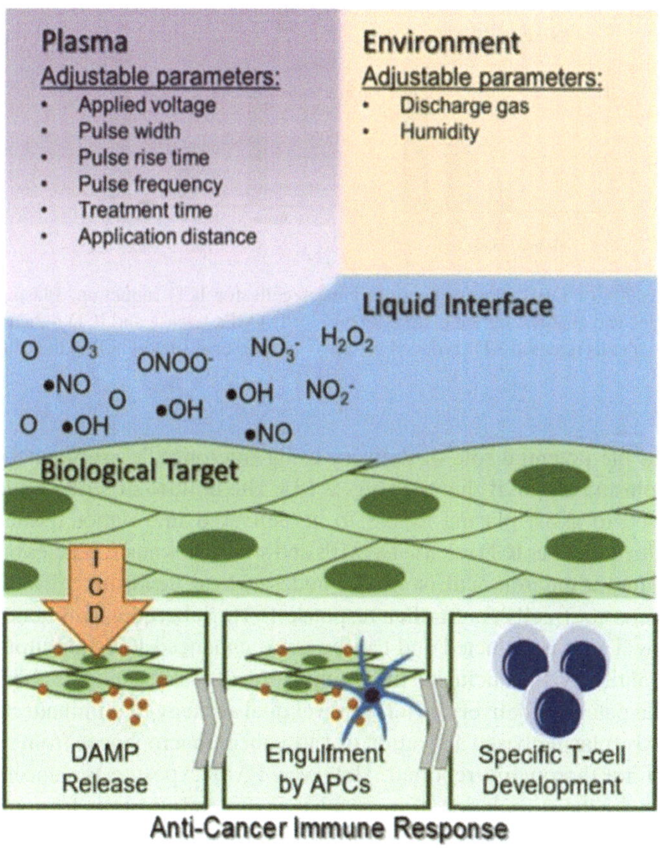

Fig. 2.11 Mechanism of nonthermal plasma interaction with cancer cells related to immunotherapy. These outcomes would offer strategic development of plasma-based medical devices for the coordinated transport of species to induce ICD [57]

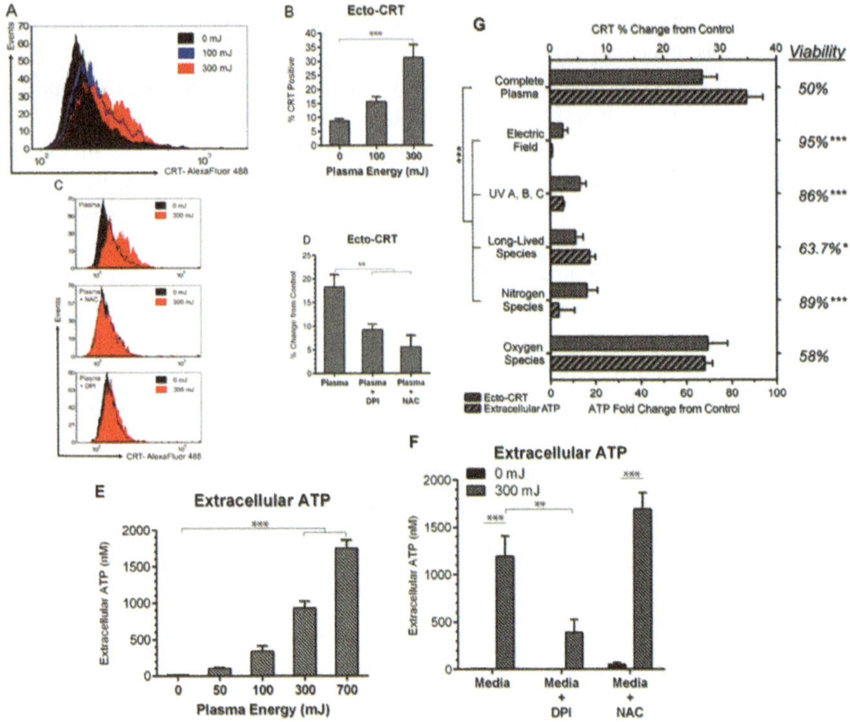

Fig. 2.12 Results of DBD plasma exposed cancer cells for ICD induction. Plasma-generated oxygen species and charges are the main effectors for DAMPs release and ICD induction. Plasma exposed cancer cells released ATP, reduced viability, and presented ecto-CRT on the cell membrane [59]

investigated the potential role of delivery using microneedle array patch [74] that combines plasma and ICB therapy (Fig. 2.13). The hollow-structured microneedle patch is used to assist plasma RONS to be delivered into tumor tissue via skin. Plasma-induced cancer ICD release DAMPs and stimulates macrophages or dendritic cell in the tumor environment or lymph nodes, where dendritic cells can present these antigens to T-cells for further response [53]. Subsequent immune response mediated by T cell is inducted and can be more enhanced ICB inhibitors, such as anti-PDL1 antibody. In conclusion the combination of plasma and ICB inhibitor with microneedle patch for delivery, offers a novel dual strategy to eliminate cancers.

Previously, plasma-based activation of anti-cancer macrophages from monocytes stimulation are thoroughly reported. However, PAM exposure to cancers still not investigated for the stimulation of macrophages or dendritic cells from monocytes. In the recent investigation, the lysate obtained from cancer cells treated with PAM is the better for stimulating dendritic cells rather than cancer lysate obtained after any other conventional procedure. PAM-A375lys-treated monocytes derived dendritic cells were highly effective in comparatively reducing the Th2 cytokine and enhancing

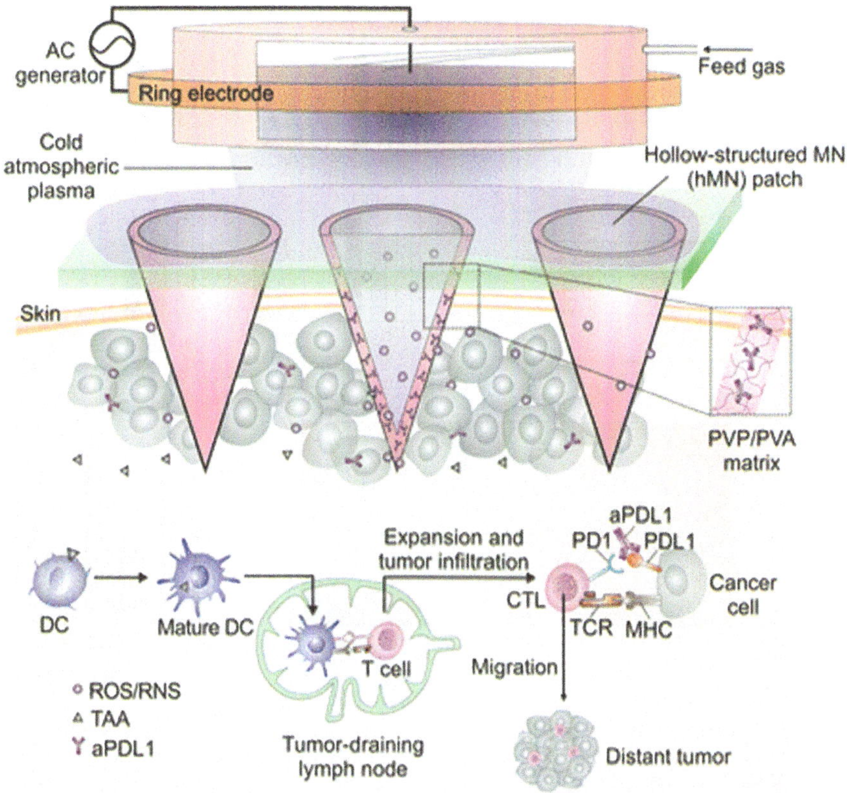

Fig. 2.13 Image of the transdermal plasma and ICB therapy with anti-PDL1 using microneedle patch [15]

the Th1 and Th17 cytokines in cell culture, validating treated media exposed cancer lysates is better for improving cancer inhibitory properties of dendritic cells. As per hypothesis, Th2 cells showed tumor favoring effects in a in vivo cancer model, but Th1 and Th17 cells exhibited the anti-tumorigenic effects. Dendritic cells treated with lysate of activated media exposed A375 cells also presented an elevated capability to promote proliferation of T cells than to unexposed counterpart (Fig. 2.14). These findings are accordance with the elevated interleukin 2 levels identified in co-cultures of stimulated dendritic cells and T cells. Differentiation and maturation of stimulated dendritic cells using activated media treated cancer cell lysate, and more IL-12 release by these cells, are crucial factors for the stimulation of T cells [91].

In another study plasma irradiation enhanced immune infiltration in the cancer microenvironment. The investigation of CD4, CRT, FOXP3, CD8, CD11c, and IL-17 was checked for all cells, due to technical limitations their individual viabilities were not taken into account. Altogether, assessment of the cancer environment showed a

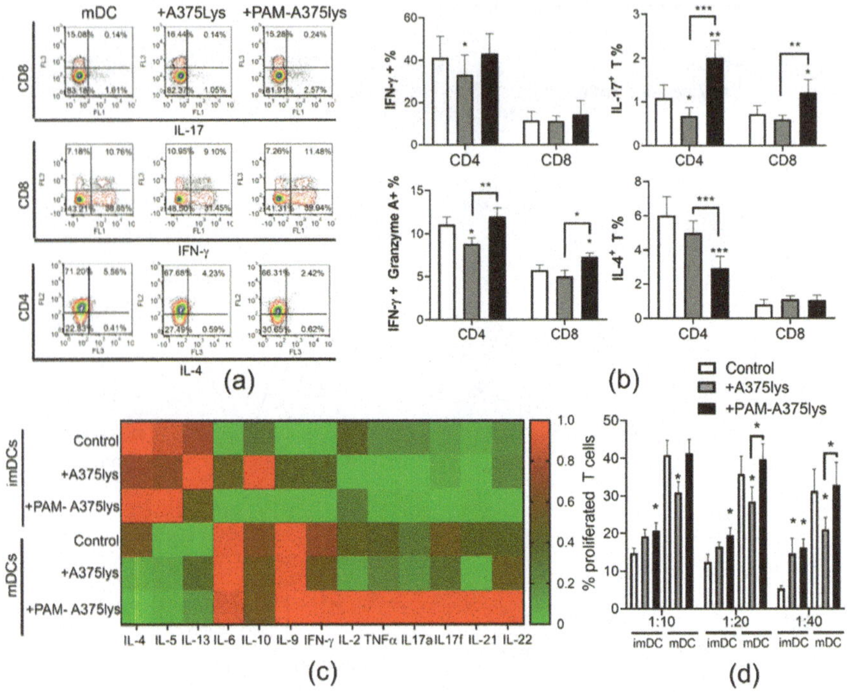

Fig. 2.14 The outcome of T cell polarization capacity dendritic cells treated with cancer cell lysates. **a** Dendritic cells, either exposed or unexposed with the untreated (+A375lys) or PAM treated A375 cancer cells lysate (+PAM-A375lys) co-cultured with T cells (1 × 10^5/well) for 5 days. **b** The percentage of cytokine/enzyme expressing immune cells. **c** cytokines levels from DC/T cell cultures detected in the supernatants. **d** The proliferation of T cells investigated after 5-day co-culture of stimulated dendritic cells and T cells [91]

substantial increase in cancer cell death, which was associated with the increase in immunogenicity and infiltration of immune cells after treatment with plasma [62].

Likewise, an immunomodulatory role of plasma treatment together with plasma induced ICD, suggested by several investigations, but research using primary immune cells is limited. A recent study showed the plasma exposed mouse skin cancer stimulates primary immune cells and changing its molecular profile. Cancer cells treated with plasma demonstrated diminished viability and motility and enhanced secretion of DAMPs such as ATP and CXCL1 release. This phenomenon induced altered molecular profile of immune factors such as cytokines. Specifically, CCL4 and IL-1β being improved in treated culture and co-cultures with immune cells. Whereas in T cells stimulated via extracellular signal-regulated Kinase phosphorylation and enhanced CD28 expression, after co-culture with cancer cells. Enhanced CD115 expression was the main indication of monocytes stimulation in this investigation. In conclusion, plasma-induced tumor cell death is accepted scientifically, and that plasma-based activation on immune cells. Lysate of plasma treated cancer cells after 24 h post treatment showed considerably enhanced levels of cytokines and chemokines such as CCL4, TNFα, IL-10, and IL-1β. DAMPs, such as CXCL1 and ATP release, were also detected at 6 h after plasma treatment in cancer cells. The response of immune cell recorded after plasma treatment for investigating immunological relevant effects of spleenocytes cultured with plasma-exposed cancers. Similarly, supernatants of spleen cells showed enhanced levels of immune signaling factors such as IL-4, IL-12, IL-10, CCL4, and IL-1β after plasma exposure (Fig. 2.15). In conclusion, plasma exposure affected cell metabolic viability and the inflammatory profile of cancer and immune cells [78].

Herein, a study recently showed ICD induction and DAMPs release from cancer cells even by plasma synthesized gold nanoparticles coated with polydopamine (Au@PDA NPs) treatment. This study demonstrated exceptional specificity towards tumor cells. The functionalized gold nanoparticles were synthesized by a plasma-based green synthesis procedure in short time and minimizing the utilization of hazardous agents. Notably, these gold nanoparticles not only stimulated ICD but also exhibited high cellular internalization in cancer cells. Furthermore, it was detected that danger molecules were secreted by exposed cells simultaneously with the process of autophagy after treatment, which acted as endogenous danger signals regulating subsequent immune response (Fig. 2.16). This investigation emphasizes the mechanism of plasma based functionalized gold nanoparticle triggered cancer

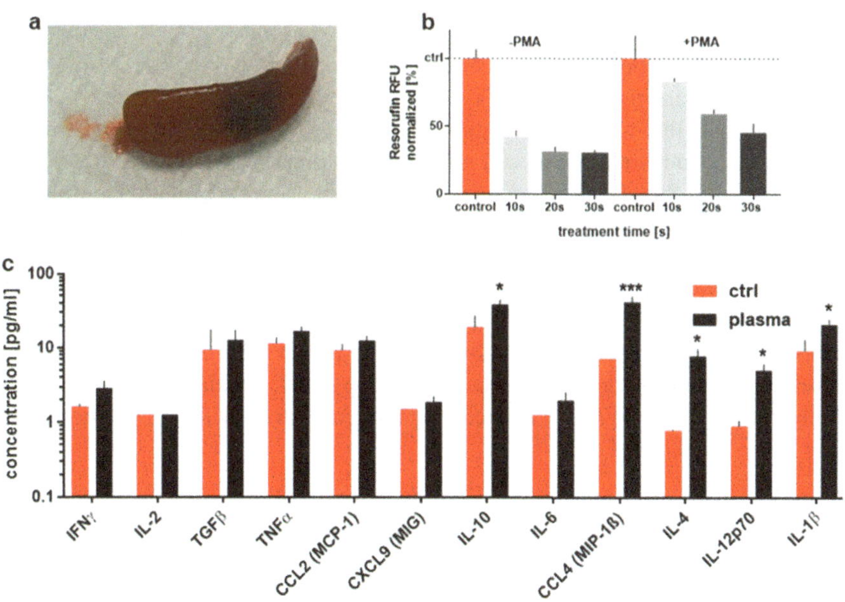

Fig. 2.15 Outcomes of murine immune cells treated by gas plasma. **a** Image of the murine spleen before homogenization; **b** viability of immature (-PMA) and stimulated (+PMA) spleen cells after exposure; **c** chemokine and cytokine concentration after 2 min of plasma exposure in immature spleen cells at 24 h [78]

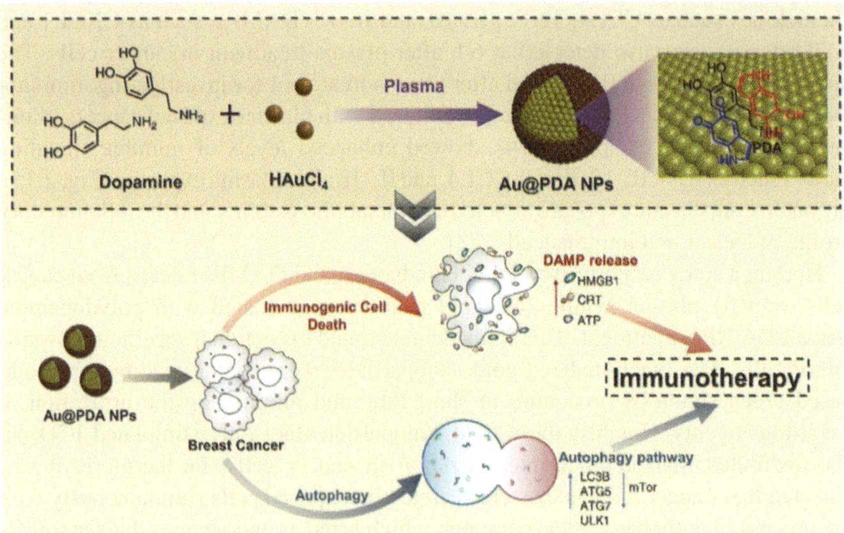

Fig. 2.16 Scheme of plasma-based green one pot preparation of dopamine coated gold nanoparticles and in vitro immunogenic potential, autophagy, and ICD induction studies against breast cancer cells [69]

targeting immunity for resistant cancers, proved the ability of these nanoparticles for immunotherapy against cancers [69].

Nonetheless, this immune-activation or immune-stimulation are of research is very new, and study on immune cell infiltrate in plasma-treated cancers are unique. Having in mind the ability of an human body immune system and green plasma treatment technology, recently several immune therapy-based strategies have been the key plasma-based revolution in this era. The immunomodulation-based concept using plasma or plasma products is established on the fact that ICD is induced, DAMPs are released and cancer targeting immune cells are activated after treatments. Plasma therapy can be peculiar or comparable to the previous treatments for anti-cancer strategies such as electrochemotherapy [14], ionizing radiation [3], and photodynamic therapy [72], ICD becoming a popular strategy for cancer therapy.

2.4 Conclusion and Future Prospective

The role of plasma generated active species and other components are associated with cancer treatment emphasized the RONS functions in plasma bioscience and medicine field. These active species are major players for several intra- and extra-cellular mechanisms or processes. Likewise, the plasma-stimulated liquid is a crucial topic to understand further as gaseous species can interact with liquids or wet tissues in actual biological conditions. Plasma-based treatment involves therapeutic applications of nonthermal plasma in various areas, such as cancers, dental, decontamination, cosmetics, control of multi-drug resistant microbes, viruses, wound healing, and neurodegenerative diseases. It is also determined that plasma can induce ICD in cancers and can produce DAMPs or antigens linked with the secretion of various cytokines/chemokines. These plasma-based immunogenic strategies lead to dendritic cells or macrophage and other immune cells stimulation and have the capability to build an advanced future technology to regulate immune response without significant side effects. The recent comparison showed that plasma has merit in the case of side effects and effects on immune cells compared with other conventional physical therapies (Table 2.2). In the future, the investigation on the synergy between plasma and nanomaterials/drugs against cancers, other conventional treatment, and plasma-based vaccine procedures should be taken into consideration. Further, the future tasks involve the standardization of plasma doses and sources for biological and medical applications.

Acknowledgements This work was funded by the NRF of Korea, funded by the Korean government (2021R1A6A1A03038785, 2021R1F1A1055694, 2021R1C1C1013875).

Table 2.2 Table depicting the comparison between plasma, radiation therapy, and photodynamic therapy [41]

	NTP	RT	PDT
Targeted	Yes	Yes	Yes
May require invasive procedure for application	Yes	No	Yes
Side effects	Unknown no major side effects reported	Skin changes, second cancer, site specific side effects caused by damage to nearby organs	Skin changes
Mechanism of action	Oxidative stress	DNA breakage	Oxidative stress, damage to tumor blood vessels
Depth of effect	Superficial	Deep	Superficial
Causes ICD	Yes	Yes	Yes
Direct effects on immune cells	Preservation/stimulatory	Suppressive	Suppressive

NTP: Non-thermal plasma, RT: radiation therapy, PDT: photodynamic therapy

References

1. T. Adachi, S. Nonomura, M. Horiba, T. Hirayama, T. Kamiya, H. Nagasawa, H. Hara, Iron stimulates plasma-activated medium-induced A549 cell injury. Sci. Rep. **6** (2016)
2. T. Adachi, H. Tanaka, S. Nonomura, H. Hara, S.-I. Kondo, M. Hori, Plasma-activated medium induces A549 cell injury via a spiral apoptotic cascade involving the mitochondrial–nuclear network. Free Radic. Biol. Med. **79**, 28–44 (2015)
3. I. Adkins, J. Fucikova, A.D. Garg, P. Agostinis, R. Špíšek, Physical modalities inducing immunogenic tumor cell death for cancer immunotherapy. OncoImmunology **3** (2014)
4. Y. Akimoto, S. Ikehara, T. Yamaguchi, J. Kim, H. Kawakami, N. Shimizu, M. Hori, H. Sakakita, Y. Ikehara, Galectin expression in healing wounded skin treated with low-temperature plasma: comparison with treatment by electronical coagulation. Arch. Biochem. Biophys. **605**, 86–94 (2016)
5. M.Y. Alkawareek, S.P. Gorman, W.G. Graham, B.F. Gilmore, Potential cellular targets and antibacterial efficacy of atmospheric pressure non-thermal plasma. Int. J. Antimicrob. Agents **43**, 154–160 (2014)
6. A.Y. Andreyev, Y.E. Kushnareva, A.A. Starkov, Mitochondrial metabolism of reactive oxygen species. Biochem. Mosc. **70**, 200–214 (2005)
7. S. Arndt, E. Wacker, Y.-F. Li, T. Shimizu, H.M. Thomas, G.E. Morfill, S. Karrer, J.L. Zimmermann, A.-K. Bosserhoff, Cold atmospheric plasma, a new strategy to induce senescence in melanoma cells. Exp. Dermatol. **22**, 284–289 (2013)
8. P. Awakowicz, N. Bibinov, M. Born, B. Busse, R. Gesche, A. Helmke, A. Kaemling, V. Kolb-Bachofen, R. Kovacs, S. Kuehn et al., Biological stimulation of the human skin applying healthpromoting light and plasma sources. Contrib. Plasma Phys. **49**, 641–647 (2009)
9. G. Bauer, Signal amplification by tumor cells: Clue to the understanding of the antitumor effects of cold atmospheric plasma and plasma-activated medium. IEEE Trans. Radiat. Plasma Med. Sci. **2**, 87–98 (2018)
10. G. Bauer, Cold atmospheric plasma and plasma-activated medium: antitumor cell effects with inherent synergistic potential. Plasma Med. **9**, 57–88 (2019)
11. L. Boeckmann, M. Schäfer, T. Bernhardt, M.L. Semmler, O. Jung, G. Ojak, T. Fischer, K. Peters, B. Nebe, B. Müller-Hilke et al., Cold atmospheric pressure plasma in wound healing and cancer treatment. Appl. Sci. **10** (2020)
12. D. Boehm, C. Heslin, P.J. Cullen, P. Bourke, Cytotoxic and mutagenic potential of solutions exposed to cold atmospheric plasma. Sci. Rep. **6** (2016)
13. A. Calugaru, L. Cremer, A. Herold, A. Lupu, G. Szegli, C. Lungu, A. Lungu, N. Georgescu, The effect of the plasma needle on tumoral cell lines apoptosis. Roum. Arch. Microbiol. Immunol. **64**, 57–64 (2005)
14. C.Y. Calvet, D. Famin, F.M. André, L.M. Mir, Electrochemotherapy with bleomycin induces hallmarks of immunogenic cell death in murine colon cancer cells. OncoImmunology **3** (2014)
15. G. Chen, Z. Chen, D. Wen, Z. Wang, H. Li, Y. Zeng, G. Dotti, R.E. Wirz, Z. Gu, Transdermal cold atmospheric plasma-mediated immune checkpoint blockade therapy. Proc. Natl. Acad. Sci. **117**, 3687–3692 (2020)
16. D.S. Chen, I. Mellman, Oncology meets immunology: the cancer-immunity cycle. Immunity **39**, 1–10 (2013)
17. D.S. Chen, I. Mellman, Elements of cancer immunity and the cancer–immune set point. Nature **541**, 321–330 (2017)
18. X. Cheng, J. Sherman, W. Murphy, E. Ratovitski, J. Canady, M. Keidar, The effect of tuning cold plasma composition on glioblastoma cell viability. PLoS ONE **9** (2014)
19. J. Condeelis, J.W. Pollard, Macrophages: obligate partners for tumor cell migration, invasion, and metastasis. Cell **124**, 263–266 (2006)
20. G. Fridman, M. Peddinghaus, M. Balasubramanian, H. Ayan, A. Fridman, A. Gutsol, A. Brooks, G. Friedman, Blood coagulation and living tissue sterilization by floating-electrode dielectric barrier discharge in air. Plasma Chem. Plasma Process. **27**, 113–114 (2006)

21. G. Fridman, A. Shereshevsky, M.M. Jost, A.D. Brooks, A. Fridman, A. Gutsol, V. Vasilets, G. Friedman, Floating electrode dielectric barrier discharge plasma in air promoting apoptotic behavior in melanoma skin cancer cell lines. Plasma Chem. Plasma Process. **27**, 163–176 (2007)
22. N. Georgescu, A.R. Lupu, Tumoral and normal cells treatment with high-voltage pulsed cold atmospheric plasma jets. IEEE Trans. Plasma Sci. **38**, 1949–1955 (2010)
23. A. Görlach, K. Bertram, S. Hudecova, O. Krizanova, Calcium and ROS: a mutual interplay. Redox Biol. **6**, 260–271 (2015)
24. D.B. Graves, The emerging role of reactive oxygen and nitrogen species in redox biology and some implications for plasma applications to medicine and biology. J. Phys. D: Appl. Phys. **45** (2012)
25. L. Guo, Y. Zhao, D. Liu, Z. Liu, C. Chen, R. Xu, M. Tian, X. Wang, H. Chen, M.G. Kong, Cold atmospheric-pressure plasma induces DNA–protein crosslinks through protein oxidation. Free Radic. Res. **52**, 783–798 (2018)
26. B. Gweon, D. Kim, D.B. Kim, H. Jung, W. Choe, J.H. Shin, Plasma effects on subcellular structures. Appl. Phys. Lett. **96** (2010)
27. J. Heinlin, G. Morfill, M. Landthaler, W. Stolz, G. Isbary, J.L. Zimmermann, T. Shimizu, S. Karrer, Plasma medicine: possible applications in dermatology. JDDG: J. der Dtsch. Dermatol. Ges. **8**, 968–976 (2010)
28. J. Huang, H. Li, W. Chen, G.-H. Lv, X.-Q. Wang, G.-P. Zhang, K. Ostrikov, P.-Y. Wang, S.-Z. Yang, Dielectric barrier discharge plasma in Ar/O_2 promoting apoptosis behavior in A549 cancer cells. Appl. Phys. Lett. **99** (2011)
29. S. Ikehara, H. Sakakita, K. Ishikawa, Y. Akimoto, T. Yamaguchi, M. Yamagishi, J. Kim, M. Ueda, J.-I. Ikeda, H. Nakanishi et al., Plasma blood coagulation without involving the activation of platelets and coagulation factors. Plasma Process. Polym. **12**, 1348–1353 (2015)
30. G. Isbary, J. Heinlin, T. Shimizu, J.L. Zimmermann, G. Morfill, H.U. Schmidt, R. Monetti, B. Steffes, W. Bunk, Y. Li et al., Successful and safe use of 2 min cold atmospheric argon plasma in chronic wounds: results of a randomized controlled trial. Br. J. Dermatol. **167**, 404–410 (2012)
31. M. Ishaq, K. Bazaka, K. Ostrikov, Intracellular effects of atmospheric-pressure plasmas on melanoma cancer cells. Phys. Plasmas **22** (2015)
32. S. Ja Kim, H. Min Joh, T.H. Chung, Production of intracellular reactive oxygen species and change of cell viability induced by atmospheric pressure plasma in normal and cancer cells. Appl. Phys. Lett. **103** (2013)
33. J.Y. Jeong, S.E. Babayan, V.J. Tu, J. Park, I. Henins, R.F. Hicks, G.S. Selwyn, Etching materials with an atmospheric-pressure plasma jet. Plasma Sour. Sci. Technol. **7**, 282–285 (1998)
34. S. Kalghatgi, G. Fridman, G. Nagaraj, M. Cooper, M. Peddinghaus, M. Balasubramanian, V. Vasilets, A. Gutsol, A. Fridman, G. Friedman, Mechanism of blood coagulation by non-thermal atmospheric pressure dielectric barrier discharge, in *2007 16th IEEE International Pulsed Power Conference* (2007), pp. 1058–1063
35. S. Kalghatgi, C.M. Kelly, E. Cerchar, B. Torabi, O. Alekseev, A. Fridman, G. Friedman, J. Azizkhan-Clifford, Effects of non-thermal plasma on mammalian cells. PLoS ONE **6** (2011)
36. N.K. Kaushik, N. Kaushik, M. Adhikari, B. Ghimire, N.N. Linh, Y.K. Mishra, S.-J. Lee, E.H. Choi, Preventing the solid cancer progression via release of anticancer-cytokines in co-culture with cold plasma-stimulated macrophages. Cancers **11** (2019)
37. N.K. Kaushik, N. Kaushik, B. Min, K.H. Choi, Y.J. Hong, V. Miller, A. Fridman, E.H. Choi, Cytotoxic macrophage-released tumour necrosis factor-alpha (TNF-alpha) as a killing mechanism for cancer cell death after cold plasma activation. J. Phys. D Appl. Phys. **49** (2016)
38. N.K. Kaushik, N. Kaushik, D. Park, E.H. Choi, Altered antioxidant system stimulates dielectric barrier discharge plasma-induced cell death for solid tumor cell treatment. PLoS ONE **9** (2014)
39. N.K. Kaushik, H. Uhm, E. Ha Choi, Micronucleus formation induced by dielectric barrier discharge plasma exposure in brain cancer cells. Appl. Phys. Lett. **100** (2012)

40. M. Keidar, R. Walk, A. Shashurin, P. Srinivasan, A. Sandler, S. Dasgupta, R. Ravi, R. Guerrero-Preston, B. Trink, Cold plasma selectivity and the possibility of a paradigm shift in cancer therapy. Br. J. Cancer **105**, 1295–1301 (2011)
41. M. Khalili, L. Daniels, A. Lin, F.C. Krebs, A.E. Snook, S. Bekeschus, W. B. Bowne, V. Miller, Non-thermal plasma-induced immunogenic cell death in cancer. J. Phys. D: Appl. Phys. **52** (2019)
42. I.E. Kieft, J.L.V. Broers, V. Caubet-Hilloutou, D.W. Slaaf, F.C.S. Ramaekers, E. Stoffels, Electric discharge plasmas influence attachment of cultured CHO K1 cells. Bioelectromagnetics **25**, 362–368 (2004)
43. C.-H. Kim, J.H. Bahn, S.-H. Lee, G.-Y. Kim, S.-I. Jun, K. Lee, S.J. Baek, Induction of cell growth arrest by atmospheric non-thermal plasma in colorectal cancer cells. J. Biotechnol. **150**, 530–538 (2010)
44. S.J. Kim, T.H. Chung, S.H. Bae, S.H. Leem, Induction of apoptosis in human breast cancer cells by a pulsed atmospheric pressure plasma jet. Appl. Phys. Lett. **97** (2010b)
45. M.G. Kong, G. Kroesen, G. Morfill, T. Nosenko, T. Shimizu, J. van Dijk, J.L. Zimmermann, Plasma medicine: an introductory review. New J. Phys. **11** (2009)
46. N. Kumar, J.H. Park, S.N. Jeon, B.S. Park, E.H. Choi, P. Attri, The action of microsecond-pulsed plasma-activated media on the inactivation of human lung cancer cells. J. Phys. D: Appl. Phys. **49** (2016)
47. O. Lademann, H. Richter, A. Patzelt, A. Alborova, D. Humme, K.D. Weltmann, B. Hartmann, P. Hinz, A. Kramer, S. Koch, Application of a plasma-jet for skin antisepsis: analysis of the thermal action of the plasma by laser scanning microscopy. Laser Phys. Lett. **7**, 458–462 (2010)
48. M. Laroussi, Low-temperature plasmas for medicine? IEEE Trans. Plasma Sci. **37**, 714–725 (2009)
49. M. Laroussi, T. Akan, Arc-free atmospheric pressure cold plasma jets: a review. Plasma Process. Polym. **4**, 777–788 (2007)
50. C.B. Lee, I.H. Seo, M.-W. Chae, J.W. Park, E.H. Choi, H.S. Uhm, K.Y. Baik, Effects of non-thermal plasma activated water on the anti-cancer immune activities of macrophages. Clin. Plasma Med. **9** (2018)
51. C.B. Lee, I.H. Seo, M.-W. Chae, J.W. Park, E.H. Choi, H.S. Uhm, K.Y. Baik, Anti-cancer activity of liquid treated with microwave plasma-generated gas through macrophage activation. Oxid. Med. Cell. Longev. **2020**, 1–13 (2020)
52. H.J. Lee, C.H. Shon, Y.S. Kim, S. Kim, G.C. Kim, M.G. Kong, Degradation of adhesion molecules of G361 melanoma cells by a non-thermal atmospheric pressure microplasma. New J. Phys. **11** (2009)
53. D.S. Leventhal, D.C. Gilmore, J.M. Berger, S. Nishi, V. Lee, S. Malchow, D.E. Kline, J. Kline, D.J. Vander, H.H. Griend et al., dendritic cells coordinate the development and homeostasis of organ-specific regulatory T cells. Immunity **44**, 847–859 (2016)
54. C.E. Lewis, J.W. Pollard, Distinct role of macrophages in different tumor microenvironments. Can. Res. **66**, 605–612 (2006)
55. A. Li, S. Dubey, M.L. Varney, B.J. Dave, R.K. Singh, IL-8 directly enhanced endothelial cell survival, proliferation, and matrix metalloproteinases production and regulated angiogenesis. J. Immunol. **170**, 3369–3376 (2003)
56. Y. Li, M. Ho Kang, H. Sup Uhm, G. Joon Lee, E. Ha Choi, I. Han, Effects of atmospheric-pressure non-thermal bio-compatible plasma and plasma activated nitric oxide water on cervical cancer cells. Sci. Rep. **7** (2017)
57. A. Lin, Y. Gorbanev, J. De Backer, J. Van Loenhout, W. Van Boxem, F. Lemière, P. Cos, S. Dewilde, E. Smits, A. Bogaerts, Non-thermal plasma as a unique delivery system of short-lived reactive oxygen and nitrogen species for immunogenic cell death in melanoma cells. Adv. Sci. **6** (2019)
58. A. Lin, B. Truong, G. Fridman, A.A. Fridman, V. Miller, Immune cells enhance selectivity of nanosecond-pulsed DBD plasma against tumor cells. Plasma Med. **7**, 85–96 (2017)

59. A. Lin, B. Truong, S. Patel, N. Kaushik, E. Choi, G. Fridman, A. Fridman, V. Miller,Nanosecond-pulsed DBD plasma-generated reactive oxygen species trigger immunogenic cell death in A549 lung carcinoma cells through intracellular oxidative stress. Int. J. Mol. Sci. **18** (2017b)
60. A.G. Lin, B. Xiang, D.J. Merlino, T.R. Baybutt, J. Sahu, A. Fridman, A.E. Snook, V. Miller, Non-thermal plasma induces immunogenic cell deathin vivoin murine CT26 colorectal tumors. OncoImmunology **7** (2018)
61. X. Lu, G.V. Naidis, M. Laroussi, S. Reuter, D.B. Graves, K. Ostrikov, Reactive species in non-equilibrium atmospheric-pressure plasmas: generation, transport, and biological effects. Phys. Rep. **630**, 1–84 (2016)
62. H. Mahdikia, F. Saadati, E. Freund, U.S. Gaipl, K. Majidzadeh-a, B. Shokri, S. Bekeschus, Gas plasma irradiation of breast cancers promotes immunogenicity, tumor reduction, and an abscopal effect in vivo. OncoImmunology **10** (2020)
63. A. Makkouk, G.J. Weiner, Cancer immunotherapy and breaking immune tolerance: new approaches to an old challenge. Can. Res. **75**, 5–10 (2015)
64. I. Mellman, G. Coukos, G. Dranoff, Cancer immunotherapy comes of age. Nature **480**, 480–489 (2011)
65. V. Miller, A. Lin, A. Fridman, Why target immune cells for plasma treatment of cancer. Plasma Chem. Plasma Process. **36**, 259–268 (2015)
66. H. Min Joh, S. Ja Kim, T.H. Chung, S.H. Leem, Reactive oxygen species-related plasma effects on the apoptosis of human bladder cancer cells in atmospheric pressure pulsed plasma jets. Appl. Phys. Lett. **101** (2012)
67. K. Miyamoto, S. Ikehara, H. Takei, Y. Akimoto, H. Sakakita, K. Ishikawa, M. Ueda, J.-I. Ikeda, M. Yamagishi, J. Kim et al., Red blood cell coagulation induced by low-temperature plasma treatment. Arch. Biochem. Biophys. **605**, 95–101 (2016)
68. S. Mohades, M. Laroussi, J. Sears, N. Barekzi, H. Razavi, Evaluation of the effects of a plasma activated medium on cancer cells. Phys. Plasmas **22** (2015)
69. L.N. Nguyen, N. Kaushik, P. Bhartiya, S.K. Gurmessa, H.-J. Kim, L.Q. Nguyen, N.K. Kaushik, E.H. Choi, Plasma-synthesized mussel-inspired gold nanoparticles promote autophagy-dependent damage-associated molecular pattern release to potentiate immunogenic cancer cell death. J. Ind. Eng. Chem. **100**, 99–111 (2021)
70. J. Van der Paal, E.C. Neyts, C.C.W. Verlackt, A. Bogaerts, Effect of lipid peroxidation on membrane permeability of cancer and normal cells subjected to oxidative stress. Chem. Sci. **7**, 489–498 (2016)
71. M.O. Palumbo, P. Kavan, W.H. Miller, L. Panasci, S. Assouline, N. Johnson, V. Cohen, F. Patenaude, M. Pollak, R.T. Jagoe et al., Systemic cancer therapy: achievements and challenges that lie ahead. Front. Pharmacol. **4** (2013)
72. E. Panzarini, V. Inguscio, L. Dini, Immunogenic cell death: can it be exploited in photodynamic therapy for cancer? Biomed. Res. Int. **2013**, 1–18 (2013)
73. L.I. Partecke, K. Evert, J. Haugk, F. Doering, L. Normann, S. Diedrich, F.-U. Weiss, M. Evert, N.O. Huebner, C. Guenther et al., Tissue Tolerable Plasma (TTP) induces apoptosis in pancreatic cancer cells in vitro and in vivo. BMC Cancer **12** (2012)
74. M.R. Prausnitz, R. Langer, Transdermal drug delivery. Nat. Biotechnol. **26**, 1261–1268 (2008)
75. S. Ptasińska, B. Bahnev, A. Stypczyńska, M. Bowden, N.J. Mason, N.S.J. Braithwaite, DNA strand scission induced by a non-thermal atmospheric pressure plasma jet. Phys. Chem. Chem. Phys. **12** (2010)
76. E.A. Ratovitski, X. Cheng, D. Yan, J.H. Sherman, J. Canady, B. Trink, M. Keidar, Anti-cancer therapies of 21st century: novel approach to treat human cancers using cold atmospheric plasma. Plasma Process. Polym. **11**, 1128–1137 (2014)
77. M.J. Regulski, Cellular senescence: what, why, and how. Wounds **29**, 168–174 (2017)
78. K. Rödder, J. Moritz, V. Miller, K.-D. Weltmann, H.-R. Metelmann, R. Gandhirajan, S. Bekeschus, Activation of murine immune cells upon co-culture with plasma-treated B16F10 melanoma cells. Appl. Sci. **9** (2019)

79. E.H. Sarsour, M.G. Kumar, L. Chaudhuri, A.L. Kalen, P.C. Goswami, Redox control of the cell cycle in health and disease. Antioxid. Redox Signal. **11**, 2985–3011 (2009)
80. T. Sato, M. Yokoyama, K. Johkura, A key inactivation factor of HeLa cell viability by a plasma flow. J. Phys. D: Appl. Phys. **44** (2011)
81. J. Schlegel, J. Köritzer, V. Boxhammer, Plasma in cancer treatment. Clin. Plasma Med. **1**, 2–7 (2013)
82. A. Schmidt, K. Wende, S. Bekeschus, L. Bundscherer, A. Barton, K. Ottmüller, K.-D. Weltmann, K. Masur, Non-thermal plasma treatment is associated with changes in transcriptome of human epithelial skin cells. Free Radic. Res. **47**, 577–592 (2013)
83. M.L. Semmler, S. Bekeschus, M. Schäfer, T. Bernhardt, T. Fischer, K. Witzke, C. Seebauer, H. Rebl, E. Grambow, B. Vollmar et al., Molecular mechanisms of the efficacy of cold atmospheric pressure plasma (CAP) in cancer treatment. Cancers **12** (2020)
84. A. Shashurin, M.A. Stepp, T.S. Hawley, S. Pal-Ghosh, L. Brieda, S. Bronnikov, R.A. Jurjus, M. Keidar, Influence of cold plasma atmospheric jet on surface integrin expression of living cells. Plasma Process. Polym. **7**, 294–300 (2010)
85. P.A. Steerenberg, J.A.J. Zonnenberg, J.A.M.A. Dormans, P.N.T. Joon, I.M. Wouters, L.V. Bree, P.T.J. Scheepers, H.V. Loveren, Diesel exhaust particles induced release of interleukin 6 and 8 by (Primed) human bronchial epithelial cells (Beas 2b) in vitro. Exp. Lung Res. **24**, 85–100
86. M.J. Steinbeck, N. Chernets, J. Zhang, D.S. Kurpad, G. Fridman, A. Fridman, T.A. Freeman, Skeletal cell differentiation is enhanced by atmospheric dielectric barrier discharge plasma treatment. PLoS ONE **8** (2013)
87. H. Tanaka, M. Mizuno, K. Ishikawa, K. Nakamura, H. Kajiyama, H. Kano, F. Kikkawa, M. Hori, Plasma-activated medium selectively kills glioblastoma brain tumor cells by downregulating a survival signaling molecule, AKT Kinase. Plasma Med. **1**, 265–277 (2011)
88. H. Tanaka, M. Mizuno, K. Ishikawa, K. Nakamura, F. Utsumi, H. Kajiyama, H. Kano, S. Maruyama, F. Kikkawa, M. Hori, Cell survival and proliferation signaling pathways are downregulated by plasma-activated medium in glioblastoma brain tumor cells. Plasma Med. **2**, 207–220 (2012)
89. H. Tanaka, M. Mizuno, K. Ishikawa, K. Takeda, K. Nakamura, F. Utsumi, H. Kajiyama, H. Kano, Y. Okazaki, S. Toyokuni et al., Plasma medical science for cancer therapy: toward cancer therapy using nonthermal atmospheric pressure plasma. IEEE Trans. Plasma Sci. **42**, 3760–3764 (2014)
90. H. Tanaka, M. Mizuno, K. Ishikawa, S. Toyokuni, H. Kajiyama, F. Kikkawa, M. Hori, Molecular mechanisms of non-thermal plasma-induced effects in cancer cells. Biol. Chem. **400**, 87–91 (2018)
91. S. Tomić, A. Petrović, N. Puač, N. Škoro, M. Bekić, Z.L. Petrović, M. Čolić, Plasma-activated medium potentiates the immunogenicity of tumor cell lysates for dendritic cell-based cancer vaccines. Cancers **13** (2021)
92. D. Trachootham, J. Alexandre, P. Huang, Targeting cancer cells by ROS-mediated mechanisms: a radical therapeutic approach? Nat. Rev. Drug Discov. **8**, 579–591 (2009)
93. E. Turrini, A. Stancampiano, E. Simoncelli, R. Laurita, E. Catanzaro, C. Calcabrini, M. Gherardi, V. Colombo, C. Fimognari, Non-thermal plasma as an innovative anticancer strategy on leukemia models. Clin. Plasma Med. **9**, 15–16 (2018)
94. F. Utsumi, H. Kajiyama, K. Nakamura, H. Tanaka, M. Mizuno, K. Ishikawa, H. Kondo, H. Kano, M. Hori et al., Effect of indirect nonequilibrium atmospheric pressure plasma on antiproliferative activity against chronic chemo-resistant ovarian cancer cells in vitro and in vivo. PLoS ONE **8** (2013)
95. M. Vandamme, E. Robert, S. Lerondel, V. Sarron, D. Ries, S. Dozias, J. Sobilo, D. Gosset, C. Kieda, B. Legrain et al., ROS implication in a new antitumor strategy based on non-thermal plasma. Int. J. Cancer **130**, 2185–2194 (2012)
96. O. Volotskova, T.S. Hawley, M.A. Stepp, M. Keidar, Targeting the cancer cell cycle by cold atmospheric plasma. Sci. Rep. **2** (2012)

97. H.E. Wagner, R. Brandenburg, K.V. Kozlov, A. Sonnenfeld, P. Michel, J.F. Behnke, The barrier discharge: basic properties and applications to surface treatment. Vacuum **71**, 417–436 (2003)
98. K.D. Weltmann, E. Kindel, T. von Woedtke, M. Hähnel, M. Stieber, R. Brandenburg, Atmospheric-pressure plasma sources: prospective tools for plasma medicine. Pure Appl. Chem. **82**, 1223–1237 (2010)
99. M. Witek, E.S. Blomain, M.S. Magee, B. Xiang, S.A. Waldman, A.E. Snook, Tumor radiation therapy creates therapeutic vaccine responses to the colorectal cancer antigen GUCY2C. Int. J. Radiat. Oncol. Biol. Phys. **88**, 1188–1195 (2014)
100. T. von Woedtke, S. Reuter, K. Masur, K.D. Weltmann, Plasmas for medicine. Phys. Rep. **530**, 291–320 (2013)
101. D. Yan, N. Nourmohammadi, K. Bian, F. Murad, J.H. Sherman, M. Keidar, Stabilizing the cold plasma-stimulated medium by regulating medium's composition. Sci. Rep. **6** (2016a)
102. D. Yan, N. Nourmohammadi, J. Milberg, J.H. Sherman, M. Keidar, Guidelines for using 3-Nitro-L-tyrosine as an antidegradation reagent of H_2O_2 in the cold atmospheric plasma-stimulated solutions. Plasma Med. **8**, 121–129 (2018)
103. D. Yan, N. Nourmohammadi, A. Talbot, J.H. Sherman, M. Keidar, The strong anti-glioblastoma capacity of the plasma-stimulated lysine-rich medium. J. Phys. D: Appl. Phys. **49** (2016b)
104. D. Yan, J.H. Sherman, X. Cheng, E. Ratovitski, J. Canady, M. Keidar, Controlling plasma stimulated media in cancer treatment application. Appl. Phys. Lett. **105** (2014)
105. D. Yan, J.H. Sherman, M. Keidar, Cold atmospheric plasma, a novel promising anti-cancer treatment modality. Oncotarget **8**, 15977–15995 (2016)
106. D. Yan, J.H. Sherman, M. Keidar, The application of the cold atmospheric plasma-activated solutions in cancer treatment. Anticancer Agents Med. Chem. **18**, 769–775 (2018)
107. D. Yan, A. Talbot, N. Nourmohammadi, J.H. Sherman, X. Cheng, M. Keidar, Toward understanding the selective anticancer capacity of cold atmospheric plasma—A model based on aquaporins (Review). Biointerphases **10** (2015)
108. X. Yan, Z. Xiong, F. Zou, S. Zhao, X. Lu, G. Yang, G. He, K.K. Ostrikov, Plasma-induced death of HepG2 cancer cells: intracellular effects of reactive species. Plasma Process. Polym. **9**, 59–66 (2012)
109. Y.J. Yoon, M.J. Suh, H.Y. Lee, H.J. Lee, E.H. Choi, I.S. Moon, K. Song, Anti-tumor effects of cold atmospheric pressure plasma on vestibular schwannoma demonstrate its feasibility as an intra-operative adjuvant treatment. Free Radic. Biol. Med. **115**, 43–56 (2018)
110. X. Zhang, M. Li, R. Zhou, K. Feng, S. Yang, Ablation of liver cancer cells in vitro by a plasma needle. Appl. Phys. Lett. **93** (2008)
111. B. Zhivotovsky, H.J. Ahn, K.I. Kim, G. Kim, E. Moon, S.S. Yang, J.-S. Lee, Atmospheric-pressure plasma jet induces apoptosis involving mitochondria via generation of free radicals. PLoS ONE **6** (2011)
112. L. Zitvogel, L. Apetoh, F. Ghiringhelli, G. Kroemer, Immunological aspects of cancer chemotherapy. Nat. Rev. Immunol. **8**, 59–73 (2008)

Chapter 3
Cold Plasma in Dentistry

Jae-Sung Kwon

3.1 Introduction on Dentistry

3.1.1 Oral Tissues

Oral tissues are unique part of our body which consists of different organs. One of main organs in oral tissues are tooth. Tooth can be further categorized into three different tissues; enamel, dentin and cementum.

Enamel is the most outer layer of the tooth at the crown which are also most highly calcified among three tissues. It provides hard surface for the efficient chewing (mastication), which would be one of the main roles of the tooth.

Meanwhile, dentine is located beneath enamel at the crown, and forms the bulk of the tooth. The junction between the dentin and enamel is called dentin-enamel junction. Dentin at the root part of the tooth is covered by cementum, which acts as layer between surrounding jaw bone, known as alveolar bone, and the dentine. Both dentine and cementum are very similar in the composition of the bone as they are mainly composed of collagen type I matrix reinforced with calcium phosphate mineral in the form of apatite. However, unlike bone tissues, regeneration of tooth is very difficult, as fracture of the part of the tooth would not result in natural healing or fusion (if broken pieces are placed together), unlike fracture of the bone which would naturally heal by casting with appropriate managements.

Within the tooth structure, a chamber is located beneath dentine, which is called pulp chamber (pulp cavity). Pulp chamber is extended into root part of the tooth as root canal, and the space is filled with nerves and vessels. Such nerves and vessels are

---

J.-S. Kwon (✉)
Department and Research Institute of Dental Biomaterials and Bioengineering, Yonsei University College of Dentistry, Seoul, Korea
e-mail: jkwon@yuhs.ac

BK21 FOUR Project, Yonsei University College of Dentistry, Seoul, Korea

E. H. Choi (ed.), *Plasma Biosciences and Medicine*, Topics in Applied Physics 148,
https://doi.org/10.1007/978-981-19-7935-4_3

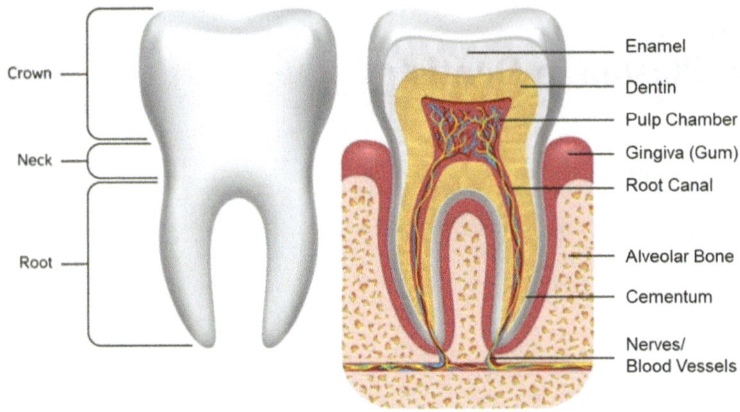

Fig. 3.1 Anatomical structure of tooth

connected to nervous and vascular system of the jaw bone (mandibular and maxillary bone), which provides important function such as providing nutrients to tooth.

Other organs of the oral cavity include gingival tissue that is commonly known as gum tissue. They are soft tissue that surrounds alveolar bone. Gingival tissues and alveolar bone can be collectively called periodontal tissue. Gingival tissues are extended to oral mucosa, which as name suggest, are mucosal tissues covering inner parts of the oral cavity including uvula etc. (Fig. 3.1).

3.1.2 Oral Environment

Oral environment is also very unique in comparison to other parts of the body. In terms of temperatures, our body temperature is maintained at relatively narrow range near 37–37.5 °C. Rise in temperatures even to 1 or 2 °C would result in difficulties of physiological functions, and therefore medications are required to lower body temperature. However, oral cavity is exposed to various conditions of temperature. Drinking or consumption of cold or hot drinks/foods would result in change of temperature between 4 to 50 °C. Still, oral cavity is capable of withstanding such radical changes of temperature.

Variations of pH can be also dramatic in oral cavity. Acidosis or alkalosis from body pH of 7.4 may result in respiratory or cardiovascular problems but oral cavity would be exposed to wide pH changes as we consume food or drink.

Changes in humidity may also have effects on oral cavity. As oral cavity is constantly occupied by saliva under normal physiological state, oral cavity maintains wet environment. Still, as with any open cavity, exposure to surrounding air by opening mouth would rapidly dry the cavity. Such problems are often associated with growth of biofilm, especially anaerobic bacteria or fungal growth on either dental tissue or dental materials.

3.1.3 Common Dental Disease

One of the most prevalent and perhaps important dental disease is dental caries. As stated above, oral environment provides good environment for bacteria to culture and consequently form biofilm. Bacteria that result in caries, also known as cariogenic bacteria, can be attached and grow on surface of tooth. One of the well-known cariogenic bacteria include *Streptococcus mutans*, which is often considered to be primary etiological agent of dental caries. As we consume carbohydrate-based food, these bacteria would then utilize such carbon and energy source molecule result in end product of acid such as lactic acid. Lactic acid would then result in erosion of enamel surface and later dentin surface. Deep erosion would result in exposure of pulp chamber or root canal to bacteria, result in endodontic infection that requires painful process of endodontic treatment (root canal treatment). As the limited regeneration of tooth structure as mentioned above, treatment of dental caries is currently based on removal of caries (eroded surface along with bacteria), and to replace with artificial dental materials known as dental prostheses or restoratives.

Gingival or periodontal disease is also common dental disease, which result from presence of bacterial biofilm on the surface of periodontal tissues, and consequent formation of calcified structure known as dental calculus to result in inflammation of gum and periodontal tissues. Although the etiology is still unclear, the main bacteria responsible for periodontal disease, *Porphyromonas gingivalis*, has been linked with development of periodontal disease and also associated with other systemic disease such as cardiovascular or neurovascular diseases. Although treatment of periodontal disease is possible with antibiotic applications or anti-inflammatory measures, best option is to prevent inflammation in the first place by regular removal of dental calculus. Severe periodontal disease would even result in loss of tooth or resorption of alveolar bone, which would require dental implant with bone filler in order to replace function.

3.2 Application of Cold Plasma on Dental Materials

3.2.1 Application of Cold Plasma on Dental Implant Surfaces

As mentioned earlier, tooth cannot be regenerated once lost or damaged. Dental implant provides functional recovery of lost tooth in relation to mastication and esthetic appearances. With aging population and increasing interest towards better quality of life, dental implant industries are growing at rapid phase, which also consequently resulted in increasing research related to the topic.

Dental implants are often divided into different parts, and screw like structure that is inserted into alveolar bone is often called dental implant though other terminologies such as dental implant fixture or dental implant body are also used. Currently, commercially pure titanium (Cp-Ti) is most commonly used as the material of choice for dental implant. Cp-Ti are graded in accordance to the contents of oxygen, where Grade 4 is the Cp-Ti that is commonly adapted for its biocompatibility and ability to form passive oxide layer for integration with surrounding bone, term called osseointegration.

Success of dental implant would depend on the bonding or integration between material and surrounding bone. This is often called osseointegration, as the process would require integration between surrounding bone and oxide layer of the titanium. Such process would require both osteoconductive and osteoinductive features of the materials. In other words, surrounding bone would require making more of bone like cells (osteoconduction) while newly supplied stem-like cells would need to differentiate into bone like cells (osteoinduction). In order to induce better and quicker osteointegration and therefore increase success of dental implant, numerous studies have been conducted. These include change in design of the material or change in topographical features of the material. Currently, the most successful implant surface in terms of topography, has been the surface known as sandblasted, large-grit, and acid-etched (SLA) surfaces. The process of SLA involves sandblasting of titanium with relatively large-grit size of particles such as alumina powder (of around 50 μm) followed by acid etching on the blasted surface with hydrochloric acid in order to from relatively nano-like roughness. The resultant surface will appear micro-nano mixed roughness which indicated to result in significantly improved osseointegration (Fig. 3.2).

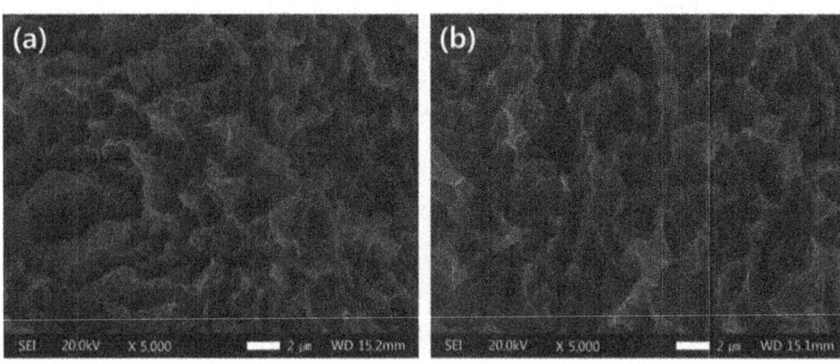

Fig. 3.2 Scanning electron microscopy image of titanium following sandblasted, large-grit, and acid-etched (SLA) process before (**a**) and after (**b**) cold plasma treatment. The results show no changes in micro-nano appearance of original SLA surface even after cold plasma treatment [9].

Many of other researchers then tried to improve such SLA treated surfaces, though most of process resulted in pressure or force that changed such favorable topographical state. Hence, method of preserving topography but enhancing chemical state of dental implant surface has been investigated. For example, ultraviolet (UV) surface treatment has indicated to preserve SLA surface while enhancing hydrophilicity and also osteoconductive and osteoinductive features, as the UV resulted in removal of hydrocarbon contaminants [11].

Similar ideas were also adapted with cold plasma. As cold plasma would provide effects similar to UV, by applying reactive oxygen species (ROS) that would react with surface chemistry of titanium and consequently remove hydrocarbon contaminant while have no effects on surface topographical features, possible use of cold plasma has been investigated.

Similar ideas were also adapted with cold plasma. As cold plasma would provide effects similar to UV, by applying reactive oxygen species (ROS) that would react with surface chemistry of titanium and consequently remove hydrocarbon contaminant while have no effects on surface topographical features, possible use of cold plasma has been investigated.

Lee et al. [9] investigated effect of applying cold plasma on SLA surface of titanium for little as 10 min, which the result indicated that the surface topographical features are preserved as above, while surface chemical analyses indicated that initial carbohydrate amount on the surface reduced with application of cold plasma, and consequently lead to improved hydrophilicity (Fig. 3.3).

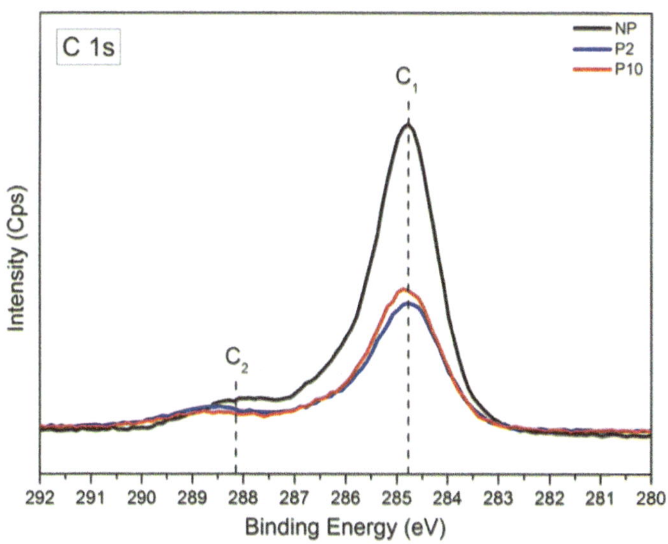

Fig. 3.3 X-ray photoelectron spectroscopy analyses of C 1 s on SLA titanium (NP) following 2 min (P2) and 10 min (P10) of cold plasma exposure. Peak of C_1 on 284.8 eV corresponds to hydrocarbon which decreased dramatically following cold plasma exposure. Also, there is small drop of C_2 peak which at 288.2 eV, which related to carbon-oxygen bond [9]

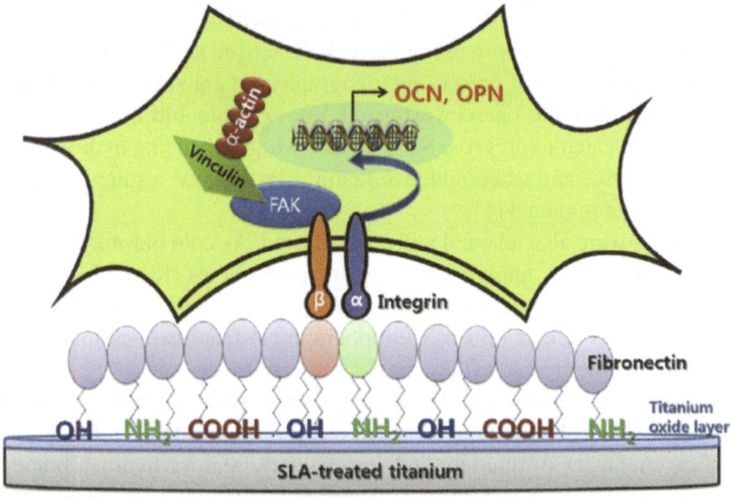

Fig. 3.4 The result of different chemical functional groups forms on SLA-treated titanium by cold plasma. Some of functional groups may be favorable to bone cell attachment via molecules such as integrin and consequently lead to bone cell differentiation by focal adhesion kinase (FAK), vinculin, actin etc. [8]

Despite may of similarities with UV treated SLA surfaces, however, there were some key differences with cold plasma treated dental implant surfaces. UV would remove hydrocarbon for improved hydrophilicity and consequent better cell reactivity, but cold plasma would not just remove hydrocarbon, but would be able to form different functional groups in accordance to ROS that is formed by cold plasma. In another paper by Lee et al. [8], authors investigated how differences in chemistry would result in cellular reaction, especially with initial bone cell attachment via integrin like molecules (Fig. 3.4).

Advantages of cold plasma in terms of forming functional groups on the surface of titanium would be favorable features of using such technique. In terms of chemical functional groups, numerous researches have been already conducted to conclude what chemical functional group would provide best results in terms of osseointegration. One of the chemical groups indicated to be linked with superior osseointegration is amine (NH_x) groups. Still, forming such chemical functional group has been difficult process as the procedures were often time consuming, expensive or even ineffective. Hence, use of cold plasma to tailor produced reactive nitrogen species (RNS) and consequently result in forming chemical functional groups such as amine like structure has been investigated. One of the methods that has been adapted was using a gas supply with humidification with water or other chemicals [4] (Figs. 3.5 and 3.6).

Along with osseointegration, other cellular and tissue activities surrounding dental implant would play important role in success of dental implant. Gingival tissues activities would be one of the key roles of soft tissue, where adequate sealing at

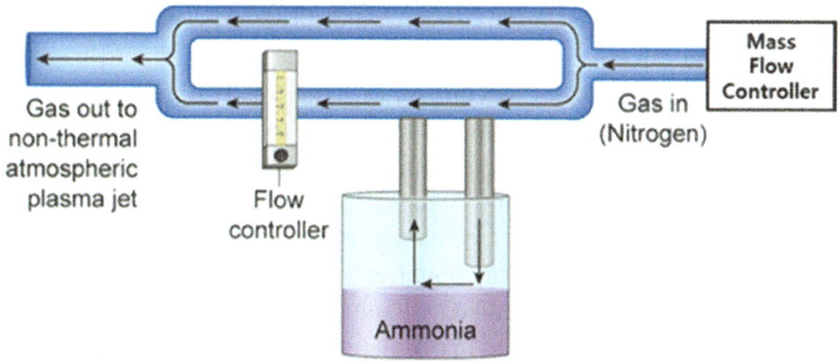

Fig. 3.5 The schematic diagram of forming tailored reactive nitrogen species from cold plasma (non-thermal atmospheric plasma jet). Nitrogen gas would pass over ammonia solution to result in humidified nitrogen with ammonia, which would then be suppled as gas source for the cold plasma [4]

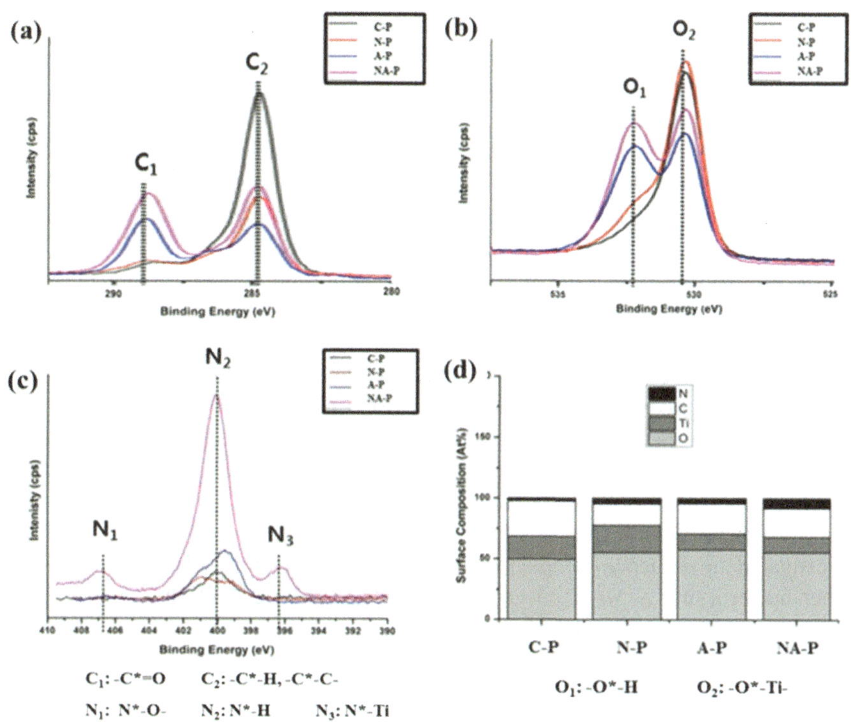

Fig. 3.6 X-ray photoelectron spectroscopy for C 1 s (**a**), O 1 s (**b**) and N1 s (**c**) of titanium before exposure to cold plasma (C-P) or following exposure to cold plasma with different gas supplies; nitrogen (N-P), air (A-P), or humidified nitrogen with ammonia (NA-P). N-P and A-P resulted in dramatic decrease in hydrocarbon as before but not much change in nitrogen related species. However, NA-P resulted in not only the reduction of hydrocarbon but formation of chemical functional groups related to nitrogen. These were also evident with chemical composition graph (**d**) [4]

the level of bone and soft tissue division would allow prevention on soft tissues to be infiltrated into alveolar bone space. Such tissue growth would be especially important during dental implant placement, as many of these patients would have inflamed periodontal tissue from chronic exposure to periodontal disease, which would have led them to undergo dental implant surgery in the first place. Hence, not only enhancing the bone cell and tissue activities, activities of gingival cells or tissues were investigated on the surface of cold plasma treated titanium. Jeong et al. [1] reported that when chemical functional groups are formed by cold plasma on titanium, cell lines such as immortalized human oral keratinocytes (IHOK) and oral fibroblasts (hTERT-hNOF) increased in cellular activities in terms of attachment and proliferations. Moreover, when these cells were induced with inflammation by LPS like molecules, interleukins and other inflammatory markers were reduced on the surface of cold plasma treated titanium (Fig. 3.7).

Formation of chemical function group that is effective while having no effects on surface topographical features are huge advantages. Understanding of how bone cells are interacting with biomaterial surface is still ongoing, and therefore, role of new chemical functional group may be presented later. In such cases, gas supply of cold plasma could be modified for provision of such favorable surfaces.

In parallel to the research on improving osseointegration of the dental implant, research related to reducing side-effects from placing dental implants also have been widely studied. One of the common side-effects would be the post-surgical infection, resulted from attachment and growth of bacteria on the implant surfaces. It has been suggested that chemical functional groups formed by ROS or RNS of cold plasma on the surface of the titanium would result in inhibition of bacteria attachment or growth. Yoo et al. [12] applied cold plasma on titanium surface and investigated attachment and growth by planktonic bacteria. Two bacteria; *Streptococcus mutans* and *Staphylococcus aureus* were investigated, which reduced number of bacteria attachments resulted from cold plasma treatment on titanium.

Such effects were not seeming to be influenced by surface topographical features, such as SLA surface treated implant. As SLA surface resulted in superior bone cell attachment and consequent improved osseointegration, it may also provide favorable surface for the bacteria to attach and therefore result in infection. Jeong et al. [2] investigated the influence of topographical features along with cold plasma treatment. When bacteria such as *Steptococcus sanguinis* were cultured on titanium, indeed the greater attachment were resulted on the rough SLA surface of titanium compared to smooth surface. However, reduction of attached bacteria was dramatically reduced to the level similar to bacteria on smooth surface following exposure to cold plasma (Fig. 3.8).

Such effects of cold plasma treated surfaces on attached bacteria were not only due to change in surface energy or surface hydrophilicity, as the appearance of individual bacteria were altered on the surface of cold plasma treated titanium (Fig. 3.9).

Why bacterial shape and structure would be altered by attaching on titanium with chemical functional group from ROS or RNS of cold plasma is still under investigation. One of the suggestions by Lee et al. [10] was linked with interactions with cell wall structure with chemical functional group, which consequently would

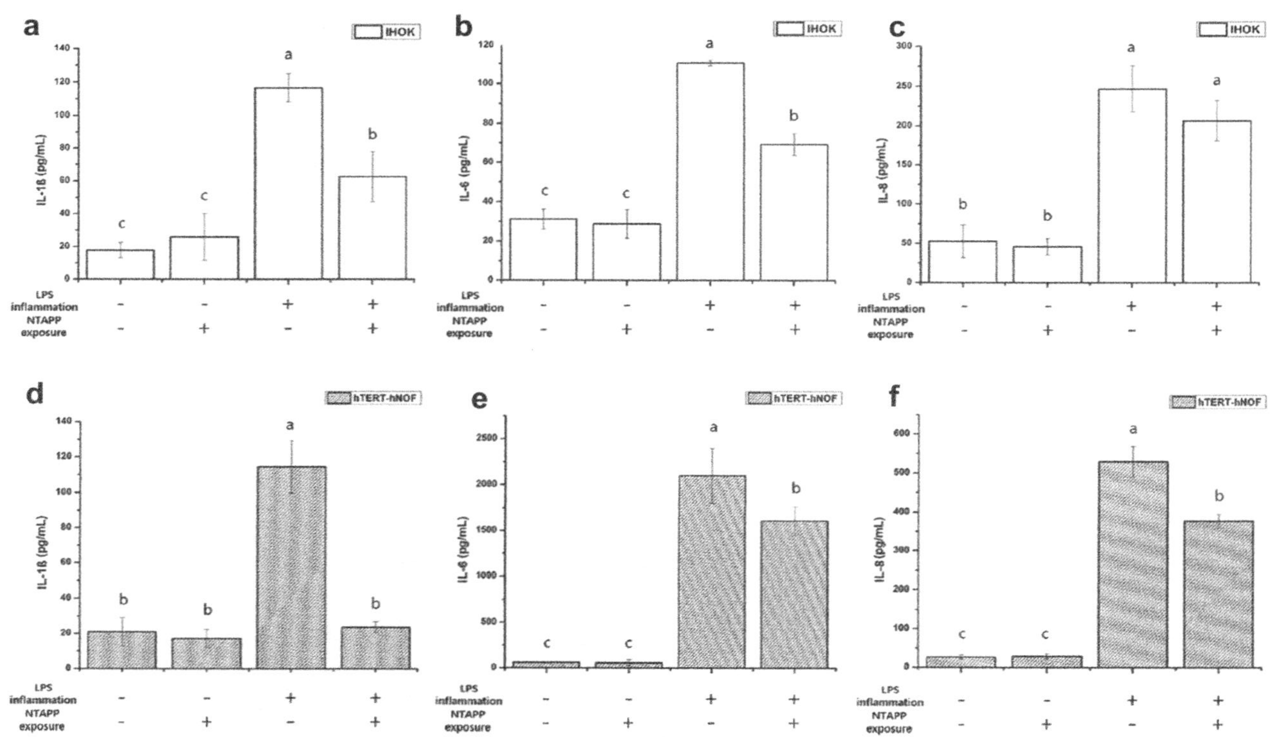

Fig. 3.7 Various inflammatory cytokines released from cell lines of immortalized human oral keratinocytes (IHOK) (**a–c**) and oral fibroblasts (hTERT-hNOF) (**d–f**), before and after induction of inflammation by LPS or placed on cold plasma (non-thermal atmospheric pressure plasma, NTAPP) exposed/unexposed titanium surface [1]

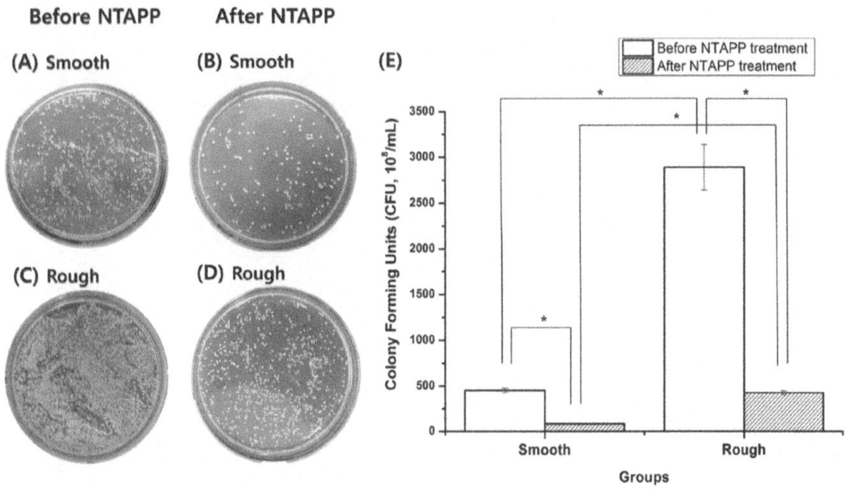

Fig. 3.8 Colony forming units (CFU) of Steptococcus *sanguinis* following cultured on; **a** before cold plasma (non-thermal atmospheric pressure plasma, NTAPP) treatment on smooth titanium, **b** after cold plasma treatment on smooth titanium, **c** before cold plasma treatment on rough titanium, and **d** after cold plasma treatment on rough titanium. **e** The bacterial attachment on each titanium surface is shown as a quantitative result [12]

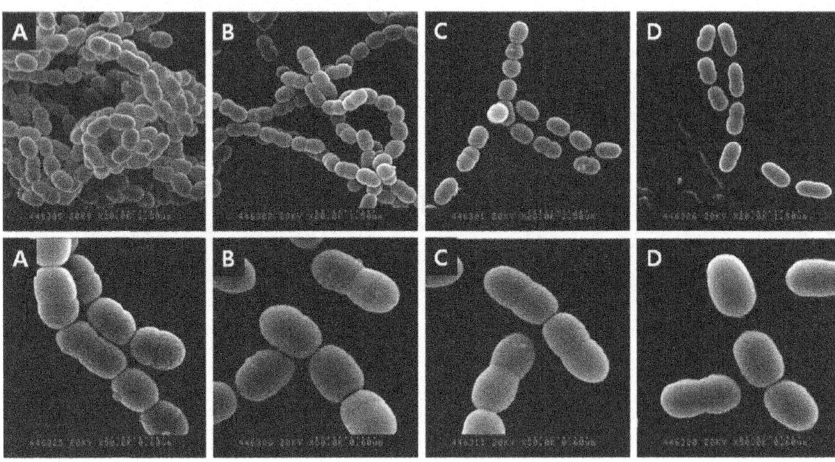

Fig. 3.9 Scanning electron microscopy images of *Streptococcus mutans* on titanium surface, before (**a**) treatment with cold plasma or after treatment with cold plasma for 30 s (**b**), 60 s (**c**) or 120 s (**d**). Not only reduced attachment growth of bacteria were evident, but also typical chained structured of *Streptococcus mutans* was altered with breakage of chains [12]

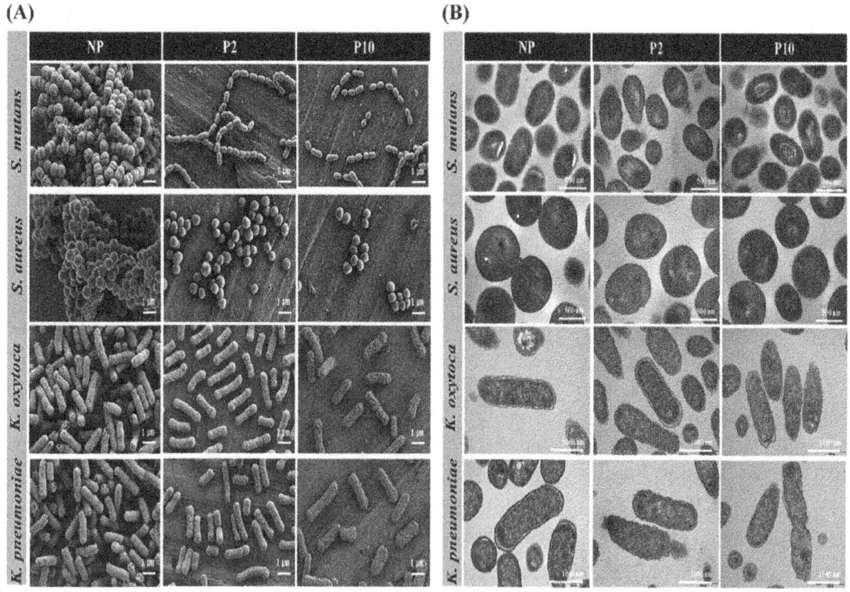

Fig. 3.10 Changes in bacterial morphology on control titanium (NP) and titanium exposed to cold plasma for 2 min (P2) or 10 min (P10 as observed by scanning electron microscope (**a**) or transmission electron microscope (**b**). Two gram-positive bacteria, *Streptococcus mutans* and *Staphylococcus aureus*, and two gram-negative bacteria, *Klebsiella oxytoca* and *Klebsiella pneumoniae* were investigated where all of them showed reduced attachment on cold plasma treated titanium but effects were less evident with gram-negative bacteria [10]

lead to different morphology of the bacteria. This was evident as the results were different between gram-positive and gram-negative bacteria which have different cell wall structure. Study concluded that exposure to cold plasma on titanium would have chemical functional groups changes leading to oxidation of bacteria, where this would be more sensitive to gram-negative bacteria as they have different cell wall structure to gram-positive bacteria (Fig. 3.10).

3.2.2 Application of Cold Plasma on Adherend for Improved Bonding

Bonding and adhesion is important in dentistry. As the lost part of tooth cannot be regenerated, materials that would act as replacement either as artificial crown, restoratives or other prosthesis would need to be boned to tooth structure. Cements or resin-like material-based adhesives are commonly used in dentistry, and improvement of adhesive force would result in increased success of long-term treatment.

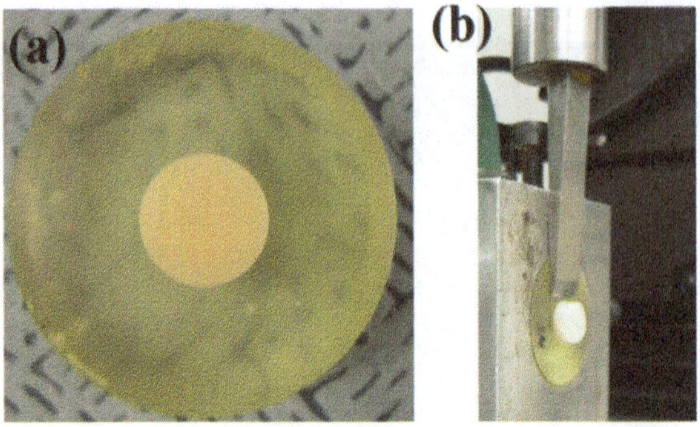

Fig. 3.11 Example of adhesion test sample (**a**) and actual test carried out (**b**) between epoxy resin and core resin used in dentistry [3]

Hence, adhesive force is often tested for dental materials between two different materials (Fig. 3.11).

As the hydrophilicity would be increased by application of cold plasma, this would be advantages for adhesives to flow freely on dental materials, aiding improved adhesions. Also, some of chemical functional groups formed by cold plasma on dental materials would aid adhesives or cements to form chemical bonding between the layer. For example, study by Kim et al. [3] demonstrated that higher shear bond strength and other related bonding test results between epoxy resin and core resin by applying cold plasma on epoxy resin. Although surface energy may be the key to the success for adhesion by cold plasma treatment, other factors such as increased salinization by cold plasma ROS or RNS, or other bonding related chemicals may have important role.

3.2.3 *Application of Cold Plasma on Dental Materials for Other Purposes*

Some of role of cold plasma that have been investigated was to possibly replace otherwise difficult or increased risk process. Conventional method of producing prostheses or restoratives involves process known as lost was technique. This first involve taking impression of prepared tooth (with removed caries) along with surrounding soft tissues. Vinyl polysiloxane based materials are commonly used for such process and hydrophilicity of these materials are important for the success as any artificial objects such as saliva would result in defect of impression and hence very low contact angle with thinned layer of saliva would be helpful. Surfactant is therefore commonly used to improve hydrophilicity but some of them may cause toxicity. Hence, alternative option to improve hydrophilicity of the dental impression material is required, which cold plasma has been suggested as one of the option [6]. As cold plasma would be portable that would allow quick treatment of dental materials before taking dental impression, this may provide alterative option to increased risk of toxicity by the surfactants (Fig. 3.12).

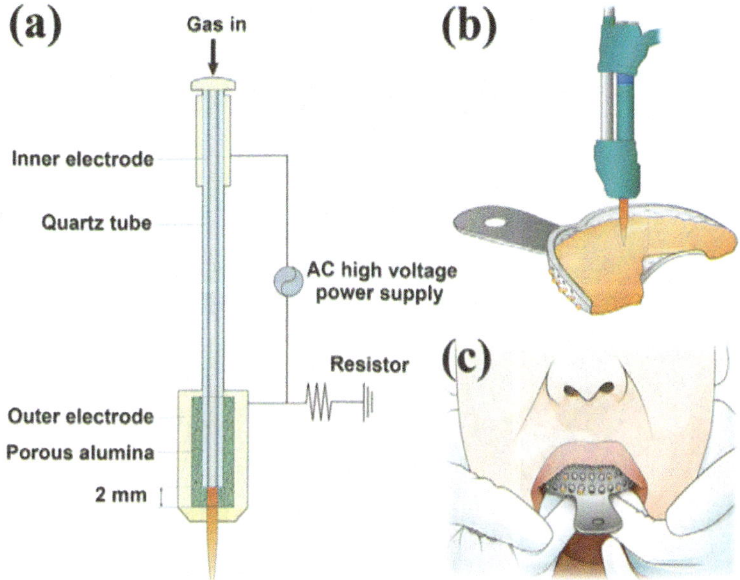

Fig. 3.12 Example of portable cold plasma in jet form (**a**) which can be applied on dental impression materials on the tray (**b**) and consequently improve process of impression taking (**c**) [6]

3.3 Application of Cold Plasma on Dental Cells or Tissues

Despite complicated structure of oral tissues, most of tissues are composed mainly of either bone like osteoblast cells or soft tissue like fibroblast cells. Although many of application of cold plasma in medical field has been focused on either killing of leading apoptosis of cancer cells or other relevant cells, studies in dentistry has been more focused on regeneration of tissues. Perhaps the reason may be due to the fact that dental tissues would be difficult to be regenerated while some of the process would require clinically long and costly process.

It has been well known that at low dose of cold plasma, cells are activated and enhanced in their growth-related cycles. When cold plasma was exposed to osteoblast, high level of actin filament was evident [7]. This, however, just required 60–240 s for the effective results (Fig. 3.13).

Like many of similar studies, effects were also evident when culture media was first exposed to cold plasma, and cells are later exposed to the culture media. Study indicated that ROS or RNS formed by cold plasma may have reaction with some of chemicals in culture media, and consequently result in change in cellular activity.

One of the commonly investigated possible ROS or RNS related chemicals linked with increased cellular activity would be nitrogen oxide (NO_x). When possible, application of cold plasma on gingival cells for gingival tissue healing following the periodontal disease, it was evident that scavengers such as c-PTIO, that removes nitrogen oxide, would result in diminished effects from the cold plasma [5] (Fig. 3.14).

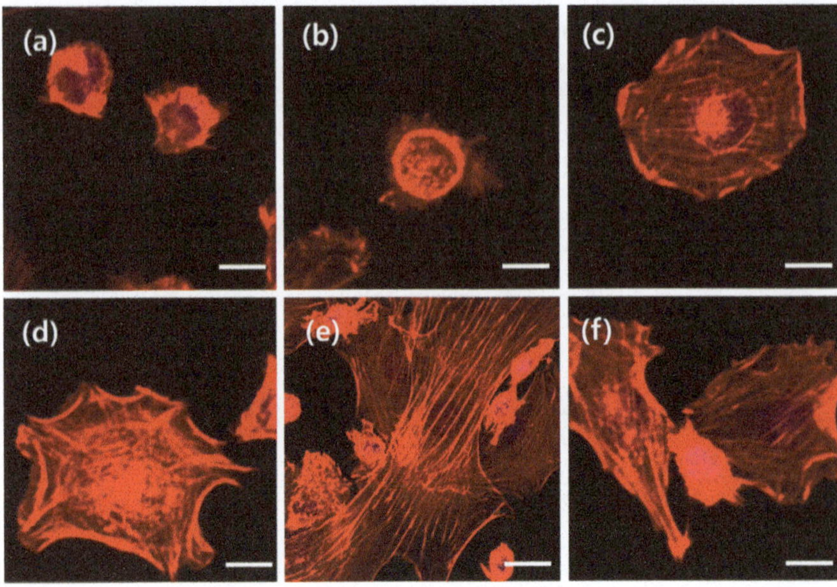

Fig. 3.13 Morphology of murine osteoblast (MC3T3-E1) observed under confocal laser microscope after 4 h of culture and staining with rhodamine phalloidin (actin, red) following direct plasma exposure to cell and culture media for; **a** 0 s (control), **b** 10 s, **c** 30 s, **d** 60 s, **e** 120 s and **f** 240 s [7]

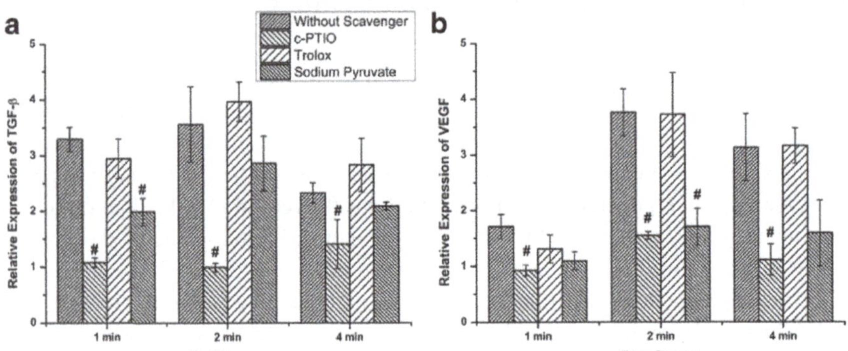

Fig. 3.14 Gene expression level of TGF-β and VEGF related to proliferation and growth of gingival cells, following exposure of human gingival cells for 1, 2 or 4 min of cold plasma. Expression levels were measured with or without scavengers, and when c-PTIO is present that is nitrogen oxide related scavenger, effect of increased gene expression as in 'Without Scavenger' seems to be diminished [5]

3.4 Others

There are many other areas where application of cold plasma has been investigated in dentistry. Esthetic application is one of the areas that is still ongoing and perhaps more interest in commercial companies than any other studies mentioned above. Current tooth or teeth whitening involve use of hydrogen peroxide or their derivatives with or without light energy application. The idea is to produce ROS from hydrogen peroxide that consequently would remove staining chemicals. However, dose of hydrogen peroxide that can be used both in home and dental clinic is limited in many parts of the world due to the danger related to chemicals.

As cold plasma would also produce ROS and perhaps safer than the high dose hydrogen peroxide, it has been naturally investigated for the purpose of tooth whitening. Many studies concluded that the cold plasma is an effective tool where higher the dose of ROS produced by altering source gas (such as use of humid gas) would result in better tooth whitening, though limitations such as ozone production and other electricity related safety issues are still to be solved.

References

1. W.S. Jeong, J.S. Kwon, E.H. Choi, K.M. Kim, The effects of non-thermal atmospheric pressure plasma treated titanium surface on behaviors of oral soft tissue cells. Sci. Rep. **8**, 15963 (2018)
2. W.S. Jeong, J.S. Kwon, J.H. Lee, S.H. Uhm, E.H. Choi, K.M. Kim, Bacterial attachment on titanium surfaces is dependent on topography and chemical changes induced by nonthermal atmospheric pressure plasma. Biomed. Mater. **12** (2017)
3. H.S. Kim, S.Y. Yang, E.H. Choi, K.M. Kim, J.S. Kwon, Adhesion between epoxy resin-based fiber post and dental core resin improved by non-thermal atmospheric pressure plasma. Appl. Sci.-Basel **10** (2020)
4. J.S. Kwon, S.H. Choi, E.H. Choi, K.M. Kim, P.K. Chu, Enhanced osteogenic differentiation of human mesenchymal stem cells on amine-functionalized titanium using humidified ammonia supplied nonthermal atmospheric pressure plasma. Int. J. Mol. Sci. **21** (2020)
5. J.S. Kwon, Y.H. Kim, E.H. Choi, C.K. Kim, K.N. Kim, K.M. Kim, Non-thermal atmospheric pressure plasma increased mRNA expression of growth factors in human gingival fibroblasts. Clin. Oral Invest. **20**, 1801–1808 (2016)
6. J.S. Kwon, Y.H. Kim, E.H. Choi, K.N. Kim, Development of ultra-hydrophilic and non-cytotoxic dental vinyl polysiloxane impression materials using a non-thermal atmospheric-pressure plasma jet. J. Phys. D-Appl. Phys. **46** (2013a)
7. J.S. Kwon, Y.H. Kim, E.H. Choi, K.N. Kim, The effects of non-thermal atmospheric pressure plasma jet on attachment of osteoblast. Curr. Appl. Phys. **13**, S42–S47 (2013)
8. E.J. Lee, J.S. Kwon, J.Y. Om, S.K. Moon, S.H. Uhm, E.H. Choi, K.N. Kim, The enhanced integrin-mediated cell attachment and osteogenic gene expression on atmospheric pressure plasma jet treated micro-structured titanium surfaces. Curr. Appl. Phys. **14**, S167–S171 (2014)
9. E.J. Lee, J.S. Kwon, S.H. Uhm, D.H. Song, Y.H. Kim, E.H. Choi, K.N. Kim, The effects of non-thermal atmospheric pressure plasma jet on cellular activity at SLA-treated titanium surfaces. Curr. Appl. Phys. **13**, S36–S41 (2013)
10. M.J. Lee, J.S. Kwon, H.B. Jiang, E.H. Choi, G. Park, K.M. Kim, The antibacterial effect of non-thermal atmospheric pressure plasma treatment of titanium surfaces according to the bacterial wall structure. Sci. Rep. **9**, 1938 (2019)

11. T. Ogawa, Ultraviolet photofunctionalization of titanium implants. Int. J. Oral Maxillofac. Implants **29**, e95–102 (2014)
12. E.M. Yoo, S.H. Uhm, J.S. Kwon, H.S. Choi, E.H. Choi, K.M. Kim, K.N. Kim, The study on inhibition of planktonic bacterial growth by non-thermal atmospheric pressure plasma jet treated surfaces for dental application. J. Biomed. Nanotechnol. **11**, 334–341 (2015)

Chapter 4
Nonthermal Plasma-Based Virus Inactivation and Sterilization

Nagendra Kumar Kaushik, Yungoh Shin, Sehoon Ki, Ihn Han, Neha Kaushik, and Eun Ha Choi

Abstract Recent reports regarding plasma and plasma-treated liquids against viruses suggested satisfactory virus inactivation or sterilization strategies to decrease contamination or treat viral diseases. In this chapter, we have discussed the role of various plasma sources on different viruses linked with viral diseases and their selectivity. Nonthermal plasma has excellent capabilities like a novel antiviral agent and has numerous benefits more than the traditional sterilization methods. Plasma can have advantages over other conventional methods since plasma application comprises the delivery of RONS that can detrimentally affect the functionality of viral pathogens including damage to nucleic acid, lipids as well as proteins. It has been also shown that plasma can selectively enhances the host cell's defense system capabilities. Moreover, plasma-based approaches for vaccine preparation against various pathogenic viruses and treatment of infected cells, immune cells, and organs have been discussed in this chapter.

4.1 Introduction of Animal Viruses

4.1.1 Definition of Virus

Viruses are described as very nanometer size small agent and capable to proliferate with in the cells of host. Viruses either consist of DNA or RNA genome and encased

N. K. Kaushik (✉) · Y. Shin · S. Ki · I. Han · E. H. Choi (✉)
Plasma Bioscience Research Center, Department of Electrical and Biological Physics, Kwangwoon University, Seoul 01897, Korea
e-mail: kaushik.nagendra@kw.ac.kr

E. H. Choi
e-mail: ehchoi@kw.ac.kr

N. Kaushik
Department of Biotechnology, College of Engineering, University of Suwon, Hwaseong 18323, Korea

E. H. Choi (ed.), *Plasma Biosciences and Medicine*, Topics in Applied Physics 148,
https://doi.org/10.1007/978-981-19-7935-4_4

by protective coat made up of protein with or without envelop. Viruses are distinguished from other microorganisms in three main properties. Viruses are smaller than other organisms and mostly are in size from 10 to 300 nm with few exceptions. In comparison, most bacteria are approximately 1000 nm and erythrocytes are 7500 nm in diameter. Secondly, the genome of viruses has either DNA or RNA. Thirdly, viruses have no metabolic activity outside susceptible host cells; they do not have any ribosomes or protein-synthesizing apparatus with some exceptions. Viruses cannot multiply in inanimate media but only inside living cells. Upon entry into a susceptible cell, the virus genome or nucleic acid is transcribed into mRNA or itself act as mRNA. Then the virus directs the replication of new virus particles and then assembles new virus particles.

4.1.2 Human Viral Epidemics of Recent Forty years

The word virus was known to be appeared in 1599 for the first time and originally meant "venom" which had been used in "病毒" with same meaning in the far East until recent time. Viral pandemics have dominated human viral diseases for a long time up to middle 19 century. However, the development and use of viral vaccines including smallpox, measles and polio had decreased viral pandemics rapidly. There were even the tendencies that many virologists and epidemiologists would expect no more serious viral pandemics except influenza and few viral respiratory diseases in the future. However, big pandemics like human immunodeficiency virus (HIV) had just prepared to begin big pandemics from macaque in 1950s. During last forty years, world faced pandemic of Human Immunodeficiency Virus (HIV) in 1981, H1N1 influenza virus in 2009, Severe Acute Respiratory Syndrome Corona Virus (SARS-CoV) in 2002~ , and most recently SARS-COV-2 in 2019 ~ till now. Approximate 75% of recent viral pandemics of forty years are known to be caused by animal originated viruses. The increase of recent viral pandemics caused by animal is considered related with changes of environment, ecosystem, and human culture including high population density and international travelling [1].

4.1.3 Structure and Function of Virus

Viruses are inactive out of the host cells and incapable to produce energy or vitality. Virus particles completely rely on the complex biological system of prokaryotic or eukaryotic host cells during replication. They used to deliver DNA or RNA genome inside the cells for transcription and translation support by the host cells. The simplest virion, complete virus particle, consists of two key elements, genome and a protective coat, the capsid. Capsid serves as a shield to protect the single stranded or double stranded RNA and DNA from nucleic acid enzymes or other factors. These proteins protective coats are coded by the virus genetic material. Virus genetic material codes for only some important proteins having structural or non-structural function which are important in virus propagation. These virus coats are consisting of protein shells made of single or double layers and few structural proteins. Thus, several protein copies need to be assembled to create 3D capsid protective structure. There are two basic patterns in virus structure, namely helical symmetry, where proteins and the genome are organized in a spiral form. The other pattern is icosahedral symmetry which is assembled into a symmetric shell.

4.1.4 Classification and Nomenclature of Animal Viruses

Classification of viruses is the system of identifying viruses for enlisting them in a taxonomic approach like complex organisms classification system [2]. These particles can be categorized by morphological properties, genome type, replication mode, type of host, and finally the viral illness they produce. There are millions of viruses are present in the earth. However, approximately 5000 viruses are widely known and studied. There are several classification models in virology depending on basis of

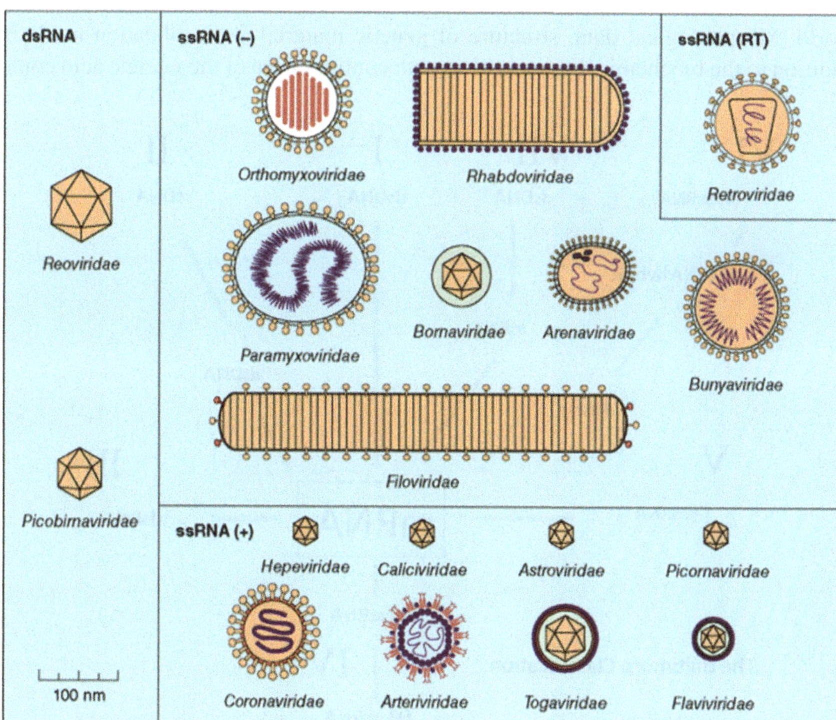

Fig. 4.1 Models of 18 RNA virus families [3]. Copyright © McGraw-Hill Education

criteria. Furthermore, virus classification is continuously being changed with time because new viruses are isolated and added for classification. Traditionally animal viruses were classified into 10 DNA virus families and 18 RNA virus families as shown in Fig. 4.1.

However, new virus families are added continuously because of new classification system being established. As in example, Hantavirus was isolated from Hantan river in South Korea was classified to Bunyaviridae family in the past. However, new classification put Hantavirus to Hantaviridae (or Orthohantaviridae).

The Baltimore classification system, one of the most important classification system to divide viruses into seven groups based on the synthesis of messenger RNA from genome (Fig. 4.2). For example, group IV include Families of Coronaviruses, Picornaviruses and Togaviruses. However, Picornaviridae family has no envelope, while Coronaviridae and Togaviridae families have envelopes. Presently, animal virus classification used multiple systems depending on virologists. In 1966, The International Committee on Taxonomy of Viruses (ICTV) was founded as the universal commission on naming viruses. This committee approves and manages the systematic categorization, and naming of viruses. The international taxonomic system for viruses has been developed by the ICTV to fittingly name, define, and categorize viruses that infect organisms.

It is true that traditional common names are very often used in real virology world. Morphological data, structure of genetic material and replication mode in addition to the biochemical composition with configuration of the nucleic acid could

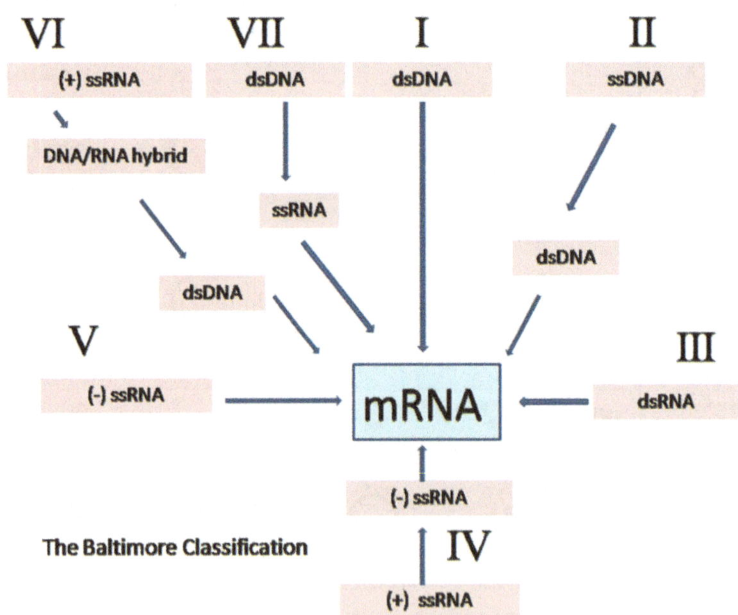

Fig. 4.2 Baltimore classification of viruses

be included in the criteria in virus nomenclature. The virus names ending in -viridae for representing virus families, -virinae is representing subfamilies and ending in -virus represent its genera has achieved worldwide recognition.

4.2 Overview of Emerging Human Coronaviruses

4.2.1 Common Cold Causing Coronaviruses in Human

So far, there are 4 species of human coronaviruses causing common cold. Alphacoronavirus genus includes 2 common cold coronavirus species; Human Corona Virus 229E (HCoV-229E) and Human Corona Virus NL63 (HCoV-NL63), while Betacoronavirus genus includes Human Corona Virus OC43 (HCoV-OC43) and Human Corona Virus HKU1 (HCoV-HKU1) [4]. On the other hand, Gamma and Delta Coronavirus genera have no common cold causing coronavirus species. HCoV-229E is a virus species which affects humans and bats. HCoV-229E virus receptor is Aminopeptidase N (APN) and is related with a range of respiratory symptoms, from common cold to pneumonia and bronchiolitis. Human coronaviruses were discovered in the 1960s by E. C. Kendall et al. from common cold patient. They could grow the virus through organ culture of human embryonic trachea. Another virus strain was isolated and grown in kidney tissue culture, designating as 229E which became worldwide reference strain of coronavirus today. HCoV-NL63 was detected in late 2004 from a child with bronchiolitis in the Netherlands. This strain was found as one of Alphacoronavirus genus and it' receptor on cell is ACE2. NL63 virus also causes common cold worldwide. Human coronavirus OC43(HCoV-OC43) infects human and cattle and belong to Betacoronavirus genus. The strain's receptor on host cell was found as the N-acetyl-9-O-acetylneuraminic acid also known as Sialic acid. HCoV-HKU1 is a coronavirus species causing common cold together with pneumonia and bronchiolitis in humans and animals with sialic acid receptors. It is one of rare coronaviruses with Hemagglutinin esterase (HE) protein in viral envelope.

4.2.2 SARS Causing Viruses

SARS-CoV-1

SARS-1 is a respiratory infection caused by virus of animal origin produced by SARS-CoV-1. The first cases occurred in 2002 in China and caused severe respiratory disorder with high mortality rate of 11.0% in 2002 outbreak. HCoV-NL63 is corona virus of bat origin, which points to the ancestry origin of HCoV-1. Asian palm civets were found as intermediatory animal of SARS-CoV-1 in Yunnan in 2017. Civet SARS-CoV showed approximately 99.8% of similarity to SARS-CoV-1, while

SARS-CoV-2 has genetic similarity of 79% to SARS-1. The outbreak of SARS-CV-1 caused 8,469 infected persons worldwide. Epidemiological and seroprevalence studies suggested zoonostic origin of SARS-CoV-1. Bats are likely to be the natural reservoir but do not show any visible signs of disease by SARS-CoV-1.

SARS-CoV-2

SARS-CoV-2 is a coronavirus species to cause severe acute respiratory syndrome. This virus shows genomic similarity with bat coronaviruses and its origin is zoonotic, indicating it might be developed from a bat coronavirus. The origin of this virus is still in debate. Researchers are speculating whether it is originated from bats or came indirectly via other intermediate. For the entry to human host, SARS-CoV-2 binds with angiotensin converting enzyme 2, a protein receptor on cell membrane. There are several variants of SARS-CoV-2 since 2019, this coronavirus and its variants causes coronavirus disease named as COVID-19. Few variants are of particularly important to study due to their capacity for enhanced transmission, enhanced virulency, or decreased efficacy of vaccines. So far, several variants causing COVID-19 have been described as variants of concern (VOC) such as Alpha, Beta, Gamma, Delta, and Omicron. A category named "variants of interest" is designated to those variants which shows some of criteria related to COVID-19 pandemic. Also "variants under investigation" are variants under investigation for validation or verification of their properties. Once validated, variants of interest or VUI may be renamed "variants of concern" by monitoring organizations. A new category is "variant of high consequence", designated by CDCs in USA or other countries worldwide in case of strong proof that the prevention efficiency and intervention procedures are significantly decreased or absent.

4.3 Plasma-Based Virus Inactivation Strategies, and Mechanisms

Mechanically, a plasma is comprised of excited electrons which produce active species that further react to make long lived species such as hydrogen peroxide, nitrates and nitrites, and many more varying on the plasma parameters and gas type. Several plasma sources are being established towards biomedical applications involving cancer, and infectious diseases treatments. Our group and other scientists have revealed that plasma has been largely developed as an promising and safe methodology for anti-tumor treatment [5–8]. Preferential selectivity of plasma towards cancer cells over their normal counterparts make hem attractive among researchers as a innovative treatment approaches. Additional plasma applications includes the elimination of cancer-resistant tumor initiating cells, stimulation of keratinocytes proliferation and wound repair process via the blockade of the gap-junction proteins [9, 10]. A mounting number of evidence suggest that plasma exposure could be utilized to eradicate infection microbes such as bacteria, viruses and fungus. Many inventors have disclosed that plasma could efficiently destroy biofilm matrices and kills bacteria, transform extracellular matrix characteristics and promote particular cellular behaviors [11–14].

Last some eras, plasma was often acknowledged to show successful antimicrobial activity towards multidrug-resistant microbes on solid surfaces in diseased and septic skins. It has been discovered that the subsequent highly plasma-generated RONS are exceptionally effective in pathogen inactivation. Hence, it could be considered as a favorable medical tool with numerous clinical issues. Reports regarding plasma outcomes on viruses suggested a satisfactory virus inactivation or sterilization strategy to decrease contamination or treat viral diseases [15, 16]. In this chapter, we discussed the role of various plasma sources on different viruses linked with viral diseases. Nonthermal plasma has great capability like a novel antiviral agent and has numerous benefits more than the traditional sterilization methods. Although plasma-based sterilization has some resemblances to some chemical methods, it could be recognized as a physical decontamination technique because it barely needs air and electricity. It is worth to mention that active species are generated in situ therefore no other additional chemical is required from outside. Distinctive plasma devices are applied by various scholars to examine the inactivation of diverse viruses (Fig. 4.3). In case of indirect treatments, the plasma is produced at a remote position to treat liquids or other substances and later these plasma-based products are used to treat viruses or experimental samples. Whereas most of the jet plasmas showed the identical operational ideologies, variances in actives species generation could be anticipated.

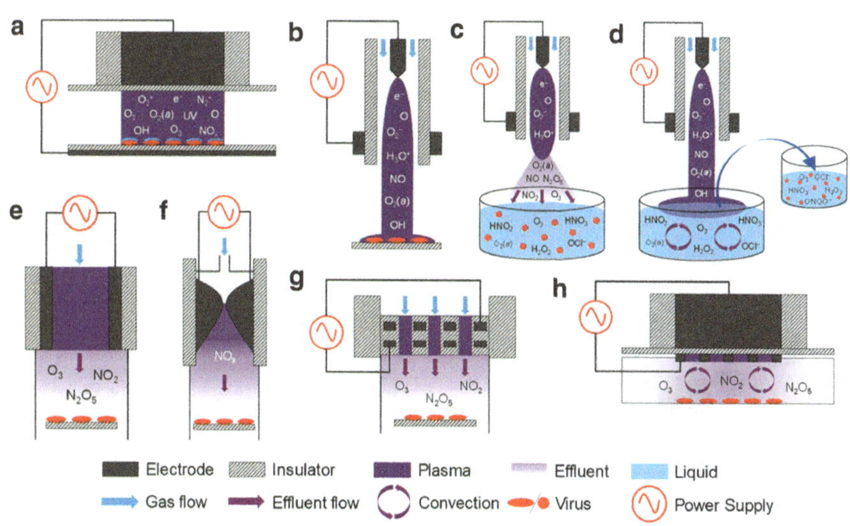

Fig. 4.3 Schematic presentation of various plasma devices can be used to sterilize viruses. **a** Direct treatment by dielectric barrier discharge, **b** jet plasma, **c** direct treatment by jet plasma, **d** indirect jet exposure using treated solutions or liquids, **e** remote treatment by volumetric dielectric barrier discharge, **f** gliding arc, **g** integrated coaxial microhollow dielectric barrier discharge, and **h** surface dielectric barrier discharge [17]

The application of plasma has been tested to both plant and animal viruses. Practically every report on plasma virus sterilization is exceptional since researchers utilized certain plasma device with distinctive characteristics such as gas, exposure time and power, gas or they interact with the exposure of various liquid capacities, substances (cells, solutions, surfaces), and virus types. Such wide-tunable range makes it challenging to compare outcomes of any plasma based strategy and to explain any comprehensive inactivation considerations (Fig. 4.4).

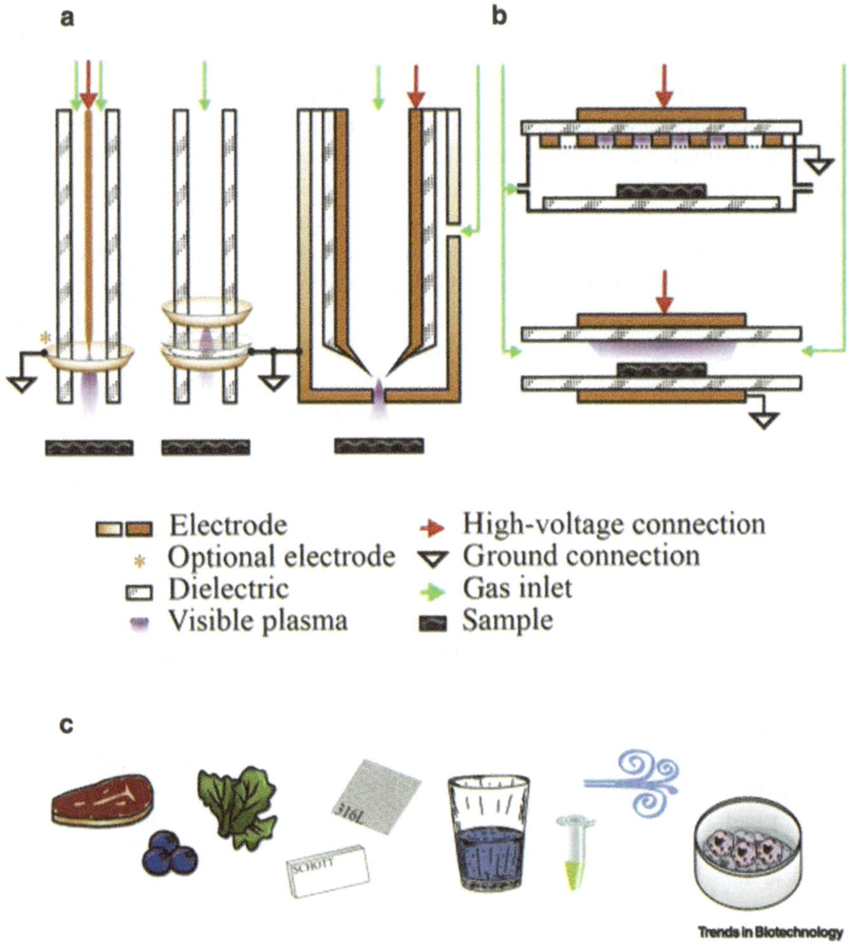

Fig. 4.4 Diagrammatic representation of few examples of plasma devices used for virus deactivation inside various substrates. **a** Various plasma jets, **b** dielectric barrier discharge and **c** different substances inoculated with virus and exposed with plasma. Reprinted from Trends Biotechnol., 38(11), Arijana Filipić et al., Cold Plasma, a New Hope in the Field of Virus Inactivation, P1278–1291. Copyright (2020), with permission from Elsevier [18]

Plasma has been extensively used for sexually transmitted, respiratory as well as enteric viruses, and reports have revealed the ability of plasma application [18]. It is already described that plasma-based virus treatment is a reasonably latest area of investigation [16], whereas certain reports over the earlier years have simply designated the plasma-induced virucidal properties alongside with its means of action [19–21]. To examine the mechanism of plasma application in viral cells, one investigation study was performed in 2018, which investigated both plasma activated solutions and DBD devices. This study determined that the plasma generated reactive species ultimately impaired the protein and nucleic acid structures to destroy virus cells [22]. Furthermore, researchers established a test that might assess DNA impairment of bacteriophage lambda viruses treated by air plasma [23]. Nitrogen gas plasma additionally deactivates influenza virus by damaging RNA and protein [24]. In another work, DBD plasma torch with feeding air gas has been fruitfully applied to deactivate feline calicivirus, surrogate of foodborne norovirus [25]. Hepatitis A and murine norovirus associated with everyday fresh meats were deactivated within few minutes' exposure using jet plasma source. Therefore, a plasma based approach can be exploited in raw meat management and transport procedures to expand safety of fresh meats [26]. These results draws a conclusion that plasma treatment deteriorates viral particles via demolishing their RNA, consequently leading to a failure in infectivity rate [27] as shown in Fig. 4.5 [18].

Direct contact among plasma and matrices can result into more efficient virus sterilization rather than remote exposures where stable species in the discharge facilitate inactivation. However, direct plasma-matrices interfaces are extremely matrices-dependent and are restricted through substrate morphologies and configuration. Therefore, many scientists have determined on the sanitization using remote plasmas strategies. Recent study displays comparable sterilization of feline calicivirus and norovirus on lettuce or steel surfaces [28]. Inactivation of pathogenic bacteria such as salmonella using the identical discharge source on food products, and steel surfaces exhibited substantial alterations in inactivation efficiencies [28]. Overall, the direct treatment strategy is more efficient towards viruses or microbes rather than distant exposure methods. Particularly, it was discovered that distant dielectric barrier discharge exposures were not effectual for dried virus condition, whereas direct exposure can sterilize virus on equally humid and dried matrices [29, 30]. Though complete inactivation, more than 4 log order reduction was accomplished for both exposure methods [31].

Plasma activated solutions (PAS) successfully inactivates bacteriophages (ϕ174, T4, and MS2) and showed similar efficacy like direct plasma exposure to these viruses [32]. Recently PAS, such as plasma activated saline, plasma activated water (PAW), and 0.3% H_2O_2 after 30 min of treatment reduced efficacy of NDV virus significantly [20]. PAW showed antimicrobial activity due to reactive species stored in water after exposure, which can control density of microbial species and can help in the treatment of virulent diseases specially in infected organs or tissue [33, 34]. This research emphasizes the significance of these activated solutions as an environmentally friendly sterilization strategy as an substitute decontamination tool in civil areas including hospitals, offices, schools, as well as reduce application of harsh chemicals

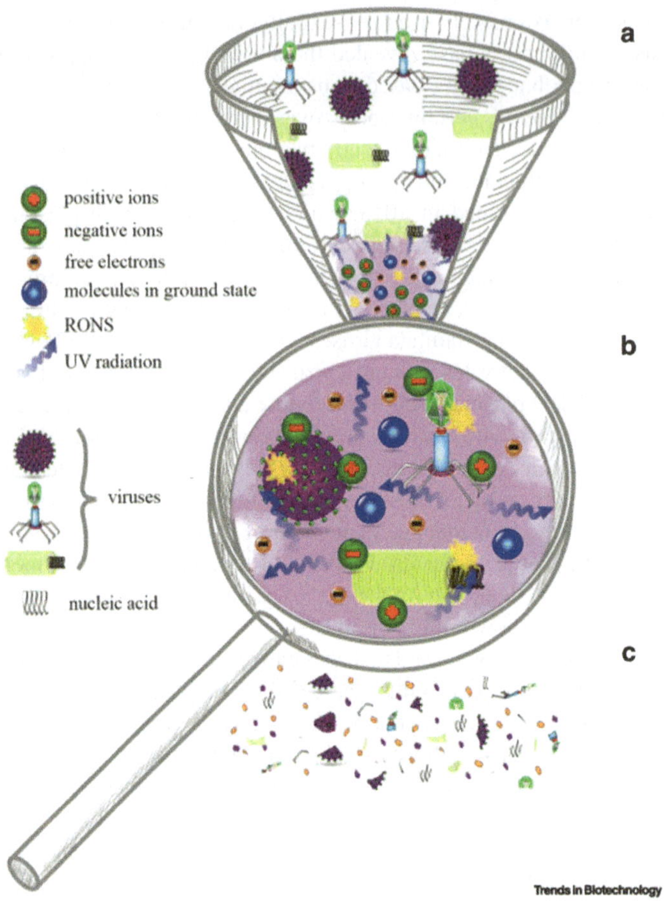

Fig. 4.5 Inactivation of viruses with the exposure of plasma. **a** Various virus particles treated by plasma. **b** Pictorial representation of characteristics such as RONS reliable decontamination. **c** After plasma treatment, viruses are partially or completely broken to noninfectious particles. Reprinted from Trends in Biotechnology, 38(11), Arijana Filipić et al., Cold Plasma, a New Hope in the Field of Virus Inactivation, P1278–1291. Copyright (2020), with permission from Elsevier [18]

such as chlorine-based disinfectants. Several zoonotic viruses are deemed as important examples for future airborne viruses which can infect humans [35, 36]. However, due to regulation of laboratory safety practices and associated risk, bacteriophages are frequently utilized as model for airborne human infecting viruses. Bacteriophages have several advantages such as they handling is easy and can be propagated in large amount and quantified by easy methods [37, 38]. Recently researchers effectively suggested and assembled a packed-bed dielectric barrier discharge source for the protection from airborne virus, and this research was relevant to air sterilization [39]. Due to current spread of SARS-COV-2 viruses and its variants pandemic [40, 41], it is of important to establish useful alleviating strategies or system that can sterilize the viruses in public places and consequently control transmission. Recently, PAW is used to demonstrated induction of structural alteration and spike (S) protein damage in a SARS-CoV-2 pseudo-virus, offering an innovative sterilization technology to combat the viruses and their variants [42]. This study specifically demonstrated the effectiveness of PAW on the receptor-binding domain and S protein on the pseudovirus of SARS-COV-2 (Fig. 4.6). However, the inactivation effects and therapeutic potential of PAW on real corona viruses or SARS-CoV-2, and their variants have not yet been reported.

Since all these reports demonstrated evidence that virus inactivation can be accomplished by various treatment strategies using nonthermal plasma source, the most important characteristic of the deactivation is not well understood except damage of virus particles by plasma reactive species. All procedures for focusing SAR-COV-2 inactivation are showed that plasma treatment have ability to induce virus

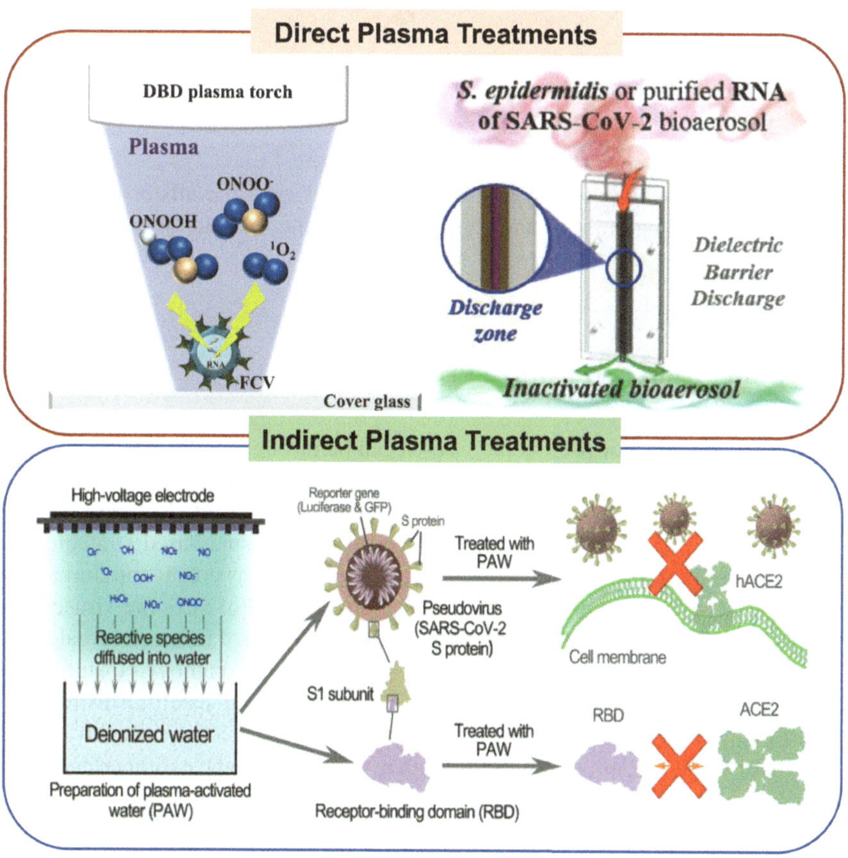

Fig. 4.6 Demonstration of FCV virus inactivation and it mechanism using DBD plasma source [25] and plasma induced decontamination of aerosolized microdroplets [43]. PAW as decontamination agent to damage spike protein specifically RBD domain to inhibit SARS-CoV-2 transmission [42]

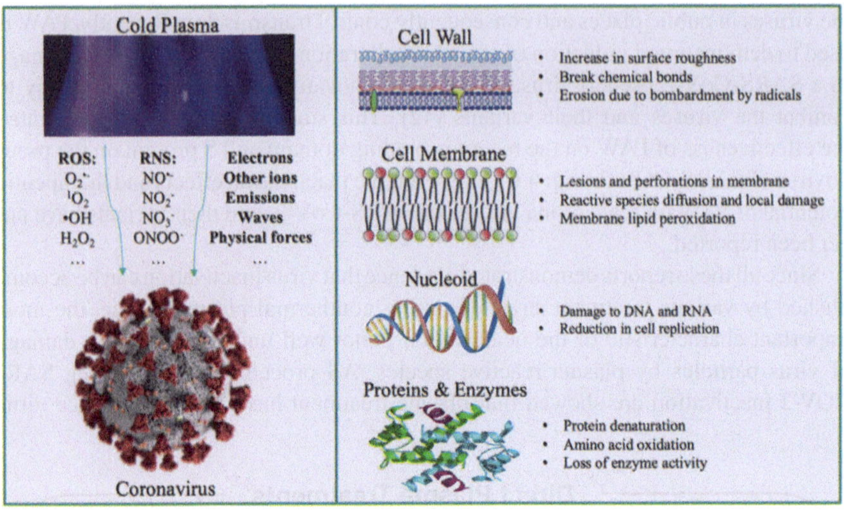

Fig. 4.7 Plasma-based mechanism of action on coronaviruses inducing functionality loss and sterilization [44]

particle damage including damaging or etching virus coat or walls or membranes, protein/enzyme denaturation, and destruction of genetic materials, as shown in Fig. 4.7.

4.4 Conclusion and Future Prospective

Mode of transmission of pathogenic viruses can be through air, liquids, or contaminated surfaces for spreading. These mode of transmission possess distinctive challenges with respect to plasma-based sterilization or inactivation of viruses. Several efforts has been focused on the advancement of new plasma-based strategies for counteract viral infections have shown that plasma can act as potent antiviral agent. Plasma induces virus inactivation via alleviating the level of airborne viruses and virus loads on infected surfaces. Several studies in this area of research revealed the effectiveness of various plasma devices for successful viral pathogen disinfection. Recent innovation emphasize the potential of plasma as a broad-spectrum disinfectant against pathogenic microorganisms including viruses. Plasma can have advantages over electrostatic precipitation [45] and other filters, highlight its importance as a cutting-edge technology. Since, plasma application comprises the delivery of RONS that can detrimentally affect the functionality of viral pathogens including damage to nucleic acid, lipids as well as proteins, it can be used as a latent phase of infection as a treatment strategy. Plasma can be used to develop approaches for vaccine preparation against various pathogenic viruses and treatment of infected cells, immune cells and organs. Plasma-based vaccine preparation strategies may include direct attenuation of

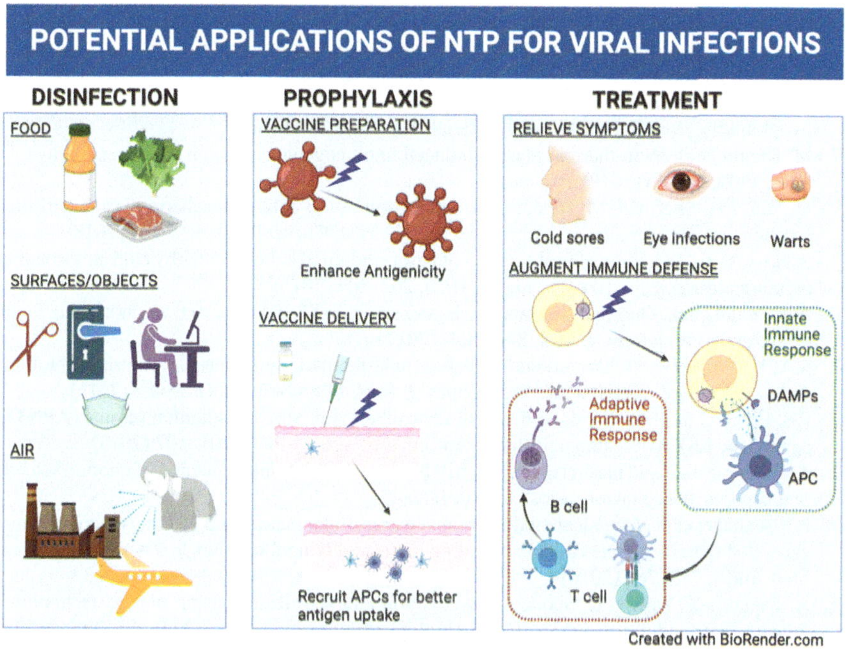

Fig. 4.8 Plasma-based treatment strategies to reduce virus transmission, vaccine development, and treatment of virus-associated diseases. Antigen presenting cell (APC); damage associated molecular pattern (DAMP) [17]

virus particles in laboratory conditions. In future, interaction of plasma inactivated viruses with immune cells should be taken into consideration for defining mechanism and further validation (Fig. 4.8). Practically, we need to address real-time strategies to treatment infected patient via targeting complex molecular or immunological systems. Ultimately, plasma could be used as a cost-effective eco-friendly tool against virus infections in humans.

Acknowledgements This work was funded by the NRF of South Korean Government (2021R1A6A1A03038785, 2021R1F1A1055694, 2021R1C1C1013875).

References

1. S. Roychoudhury et al., Viral pandemics of the last four decades: pathophysiology, health impacts and perspectives. Int. J. Environ. Res. Public Health **17**(24) (2020)
2. M.J. Adams et al., Recently agreed changes to the international code of virus classification and nomenclature. Arch. Virol. **158**(12), 2633–2639 (2013)
3. M.M. Pierce, Virus classification. AccessScience (McGraw-Hill Education, 2020)
4. Y. Jin et al., Virology, epidemiology, pathogenesis, and control of COVID-19. Viruses **12**(4) (2020)

5. N.K. Kaushik et al., Glycolytic inhibitor induces metabolic crisis in solid cancer cells to enhance cold plasma-induced cell death. Plasma Process. Polym. **18**(5) (2021)
6. N.K. Kaushik et al., Low doses of PEG-coated gold nanoparticles sensitize solid tumors to cold plasma by blocking the PI3K/AKT-driven signaling axis to suppress cellular transformation by inhibiting growth and EMT. Biomaterials **87**, 118–130 (2016)
7. M. Khalili et al., Non-thermal plasma-induced immunogenic cell death in cancer. J. Phys. D: Appl. Phys. **52**(42) (2019)
8. D. Yan, J.H. Sherman, M. Keidar, The application of the cold atmospheric plasma-activated solutions in cancer treatment. Anticancer Agents Med. Chem. **18**(6), 769–775 (2018)
9. S. Hasse et al., Induction of proliferation of basal epidermal keratinocytes by cold atmospheric-pressure plasma. Clin. Exp. Dermatol. **41**(2), 202–209 (2016)
10. A. Schmidt et al., One year follow-up risk assessment in SKH-1 mice and wounds treated with an argon plasma jet. Int. J. Mol. Sci. **18**(4) (2017)
11. S.A. Ermolaeva et al., Bactericidal effects of non-thermal argon plasma in vitro, in biofilms and in the animal model of infected wounds. J. Med. Microbiol. **60**(1), 75–83 (2011)
12. P.B. Flynn et al., Bactericidal efficacy of atmospheric pressure non-thermal plasma (APNTP) against the ESKAPE pathogens. Int. J. Antimicrob. Agents **46**(1), 101–107 (2015)
13. T. Coenye et al., Cold plasma inactivation of bacterial biofilms and reduction of quorum sensing regulated virulence factors. Plos One **10**(9) (2015)
14. P. Eisenhauer et al., Chemical modification of extracellular matrix by cold atmospheric plasma-generated reactive species affects chondrogenesis and bone formation. J. Tissue Eng. Regen. Med. **10**(9), 772–782 (2016)
15. A. Sakudo, Y. Yagyu, T. Onodera, Disinfection and sterilization using plasma technology: fundamentals and future perspectives for biological applications. Int. J. Mol. Sci. **20**(20), 5216 (2019)
16. M. Weiss et al., Virucide properties of cold atmospheric plasma for future clinical applications. J. Med. Virol. **89**(6), 952–959 (2017)
17. H. Mohamed et al., Non-thermal plasma as a novel strategy for treating or preventing viral infection and associated disease. Front. Phys. **9** (2021)
18. A. Filipić et al., Cold plasma, a new hope in the field of virus inactivation. Trends Biotechnol. **38**(11), 1278–1291 (2020)
19. T. Xia et al., Inactivation of airborne porcine reproductive and respiratory syndrome virus (PRRSv) by a packed bed dielectric barrier discharge non-thermal plasma. J. Hazard. Mater. **393** (2020)
20. X. Su et al., Inactivation efficacy of nonthermal plasma-activated solutions against Newcastle disease virus. Appl. Environ. Microbiol. **84**(9) (2018)
21. A. Filipić et al., Inactivation of pepper mild mottle virus in water by cold atmospheric plasma. Front. Microbiol. **12** (2021)
22. L. Guo et al., Mechanism of virus inactivation by cold atmospheric-pressure plasma and plasma-activated water. Appl. Environ. Microbiol. **84**(17), e00726-e818 (2018)
23. H. Yasuda et al., Biological evaluation of DNA damage in bacteriophages inactivated by atmospheric pressure cold plasma. Plasma Process. Polym. **7**(3–4), 301–308 (2010)
24. A. Sakudo et al., N2 gas plasma inactivates influenza virus by inducing changes in viral surface morphology, protein, and genomic RNA. Biomed. Res. Int. **2013**, 694269 (2013)
25. R. Yamashiro, T. Misawa, A. Sakudo, Key role of singlet oxygen and peroxynitrite in viral RNA damage during virucidal effect of plasma torch on feline calicivirus. Sci. Rep. **8**(1), 17947 (2018)
26. S.C. Bae et al., Inactivation of murine norovirus-1 and hepatitis A virus on fresh meats by atmospheric pressure plasma jets. Food Res. Int. **76**, 342–347 (2015)
27. S.E. Hanbal et al., Atmospheric-pressure plasma irradiation can disrupt tobacco mosaic virus particles and RNAs to inactivate their infectivity. Adv. Virol. **163**(10), 2835–2840 (2018)
28. H.A. Aboubakr et al., Ìn situ inactivation of human norovirus GII.4 by cold plasma: ethidium monoazide (EMA)-coupled RT-qPCR underestimates virus reduction and fecal material suppresses inactivation. Food Microbiol. **85** (2020)

29. A. Moldgy et al., Comparative evaluation of the virucidal effect of remote and direct cold air plasmas with UV-C. Plasma Process. Polym. **17**(4) (2020)
30. G. Nayak et al., Reactive species responsible for the inactivation of feline calicivirus by a two-dimensional array of integrated coaxial microhollow dielectric barrier discharges in air. Plasma Process. Polym. **15**(1) (2017)
31. A. Moldgy et al., Inactivation of virus and bacteria using cold atmospheric pressure air plasmas and the role of reactive nitrogen species. J. Phys. D: Appl. Phys. **53**(43) (2020)
32. L. Guo et al., Mechanism of virus inactivation by cold atmospheric-pressure plasma and plasma-activated water. Appl. Environ. Microbiol. **84**(17) (2018)
33. J. Zhang et al., Discharge plasma-activated saline protects against abdominal sepsis by promoting bacterial clearance. Shock **52**(1), 92–101 (2019)
34. R. Zhou et al., Plasma-activated water: generation, origin of reactive species and biological applications. J. Phys. D: Appl. Phys. **53**(30) (2020)
35. N. Cimolai, Environmental and decontamination issues for human coronaviruses and their potential surrogates. J. Med. Virol. **92**(11), 2498–2510 (2020)
36. J.A. Otter et al., Transmission of SARS and MERS coronaviruses and influenza virus in healthcare settings: the possible role of dry surface contamination. J. Hosp. Infect. **92**(3), 235–250 (2016)
37. B.R. McMinn, N.J. Ashbolt, A. Korajkic, Bacteriophages as indicators of faecal pollution and enteric virus removal. Lett. Appl. Microbiol. **65**(1), 11–26 (2017)
38. D. Verreault, S. Moineau, C. Duchaine, Methods for sampling of airborne viruses. Microbiol. Mol. Biol. Rev. **72**(3), 413–444 (2008)
39. T. Xia et al., Inactivation of airborne viruses using a packed bed non-thermal plasma reactor. J. Phys. D: Appl. Phys. **52**(25) (2019)
40. N. van Doremalen et al., Aerosol and surface stability of SARS-CoV-2 as compared with SARS-CoV-1. N. Engl. J. Med. **382**(16), 1564–1567 (2020)
41. Y. Bai et al., Presumed asymptomatic carrier transmission of COVID-19. JAMA **323**(14) (2020)
42. L. Guo et al., Plasma-activated water: an alternative disinfectant for S protein inactivation to prevent SARS-CoV-2 infection. Chem. Eng. J. **421** (2021)
43. A. Bisag et al., Cold atmospheric plasma inactivation of aerosolized microdroplets containing bacteria and purified SARS-CoV-2 RNA to contrast airborne indoor transmission. Plasma Process. Polym. **17**(10) (2020)
44. Z. Chen, R. Wirz, Cold atmospheric plasma for COVID-19. Preprints (2020)
45. E.M. Kettleson et al., Airborne virus capture and inactivation by an electrostatic particle collector. Environ. Sci. Technol. **43**(15), 5940–5946 (2009)

Chapter 5
Cold Plasma Based Wound Healing Application

Kai Masur

Abstract The healing of wounds displays a complex process of consecutive steps: homeostasis (stop the bleeding by forming a clot), followed by an inflammation phase (to kill micro-organisms), subsequently followed by a phase of cell proliferation and finally a phase of tissue remodeling. Usually human body is capable to heal acute wounds within a few days. However, these well orchestrated phases can be interrupted at any or multiple stages—causing a chronification of wounds. In most cases an underlying disease such as diabetes or the metabolic syndrome leading to such delays in wound healing—often accompanied by infections. Those wounds display a decreased support of nutrition and oxygen—weakening the body´s defense and repair abilities. This chapter will highlight the effects of cold atmospheric pressure plasmas (CAP) on the modulation cell activities and the support of wound healing. The complex mixture of reactive species, electric fields in combination with mild heat and various kinds of radiation—ranging from UV, over visible to near infrared light interacts with micro-organisms and human cells. CAP displays antimicrobial efficacy, support cell proliferation and migration by modulation cellular redox balance. And finally, clinical trials could show that CAP leads to an increased micro-circulation and therefore an elevated tissue oxygenation.

5.1 Background/Introduction

The development of life on our planet was influenced by many factors—one of which is oxygen. During the first billions of years, only single-celled life existed in the primordial oceans—and all developments initially took place under anaerobic conditions … the exclusion of oxygen. Due to the facts, that the first cyanobacteria began to photosynthesize, oxygen was released for the first time. However, even this oxygen did not reach the earth's atmosphere, but first oxidized everything that was freely available in the water. This is how most of today's iron ore deposits were

K. Masur (✉)
Leibniz-Institute for Plasma Science and Technology, Greifswald, Germany
e-mail: kai.masur@inp-greifwald.de

E. H. Choi (ed.), *Plasma Biosciences and Medicine*, Topics in Applied Physics 148,
https://doi.org/10.1007/978-981-19-7935-4_5

formed—by freely available iron being oxidized to rust by the oxygen generated by cyanobacteria and deposited on the sea floor.

Only after this and similar oxidation processes had saturated over millions of years, the concentration of free oxygen did increase, first in the ocean and later in the Earth's atmosphere. This laid the foundation of life as we know it today, but initially also led to one of the greatest mass extinctions in the history of the Earth [1]. The background is that free oxygen—and the reactive oxygen species (ROS) derived from it—are highly reactive, and oxidize everything … including organic matter. Therefore, life had to adapt to the new conditions, and some organisms also managed this step. Numerous adaptations enabled a manageable number of single-celled organisms to come to terms with the highly reactive oxygen. They developed defence mechanisms to capture the oxygen—or the oxygen radicals, to make them harmless, or to repair the damage caused. The advantage of breathing air—containing free oxygen—led to development of multi-cellular organisms enabled to consume organic materials, which are metabolised by applying oxidative processes. In turn, they had to build up mechanisms to control/convert and repair the damages caused by oxygen and its reactive species.

In this context, aerobic cellular respiration also developed, and with it life as we know it today, which incorporates oxygen as an energy-rich compound into the processes of life. Thus, oxygen not only plays an essential role in the mitochondria in the cellular respiration of all eukaryotic organisms, these organisms also developed strategies to use oxygen as a weapon[2]. Besides this, ROS play a central in many physiological processes, orchestrating several signalling pathways and even influence tissue regeneration. However, the major function of reactive species in the human body this so-called oxidative or respiratory burst. The respiratory burst of phagocytes is required for the optimal killing of a wide variety of bacteria and fungi. The burst of 0_2 consumption is utilized by an NADPH-oxidase to generate highly-reactive oxygen species (ROS) starting with one and two electron reductions to generate superoxide anion $\left(O_2^-\right)$ and finally hydrogen peroxide (H_2O_2), respectively [3].

5.2 Wound Healing

5.2.1 Acute Wounds

Physiological wound healing is divided into four phases: hemostasis, inflammatory reaction, proliferation and remodelling phase—whose transitions are fluid [4]. In the first hours after wound formation, hemostasis begins, with constriction of blood vessels, platelet aggregation and finally thrombus formation. So the initial wound will be sealed to ensure that no further microorganisms might intrude the body. In the subsequent inflammatory phase the immune systems becomes activated and especially leukocytes such as monocytes, neutrophil granulocytes and lymphocytes infiltrate the wound. Usually this phase lasts between a few hours up to two or three days. As a result, the wound has been cleared of the microorganisms and inflammation processes are turned down.

After the infection is deminished, the proliferation phase follows. Usually this phase lasts days to weeks, characterized by the re-epithelialization processes. This involves keratinocytes and fibroblasts, which are predominantly responsible for fibronectin, keratin, and collagen synthesis to ensure production of extracellular matrix (ECM) as well matrix contraction—necessary for correct wound closure. The proliferation phase also includes the beginning of angiogenesis: the formation of new blood vessels. However, the major part is the formation of granulation tissue in order to reduce the wound size and finally to close the wound. The proliferation phase transitions seamlessly into the remodelling phase, where vascularization of blood vessels and collagen formation is completed and reduction of scarring occurs.

5.2.2 Chronic Wounds

Alterations in any of these above mentioned phases can promote chronic wound development and may impede wound healing [5]. The pathological wound healing for example of the diabetic foot has to be distinguished from the physiological one. The dysfunction of the granulocytes is of central importance, but also fibroblasts are impaired in their function. The cause of this is often a disease-related shortage of tissue oxygenation and a reduction of nutrients. This is often accompanied by a permanent colonisation with microorganisms—causing a chronic infection. As a consequence, there is a reduced level of growth factors, e.g. Platelet-Derived Growth-Factor (PDGF), an increased level of proteases, especially an increased protein expression of matrix-metallo proteases (MMPs) [6]. The elevated level of proteases prevents coordinated wound healing by constantly degrading extracellular matrix, wound healing-promoting growth factors, and their receptors, and thereby preventing cell migration processes of fibroblasts and keratinocytes. As a result, subsequent re-epithelialization processes are prevented and immune cell persistence is increased.

However, physiological wound healing and a rapidly subsiding inflammatory response require a balanced concentration of growth factors and cytokines, as well as of proteases and extracellular matrix turnover, which interact with each other during the complex repair mechanisms (Fig. 5.1).

A major difference of wound healing in healthy individuals, in addition to the altered expression pattern of various endogenous factors, is the prolonged inflammatory phase in chronic wounds. Especially in diabetic foot, endotoxins released by pathogenic agents prolong inflammatory responses and worsen wound healing. In those wounds, the neutrophil count is upregulated, which maintains inflammation via the secretion of growth factors, such as TNF-alpha and interleukins—such as IL-6, and causes an insufficient supply of oxygen (ischemia and hypoxia).

Therefore, the development of innovative therapy options which significantly contribute to the healing of chronic wounds, by reducing microbial load, improving

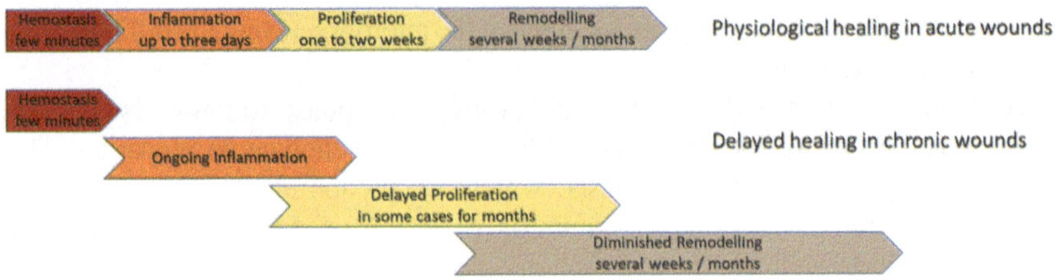

Fig. 5.1 Comparison of acute and chronic wound healing—delays in either inflammatory phase/proliferation phase and/or remodelling phase lead to chronification

tissue regeneration by enhanced cell proliferation and cell migration would be of tremendous importance.

5.3 Cold Atmospheric Pressure Plasma and Chronic Infected Wounds

In the recent years, a new field of research has been established: plasma medicine. Plasma medicine is an innovative research field combining plasma physics, life science, and clinical medicine. The main aspect in plasma medicine is the biological and clinical application of cold atmospheric pressure plasma (CAP). Applying energy to molecular or noble gases—can generate partially ionized gases that modulate biological response mediated by reactive oxygen and nitrogen species (ROS and RNS) in combination with electric fields and mild UV radiation [7]. Thereby these partially ionized gas plasmas contribute to wound healing by modulating several processes: reduction of microbial load, modulation of cell proliferation and cell migration and improving tissue oxygenation. In Europe, several plasma sources are already certified as class 2A medical devices. Based on their ability to generate cold atmospheric plasmas, which in turn are capable to inactivate microorganisms but also to stimulate tissue regeneration, current medical applications are focused on the treatment of wounds and skin diseases [7].

5.3.1 Anti-Microbial Effects

During the early years of plasma medicine the focus was led on the anti-microbial efficacy of CAP—which could be demonstrated for several plasma sources. Especially with focus on novel concepts to limit the spread of multidrug-resistant bacteria (MDR) the application of CAP was investigated intensively.

Usually, for each plasma source the appropriate treatment conditions have to be evaluated. Typically the distance and treatment time are crucial for a proper

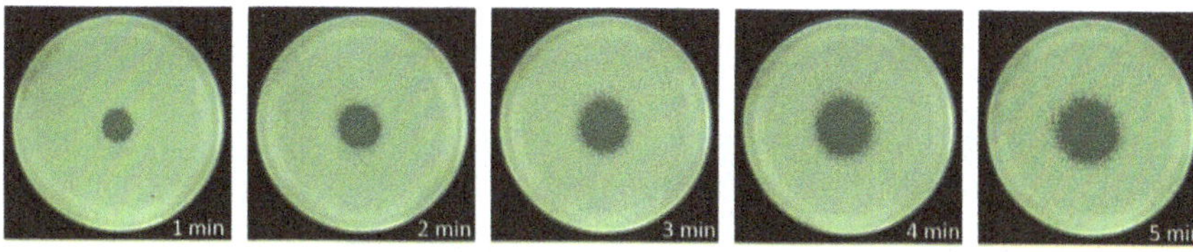

Fig. 5.2 Treatment time dependent increase of inhibition zones of Staphylococcus aureus seeded on agar plates and treated with argon plasma by kINPenMed ® for 1, 2, 3, 4 and 5 min per square centimetre

reduction of microorganisms. Most of the known plasma sources are capable to reduce bacteria within a few minutes or even seconds. A well established method to visualize the anti-microbial effects is to seed a distinct amount of microorganisms on an agar plate/perform the plasma treatment/incubate over-night and finally observe the inhibition zone—the area free of microorganisms (Fig. 5.2).

Daeschlein et al. evaluated the ability of CAP to eliminate MDR- compared to non-MDR-pathogens in chronic wounds. They could proof that a single CAP treatment reduced MDR in all wounds. In 14 treatments (63.6%) and for 16 pathogens (66.7%), a 100% reduction of the bacterial load was observed. For 11 of 17 (64.7%) MDR-pathogens a complete eradication was achieved [8].

However, there seems to be a discrepancy between the reduction rate of micro-organisms cultured and plasma treated on agar plates and micro-organisms treated on real wounds. While 5–6 log reduction (five to six orders of magnitude) are commonly found for plasma treated micro-organisms on agar plates—the same plasma sources used for wound treatment typically show reduction rates of one or two orders of magnitude (1–2 log reduction). This is mainly due to the fact, that on agar plates or generally in cell culture optimized conditions provide a better micro-environment compared to the situation found in real wounds. The wound fluid contains a huge amount of partially unknown proteins, buffering substances and contamination—ranging from bacteria and fungi to dead cells and debris of all kinds. This complex mixture of organic molecules will interact with plasma components and therefore diminish the effects on the micro-organisms themselves—finally leading to a smaller reduction rate (Fig. 5.3).

Besides the fact that reduction of bacterial load in patients wounds displays lower efficacy compared to clean room cultured organisms on agar plate—plasma treatment of wounds is most effective when combined with proper debridement. For well-orchestrated wound care, each wound needs to be assessed by a professional nurse or physician—including management of any complications and comorbidities. Therefore, plasma is a very helpful supporting technology and has the enormous potential to optimize a professional cleaning of wounds regarding the reduction of bacterial loads of chronic infected wounds.

However, this issue is of even greater importance, when patients are released from clinics and transferred to home care. In this case, a similar standard of wound care has to be ensured for a return to clinic for an ambulant plasma treatment. Figure 5.4 depicts the bacterial load of "patient 37" over several weeks of plasma treatment—with first nine treatments when "patient 37" was hospitalized followed by wound care

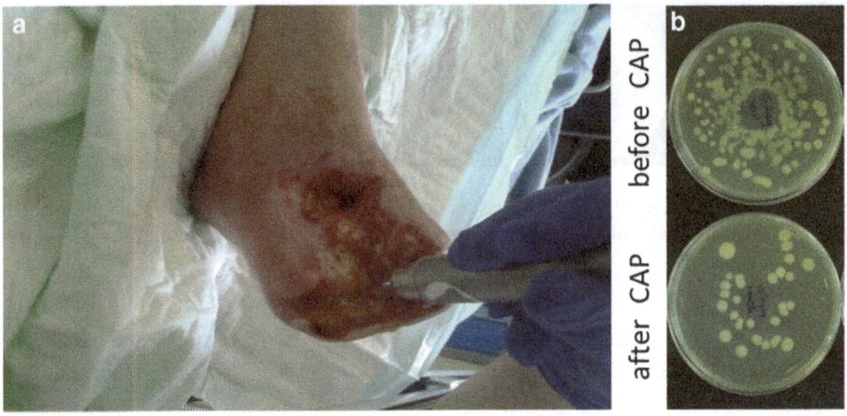

Fig. 5.3 Plasma treatment of a diabetic foot ulcer applying argon plasma of the kINPen Med ® for 30 s per square centimetre. **a** Over-night culture of micro-organisms **b** of a sample taken from the wound before (top picture) and after (bottom picture) argon plasma treatment

at home performed by day care and a accompanied ambulant plasma treatment at the clinic for several weeks. A less professional wound management led to an increase of bacterial load reaching levels higher that before the initial plasma treatment. Each follow-up plasma treatment in ambulant clinic procedure was able to reduce the bacterial load, but efforts were disrupted by bacterial re-growth in between the weekly plasma treatments.

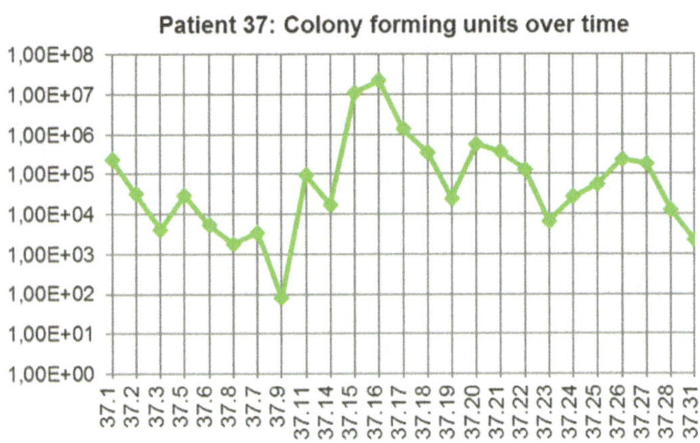

Fig. 5.4 Up and down of bacterial load (colony forming units) of "patient 37"—starting with stationary plasma treatments until treatment number nine—at the day of release from stationary wound care to day care. The following records of wound colonisation were dependent on the professionalism of wound care and the additional effects of cold plasma to diminish micro-organisms in wounds

In summary, plasma treatment is able to reduce bacterial load in chronic infected wounds—but success strongly depends on proper debridement and a state-of-the-art wound care management. There are differences between findings in the laboratory, with microorganisms cultured and treated under optimal conditions—and the real world treatment on the patient side. The ability of cold plasma to reduce bacterial load is affected by many factors—such as dead cells, components of extracellular matrix, body fluids (blood/lymph) containing large amounts of organic molecules, which scavenge reactive species generated in cold plasma. Therefore, cleaning wounds, removing necrotic tissue and visible bacterial films strongly supports an additive cold plasma treatment—and finally ensures a successful wound healing. The major advantage—from a antiseptic point of view—is the ability of CAP to kill antibiotic resistant bacteria in a same manner as non-resistant strains. So far, no report of any plasma-resistance has been reported.

5.3.2 Cold Plasma in Cell Culture

Besides its antimicrobial effects, CAP also modulates cell activities in dermis and epidermis. However, before starting applications on human beings, cell culture experiments have been performed in order to identify the mode of action. Applying CAP in a similar procedure compared to the tests for its capabilities on microorganisms, the focus now is on its stimulating effects. In most experiments keratinocytes and fibroblasts—in rare cases also immune cells are the focus of investigations. In first sight cell viability/cytotoxicity tests have been performed in order to find optimal treatment times and conditions for each plasma source. Usually a few seconds up to one or two minutes per square centimetre of CAP treatment are sufficient to modulate cellular activities. One major player identified is the NRF-2/Keap-1 system—which are sensors for oxidative stress in cells. Several authors showed that expression and cellular translocation of NRF-2 from cytosolic fraction into the nucleus could be detected after CAP treatment. Schmidt et al. could show that the cellular redox homeostasis was maintained and cells were defended from damage by a strong modulation of the nuclear E2-related factor (NRF-2) pathway [9]. As a transcription factor, NRF-2 binds to antioxidant response elements (AREs) in the nucleus leading to transcription of ARE genes [10]. In general, cells can overcome chronic oxidative stress by enhancing activities of anti-oxidant enzymes, thereby protecting cells from DNA damage [11]. While NRF-2/Keap-1 act as intra-cellular sensors for oxidative stress in order to prevent damage and to start repair mechanisms, when ROS and RNS are applied at higher concentrations—a short term treatment with cold plasma is capable to induce another transcription factor system: YAP/TAZ. This transcription factor is associated with the HIPPO pathway, known to activate genes for regeneration and proliferation [12, 13]. Shome et al. already demonstrated successfully that also short term plasma treatment of fibroblasts and keratinocytes led to an activation and translocation of YAP to the nucleus—leading to gene activation [14] (Fig. 5.5).

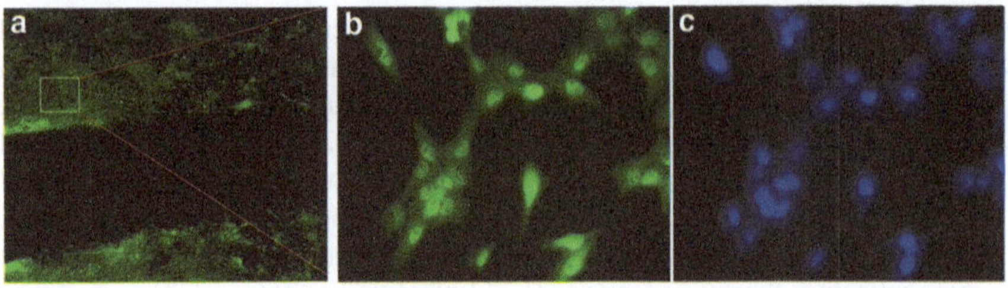

Fig. 5.5 Yap is translocated to nucleus of fibroblasts upon scratch wound and CAP treatment—indicated by green fluorescence in left figure (**a**) and in magnification in middle part (**b**). For comparison the DAPI counterstaining in the right figure (**c**) depicting nuclear staining in both cases

CAP treatment resulted in an upregulation of the HIPPO transcription factor YAP in both keratinocytes and fibroblasts. Downstream effectors of the HIPPO signalling pathway (CTGF and Cyr61) were upregulated mainly in fibroblast fraction. In addition, the administration of antioxidants such as N-acetyl-cysteine could inhibit CAP-mediated wound healing and abrogate the gene expression of the HIPPO downstream effectors. Furthermore, Shome et al. could also prove a paracrine signalling between fibroblasts and keratinocytes when co-cultured and plasma treated under this condition. In this case, a CAP treatment led to an activation of dermal fibroblasts—resulting in secretion of Cyr61 and CTGF—which in turn led to the paracrine stimulation of co-cultured keratinocytes (Fig. 5.6).

In addition, they could show that keratinocytes revealed an elevated cell migration when incubated with CAP-treated fibroblast-conditioned media compared to an incubation with untreated cell culture media—leading to a significantly improved cell migration of keratinocytes [14]. Taken these facts together, they could show an improved keratinocyte wound healing in co-culture—with CAP treated fibroblasts as central players. This also proves that experimental set-ups should be carefully

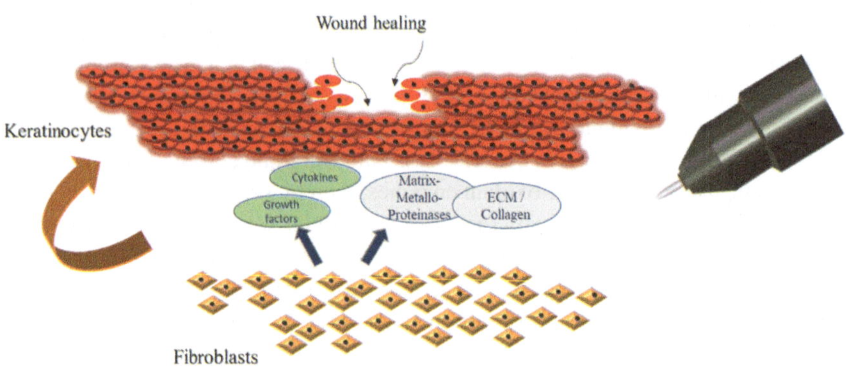

Fig. 5.6 Paracrine activation of keratinocytes by CAP modulated fibroblasts. Adapted from Shome et al. [14]. Oxidative Medicine and Cellular Longevity Plasma stimulated fibroblasts secret cytokines and growth factors, leading to increased keratinocyte migration and proliferation

planned and in best case should mimic real wound conditions. Co-cultured fibroblasts and keratinocytes were capable to tolerate a CAP treatment much better than the mono-cultured cells at same density. Cell communication after plasma treatment is more complex in co-cultured cells and therefore reflects real skin situation more close than mono-culture cell experiments.

5.4 Animal Studies Applying Cold Plasma

Another aspect in the CAP mediated effects on wound healing is reflected by animal experiments—mostly in mice. There are several groups showing an improved healing of acute wounds. In early studies Schmidt et al. examined the cold plasma's efficacy on dermal regeneration in a murine model of dermal full-thickness ear wound [15]. Within a period of two weeks, female mice received daily plasma treatment. Their results showed a significantly accelerated wound re-epithelialization at days 3–9 in comparison with untreated controls. But also cell communication, cell migration and cell attachment is influenced by cold plasma treatment. By combining in vitro analyses in primary dermal fibroblasts isolated from murine skin with in vivo studies in another murine wound model Schmidt et al. could demonstrate that plasma treatment changed phosphorylation of signalling molecules such as focal adhesion kinase and paxillin alpha in adhesion-associated complexes [16]. The same group also investigated the integrity of healthy skin of plasma treated mice by analysing tissue oxygenation, perfusion, hemoglobin, and water index by applying hyperspectral imaging. In this study Schmidt et al. could show a plasma based modification of the junctional network in skin, which promoted tissue oxygenation, and restricted penetration, implicating that plasma may provide a novel and sensitive tool of skin barrier regulation [17]. Furthermore, animal experiments also showed a direct activation of immune cells as another important fact in plasma mediated wound healing. Kupke et al. could prove a CAP-related induction of neutrophils in wound tissue from mice by investigating the functionality of human polymorphonuclear cells (PMN)/granulocytes through either a plasma-treated solution (PTS) or the direct CAP treatment [18]. They stated that the modification of PMN immunoreactivity by direct plasma treatment might be a main supporting mechanism for CAP-induced improvement in wound healing.

Besides the confirmation that CAP treatment led to significant improvements in wound healing—several studies also investigated the safety issues in long-term animal studies [19, 20]. Schmidt et al. studied in an one year follow-up risk assessment in SKH-1 mice the possible side effects of a CAP treatment applying an argon jet plasma. They applied quantitative PCR, to investigate expression levels of several cytokines and tumour markers in liver, lung, and skin. In addition, also histological and immune-histochemical analysis failed to detect abnormal morphological changes and the presence of tumour markers. Also magnetic resonance imaging and positron emission tomography confirmed the absence of neoplastic lesions in these mice [20]. Evert et al. investigated in their study the long-term risk assessment of

CAP treatment in the oral cavity. Histological analysis of 406 animals revealed that repeated CAP exposure did not foster non-invasive lesions or squamous cell carcinoma. In conclusion Evert et al. stated that a repeated CAP exposure of murine oral mucosa was well tolerated, and carcinogenic effects did not occur, motivating CAP applications in patients for dental and implant treatments in the future [19].

Similar results from cell culture experiments of various groups working with different plasma sources could confirm that none of the CE-certified plasma sources showed any mutagenic potential [20–23]. In summary, all certified plasma sources are safe in medical application, if the devices are handled according the manufactures advices. In the recent years, there are first efforts for a standardisation of plasma sources—with a German pre-standard DIN Spec91315. A first application of the DIN Spec91315 is published by Mann et al. and provides a basic approach how to test plasma sources for safety and efficacy [24]. Ongoing efforts recently started to transform this German pre-standard into a DIN norm, and parallel on an international mutual project the preparation of an IEC standard started in 2021.

Further information on safety aspects and standardisation can be found in chapter nine.

5.5 Clinical Application of Cold Plasma for Wound Healing

Based on very promising results from cell culture and microbiology first case reports were started in early 2012. Metelmann et al. investigated in a case report including five individuals with identical settings how a argon plasma can stimulate the healing of skin lesions of CO_2 laser and observed the recovery of these artificial acute wounds [25]. Those 20 laser lesions have been treated with argon plasma for 10, 30 s or three-times for 10 s, and compared to untreated control laser lesions. While in first approach, the scar formation was observed for 10 days, in a second follow-up study further evaluations of those lesions were evaluated after six and 12 months. As a result, Metelmann et al. stated, that plasma treatment shows superior aesthetics during scar formation. No precancerous skin features occurred up to 12 months [26].

Stratmann et al. started a clinical trial in order to determine whether the application of CAP accelerates wound healing in diabetic foot ulcers compared with standard care therapy [27]. This prospective, randomized, placebo-controlled, patient-blinded clinical trial was conducted at two clinics—and was the first study which could prove a positive influence of CAP on wound size reduction. Therefore, standard care treatment with eight applications of either CAP generated from argon gas of an atmospheric pressure plasma jet was compared to eight applications of placebo treatment in a patient-blinded manner. CAP therapy yielded a significant increase in wound healing, both in total mean (SD) area reduction and mean (SD) time to relevant wound area reduction [27]. Moreover, cutaneous blood flow and oxygen saturation can be improved in human skin—due to a clinical application of cold plasma [28]. As stated by Kisch et al., these effects are mostly explained by reactive oxygen species

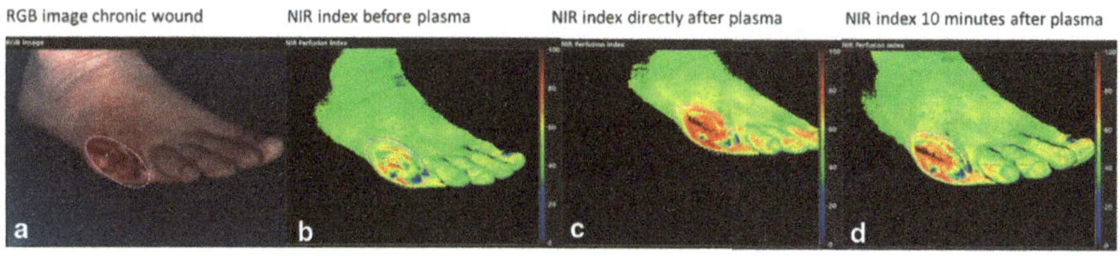

Fig. 5.7 Hyperspectral imaging of a plasma treated chronic wound **a** RGB coloured image—and **b** false colours indicated an increase in microcirculation indicated by shift to orange and red colour after plasma **c** which stayed elevated several minutes after CAP **d**

(ROS), but electric fields, currents and ultraviolet radiation may also have an impact on cells in the treated area [28].

Applying hyperspectral imaging we also could show that a treatment with the argon plasma jet improved tissue oxygenation and microcirculation in a chronic foot ulcer. Analysing the NIR perfusion index we could show a significant increased micro-circulation directly after plasma treatment (indicated by increase in orange and red colour—which stayed elevated after the procedure (Fig. 5.7).

In another clinical trial applying the argon jet plasma jet, wound exudate was investigated within a prospective, randomised, patient-blinded clinical trial. Hiller et al. recently published that those CAP-treated wounds showed increased levels of tumour necrosis factor-alpha, interleukins 1alpha and 8. They also found an induction of crucial growth factors, like FGF-2 and VEGF-A, and interleukins appears to be an important component of CAP-mediated promotion of granulation, vascularisation and reepithelialisation in the diabetic foot [29].

Taken exudate samples on a regular base allows a permanent control of wound progression. Besides the detection of inflammation markers such as cytokines, the analysis of matrix-metallo-proteinases (MMP) could be an hallmark for an evaluation of the stimulation by cold plasma. We could show, that CAP treatment led to a significant reduction of MMP-2, MMP-8 and MMP-9 levels over a duration of a 10 week plasma treatment (Fig. 5.8).

In summary, several case reports and clinical trails proved the positive effects of cold plasmas generated by different devises. Besides its antimicrobial efficacy cold plasma significantly induced wound healing—especially in diabetic patients often

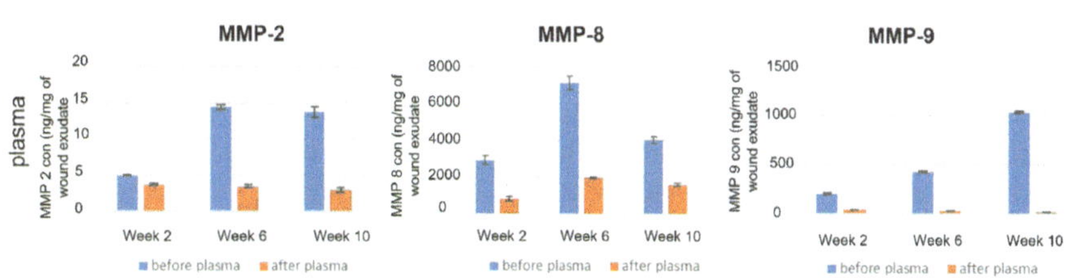

Fig. 5.8 Results from wound exudate analysis—CAP treatment reduces expression of matrix-metallo proteinases

suffering from chronic wounds for years. There is a clear tendency that CAP treatment is mediating growth factor induction and modulation of matrix-composition—resulting in a faster wound closure. A further surplus is the plasma modulation of tissue oxygenation and elevated levels of micro-circulation leading to an improved support with nutrients and oxygen. These data are in concurrence with results previously described in several in vitro and rodent experiments.

In chapter eight more detailed information are summarized about clinic trails and clinical applications.

5.6 Cellular Redox Balance Modulated by Cold Plasma

In order to heal, a wound is in need for energy—in form of nutrients (carbohydrates, amino acids etc.) and oxygen for an active metabolism to finally close the wound and regenerate the tissue. In chronic wounds, these processes are diminished—often caused by diseases like the metabolic syndrome—with all its side effects. Often reduced blood flow and dimished oxygenation are the reason for an insufficient supply with nourishment and oxygen, which have to be restored in first place. A lack of oxygen and nutrients not only hampers tissue regeneration, but also reduces the ability of immune cells to defend invading microorganisms finally leading to infected chronic wounds. Those missing natural ingredients—especially the reactive species (ROS/RNS) formed during immune defences but also tissue regeneration can be replaced by identical species generated in cold plasmas. Most reactive species generated in plasma are also known in biology—functioning in redox signalling and mammalian cells are equipped to interpret the plasma derived redox signal [30] (Fig. 5.9).

The hypothesis that a single plasma component could be traced to a specific cellular event or effect is somewhat difficult to disentangle in the case of cold plasma. Due to the fact, that a complex cocktail of reactive species and various kinds of radiation are formed during the plasma generation, and the fact that most of these species are very short lived—a lot of intermediates formed in the gas phase never will reach the cells or their liquid environment. To our knowledge, the long-lived redox-active species in combination with energy from various electromagnetic radiations dominate the biological effects, while moreover, plasma biological effects are significantly modulated by plasma modifications of the liquid environment. Furthermore, there is no such single receptor of plasma species inside (or outside) the cells—nothing like a drug that binds to a single receptor. As mentioned above, ROS and RNS are natural species employed in many cellular processes. There is a constant flux of reactive species formed by diverse occasions—such as respiratory chain or immune defence. Each cell type has—based on its function and metabolic state—its own redox balance. This means all cells produce their own reactive species—(roughly two percent of oxygen consumed for cellular respiration escape within mitochondria during the processes of respiratory chain). Hence, the need for counter reactions or repair mechanisms has to be active in all oxygen consuming cells in order to ensure a

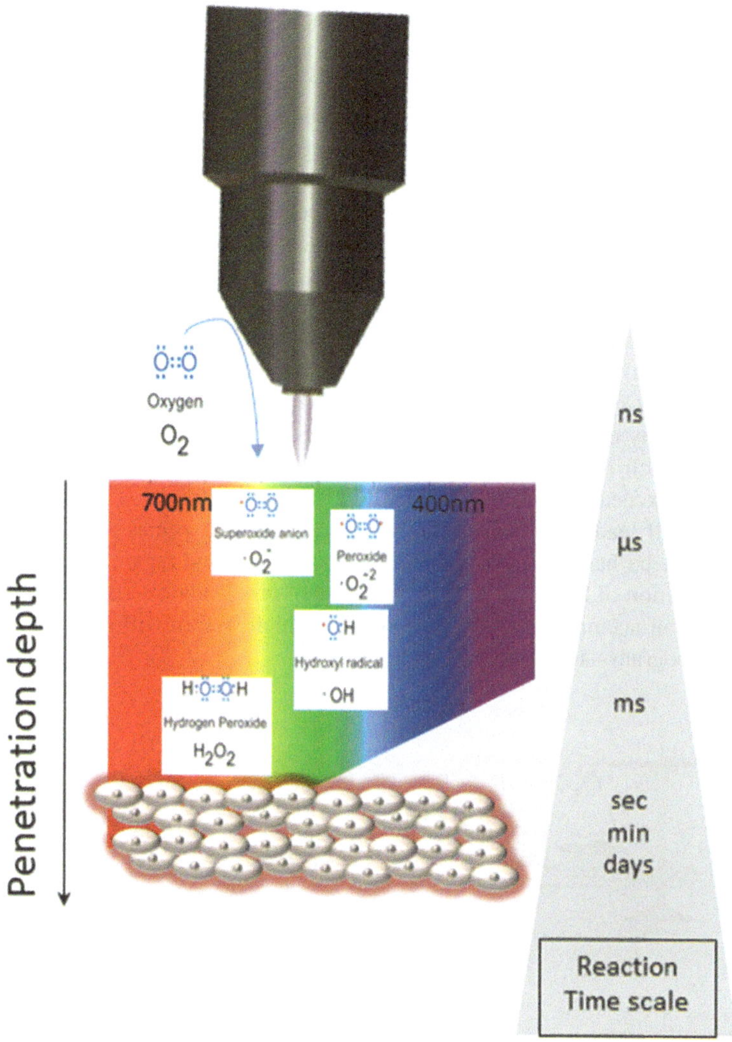

Fig. 5.9 The hub of plasma medicine. The addition of energy to gases leads to the formation of (partially) ionized gases—cold plasma—which are used to modulate the redox balance of living cells. Very short-lived radicals are formed in plasma—with half-lives ranging from nanoseconds to microseconds. Long-lived redox-active species (with half-lives up to several seconds) dominate the biological effects, while biological plasma effects are essentially modulated by plasma changes in the liquid environment, followed by secondary reactions within cells. The overall chain reaction—triggered by the plasma—can last for days and includes triggered effects such as cell proliferation and migration. Over time, the availability and penetration depth of the reactive species also change. The penetration depth of light—produced by the previously excited species—also depends on the wavelength, while UV light is absorbed by the air and fluids above the treated tissue—and only red and near IR light can penetrate multiple cell layers. Plasma reactions are very complex and cover a wide range in time. There are energy transfers occurring in the very first nanoseconds, triggering the generation of new reactive species. Some of them will react with an half-life of a few µs and milliseconds—finally leading to more stable species such as hydrogen peroxide, which is able to interact with living matter. The reaction triggered directly and indirectly at the cells are causing the next stage of plasma mediated reactions, such as the activation of transcription factors. Once transcription factors caused the activation of signalling cascades, and genes are transcribed, cell proliferation and cell migration are triggered within a time frame of hours. Finally the plasma directly and indirectly activates cellular activities which can last for up to a few days

proper cell cycle. However, there is a limit on the amount of reactive species, which can be handled by a single cell. For each cell type (based on metabolic state and function) there is a certain threshold—the **AOP anti-oxidative potential** of a cell, encircling all anti-oxidative possibilities and counter reactions (including repair).

For each oxidative influence (endogenous e.g. from respiratory chain/or exogenous like UV radiation, oxidants) a cellular counter reaction is induced by effectors such as NRF-2/TAZ—leading to a restoring of cellular redox balance. These modulations can be triggered by cold plasma—as an exogenous influence of the cellular redox balance like any other source from outside. Therefore, CAP can induce activation of the transcription factors NRF-2 and YAP and finally can lead to an activation of tissue repair and regeneration too. Those processes (as a sum of internal and external triggers) will lead to a cellular activation. By modulating the ROS/RNS based processes of the cells tissue regeneration but also tissue inflammation can be influenced by cold plasma. However, there is a certain limit for each cell type or tissue—which is defined by the anti-oxidative potential of the cells. Once this threshold is exceeded, the cellular redox balance is destroyed (Fig. 5.10d). This imbalance will lead to an accumulation of redox active species and compounds, which cannot be handled by the cellular anti-oxidants and repair mechanisms. A chronic elevation of ROS will lead to an excess of pro-oxidative effects, and finally will lead to an accumulation of cell damages, which in turn will start the cellular apoptosis program—the programmed suicide.

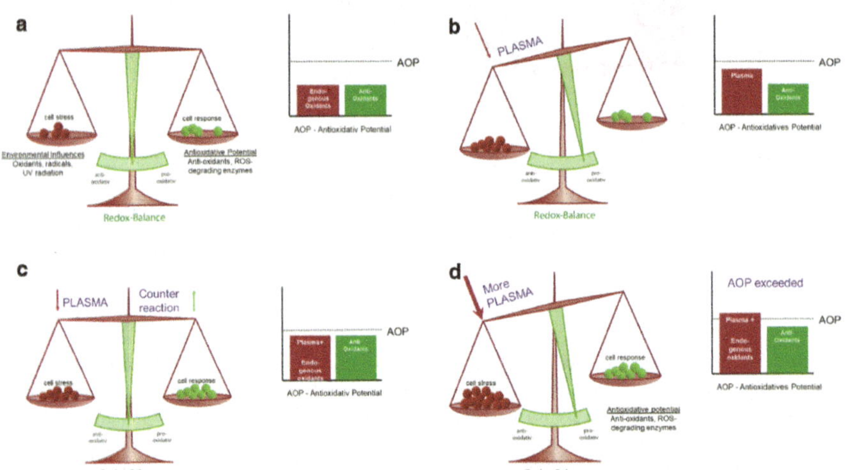

Fig. 5.10 The cellular redox balance: Each cell type has its own redox balance of exogenous and endogenous oxidative processes and its counter reactions of the cell. **a** Each increase on the oxidative side, **b** will lead to a redox balance restoration by cellular counter reaction. **c** This balance and counter-balance effects will ensure cell survival, as long the anti-oxidative potential is not exceeded. Once this threshold is exceeded, **d** cellular damage will occur—finally leading to the programmed cell death apoptosis

Therefore, each plasma source needs to be evaluated in detail, before an application on human beings is planned. Again, the German pre-standard DIN Spec 91,315 contains a selection of tests in physics, cell- and micro-biology in order to ensure a safe and efficient plasma treatment with a clinical focus. (more details in Chap. 9 on safety and standardisation.)

5.7 Summary

About two decades ago, new devices have been developed, capable to generate partially ionised gases—so-called cold plasmas. These energy-rich mixtures of reactive species (mainly ROS and RNS) in combination with mild heat and UV radiation are tissue-tolerable. These well prepared cocktails display several properties for a medical application with focus on wound healing: they show a high anti-microbial efficacy—with the surplus that they are effective on anti-biotic resistant strains as they are on normal skin microbes. Especially in chronic infected wounds with permanent bacterial load in combination with co-existing fungi an advantage to most antiseptics. However, besides anti-microbial effects, cold plasmas are capable to stimulate human tissue, by modulation the cellular redox balance. This well orchestrated signalling cascade of redox sensors (such as NRF-2/Keap-1) also other transcription factors will be activated upon CAP treatment. For example, the YAP/TAZ system which belongs to the HIPPO pathway is an important activator for tissue regeneration. Following an activation via NRF-2 or YAP cells start to synthesize cytokines and growth factors, enabling the plasma treated cells and surrounding tissues to elevate proliferation and cell migration activities. Both, animal and human trials proved that a third hallmark of cold plasma treatment is an increased tissue oxygenation and micro-circulation. This further supports a proper wound healing due to an improved support with nutrients and oxygen. All three facts: anti-microbial efficacy, modulation of cellular redox balance and thereby stimulation of cell migration and proliferation and third an elevated tissue oxygenation led to the superiority of cold plasma in healing infected chronic wounds. This could be proven in several case studies and clinical trials (Fig. 5.11).

CAP components are supporting cellular processes, which were reduced due to underlying diseases and limited supply of energy. Generating ROS/RNS as physiological components of immune defence and signalling cascades, plasma supports the diminished activities of a weakened immune system or tissue regeneration. There are hints that a plasma treatment of chronic wounds turns them back into acute wounds—where normal wound healing can start on its own. However, this plasma mediated support of wound healing can only be accomplished when a proper wound management is performed, and all other side effects of the underlying disease are corrected. Once the debridement is done, and blocked blood vessels are re-opened, an additional plasma treatment of such chronic wounds will be a great support for

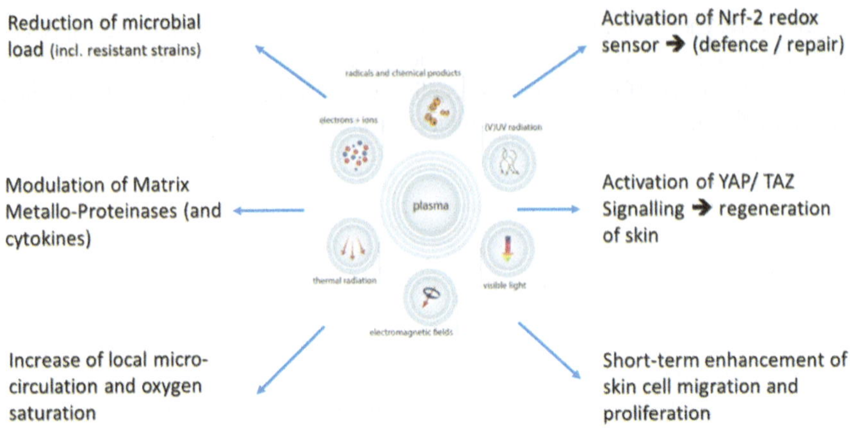

Fig. 5.11 Summary of the three hallmarks of cold plasma for wound healing: (1) anti-microbial efficacy; (2) modulation of cellular redox balance with subsequent stimulation of cell growth and migration and (3) increased tissue oxygenation and micro-circulation

most patients. Further studies and clinical trials will help to understand the underlying mechanisms in more detail, so that a more personalized CAP treatment will be even more effective.

References

1. T.W. Lyons, C.T. Reinhard, N.J. Planavsky, The rise of oxygen in Earth's early ocean and atmosphere. Nature **506**(7488), 307–315 (2014)
2. P. Yang, S. Huang, A. Xu, Chapter 8—The oxidative burst system in amphioxus, in *Amphioxus Immunity*, ed. by A. Xu. (Academic Press, 2016), pp. 153–165
3. C. Dahlgren, A. Karlsson, J. Bylund, Measurement of respiratory burst products generated by professional phagocytes. Methods Mol. Biol. **412**, 349–363 (2007)
4. J.M. Reinke, H. Sorg, Wound repair and regeneration. Eur. Surg. Res. **49**(1), 35–43 (2012)
5. Y.H. Almadani et al., Wound healing: a comprehensive review. Semin. Plast. Surg. **35**(3), 141–144 (2021)
6. R. Lobmann, G. Schultz, H. Lehnert, Molecular fundamentals of wound healing in diabetic foot syndrome. Med. Klin. (Munich) **98**(5), 292–301 (2003)
7. T. von Woedtke et al., Perspectives on cold atmospheric plasma (CAP) applications in medicine. Phys. Plasmas **27**(7) (2020)
8. G. Daeschlein et al., Skin and wound decontamination of multidrug-resistant bacteria by cold atmospheric plasma coagulation. J. Dtsch. Dermatol. Ges. **13**(2), 143–150 (2015)
9. A. Schmidt et al., Nrf2 signaling and inflammation are key events in physical plasma-spurred wound healing. Theranostics **9**(4), 1066–1084 (2019)
10. A.P. Gureev, V.N. Popov, A.A. Starkov, Crosstalk between the mTOR and Nrf2/ARE signaling pathways as a target in the improvement of long-term potentiation. Exp. Neurol. **328**, 113285–113285 (2020)
11. A. Yanaka et al., Role of the nrf-2 gene in protection and repair of gastric mucosa against oxidative stress. Inflammopharmacology **13**(1–3), 83–90 (2005)

12. A.T. Nguyen-Lefebvre et al., The hippo pathway: a master regulator of liver metabolism, regeneration, and disease. FASEB J. **35**(5), e21570 (2021)
13. D. Elster, B. von Eyss, Hippo signaling in regeneration and aging. Mech. Ageing Dev. **189**, 111280 (2020)
14. D. Shome et al., The HIPPO transducer YAP and its targets CTGF and Cyr61 drive a paracrine signalling in cold atmospheric plasma-mediated wound healing. Oxid. Med. Cell Longev. **2020**, 4910280 (2020)
15. A. Schmidt et al., A cold plasma jet accelerates wound healing in a murine model of full-thickness skin wounds. Exp. Dermatol. **26**(2), 156–162 (2017)
16. A. Schmidt et al., Gas plasma-spurred wound healing is accompanied by regulation of focal adhesion, matrix remodeling, and tissue oxygenation. Redox Biol. **38**, 101809 (2021)
17. A. Schmidt et al., The molecular and physiological consequences of cold plasma treatment in murine skin and its barrier function. Free Radic. Biol. Med. **161**, 32–49 (2020)
18. L.S. Kupke et al., Cold atmospheric plasma promotes the immunoreactivity of granulocytes in vitro. Biomolecules **11**(6) (2021)
19. K. Evert et al., Repeated exposure of the oral mucosa over 12 months with cold plasma is not carcinogenic in mice. Sci. Rep. **11**(1), 20672 (2021)
20. A. Schmidt et al., One year follow-up risk assessment in SKH-1 mice and wounds treated with an argon plasma jet. Int. J. Mol. Sci. **18**(4) (2017)
21. S. Bekeschus et al., Risk assessment of kINPen plasma treatment of four human pancreatic cancer cell lines with respect to metastasis. Cancers (Basel) **11**(9) (2019)
22. K. Wende et al., Risk assessment of a cold argon plasma jet in respect to its mutagenicity. Mutat. Res.-Genet. Toxicol. Environ. Mutagen. **798**, 48–54 (2016)
23. T. Maisch et al., Investigation of toxicity and mutagenicity of cold atmospheric argon plasma. Environ. Mol. Mutagen. **58**(3), 172–177 (2017)
24. M.S. Mann et al., Introduction to DIN-specification 91315 based on the characterization of the plasma jet kINPen® MED. Clin. Plasma Med. **4**, 35–45 (2016)
25. H.-R. Metelmann et al., Experimental recovery of CO_2-laser skin lesions by plasma stimulation. Am. J. Cosmet. Surg. **29**(1), 52–56 (2012)
26. H.-R. Metelmann et al., Scar formation of laser skin lesions after cold atmospheric pressure plasma (CAP) treatment: a clinical long term observation. Clin. Plasma Med. **1**(1), 30–35 (2013)
27. B. Stratmann et al., Effect of cold atmospheric plasma therapy versus standard therapy placebo on wound healing in patients with diabetic foot ulcers: a randomized clinical trial. JAMA Netw. Open **3**(7), e2010411 (2020)
28. T. Kisch et al., The repetitive use of non-thermal dielectric barrier discharge plasma boosts cutaneous microcirculatory effects. Microvasc. Res. **106**, 8–13 (2016)
29. J. Hiller et al., Enhanced growth factor expression in chronic diabetic wounds treated by cold atmospheric plasma. Diabet. Med. e14787 (2022)
30. K. Wende et al., Chemistry and biochemistry of cold physical plasma derived reactive species in liquids. Biol. Chem. **400**(1), 19–38 (2018)

Chapter 6
Agriculture and Food Processing Applications

Henrike Brust, Nicola Wannicke, and Gyungsoon Park

Abstract Non-thermal plasma produced at atmospheric or low pressure has the potential to solve problems in modern agriculture and the food industry and may specifically address challenges resulting from the current climate crisis and environmental changes. In laboratory conditions, non-thermal plasma has shown promising results in applications such as plant disease control, seed germination, plant growth, food sanitation, and improvement of food quality and functionality. In particular, the improvement of plant vitality under stress conditions and storage time suggests that plasma can play a pivotal role in sustainable agriculture and the food industry. Advances in field- and industrial-scale applications are currently underway, as reported by an increasing number of studies. In this chapter, we summarize and discuss studies on the application of low-and atmospheric-pressure plasma to agriculture and food production.

6.1 Background

The agriculture and food industries face many challenges, including some resulting from climate change and environmental pollution. Climatic change has caused the emergence of new diseases and changes in plant susceptibility to diseases [76]. Some estimate a 5–50% reduction in crop yield as a result of climatic change by 2100 [10]. Thus, climate change, sustainable agriculture, food preservation, and improved storage of fresh produce have become important issues to be resolved.

H. Brust · N. Wannicke
Leibniz Institute for Plasma Science and Technology, Felix-Hausdorff-Str. 2, 17489 Greifswald, Germany
e-mail: henrike.brust@inp-greifswald.de

N. Wannicke
e-mail: nicola.wannicke@inp-greifswald.de

G. Park (✉)
Department of Plasma-Bio Display and Department of Electrical and Biological Physics, Kwangwoon University, Seoul 01897, Republic of Korea
e-mail: gyungp@kw.ac.kr

E. H. Choi (ed.), *Plasma Biosciences and Medicine*, Topics in Applied Physics 148,
https://doi.org/10.1007/978-981-19-7935-4_6

Conventional approaches to solve problems in agriculture and food processing are mostly focused on the use of chemicals. However, chemical-based techniques have frequently shown their limitations with respect to safety and emergence of resistance. Alternative technologies may be needed to overcome these challenges.

Non-thermal plasma is a promising technology to solve problems in the agriculture and food industries. Although intensive studies of plasma application to agriculture and foods started relatively later compared to medical applications, enormous advances in knowledge and technical development have been made in the last decade. Non-thermal plasma generated under atmospheric, low, or medium pressure has been examined for its potential in food sanitation, food storage, plant disinfection, and enhancement of seed germination and growth [12, 36, 106, 187, 278]. Although most investigations have been performed under laboratory conditions, the application of plasma at industrial and field scales is currently increasing. This chapter presents a synthesis of studies performed on plasma applications in agriculture and food production. We attempted to include as many studies as possible, and any omissions were unintentional.

6.2 Application of Non-thermal Atmospheric Pressure Plasma to Prevent Seed Borne Infections

6.2.1 General Treatment of Seeds

Seed is a basic and vital input for agricultural productivity considering that ninety percent of food crops are grown from seed. To guarantee health and quality (maximum germination above 80%) of seeds, they are generally subjected to pre-harvest manipulation directed towards improving germination and to deliver protection against pathogens and related pests and diseases. Factors that can impair seed quality are related to: biological factors (pathological, entomological, animal grub); physiological factors (physiological disorders, nutritional imbalances, maturity); environmental/cultural factors (e.g. climate, weather, soils, water relations, light intensity); mechanical damage during processing; extraneous matter (growing medium, vegetable matter, chemical residues); and genetic variation and aberrations [148]. The chosen seed treatment should be functional across a wide variety of soil types, cultural practices and environmental conditions. They aim at changing physical form to facilitate sowing (pelleting), enhance germination by improving physiological performance (priming) and extend longevity by removing pathogens. Commonly applied modes of seed treatment can be categorized into: (1) mechanical methods (scarification, separation from infectious agents), (2) physical methods (electron beam treatment, hot water treatment: dry heat treatment: aerated heat treatment, radiation treatment, microwave, ultrasound), (3) biological methods (treatment with beneficial microorganisms including fungi and bacteria e.g. species of *Trichoderma*, *Pseudomonas*, *Bacillus*, *Rhizobia*), and (4) chemical methods (organic or inorganic, metallic or non-metallic, insecticides, fungicides, bactericides using coating to form pellets or entrustments). Methods can also be combined to ensure pathogen inactivation, next to addition of auxiliary materials like nutrients or other growth promoting agents. In the past years, the application of chemical seed dressing has declined due to suspected negative effects on diversity of organisms agricultural landscapes [102, 261]. Growing concerns has led to a banned of most insecticides in the European Union, as well as chemical seed dressing using the agent Thiram (TMTD), widely applied as a fungicide in rape and leguminous seed treatment to prevent soil-borne infections [52].

Application of alternative non-chemical seed treatment methods are propagated, like electron beam treatment or are under development like cold plasma application in pre-harvest. Most non-chemical alternatives are not functional against soil borne pathogens or insect or animal grub, because no long-lasting reservoir of agents is formed. Commonly found soil borne pathogens, which often can also be spread by invested seeds, include the fungal genus *Fusarium, Pythium, Rhizoctonia, Phytophthora, Verticillium, Rhizopus, Thielaviopsis,* and *Sclerotia* [235]. In addition to being soil borne, some pathogenic bacteria like *Ralstonia solanacearum, Streptomyces scabies, Clavibacter michiganensis* subsp. *sepedonicum, Pectobacterium* spp., *Dickeya* spp., *and Agrobacterium tumefaciens* are also transmitted through infected planting materials such as tubers and cuttings. Some soil borne viruses, such as Tomato mosaic virus (tomato), Tobacco ringspot virus (tobacco), and Indian peanut clump virus can also be transmitted by nematode vectors or via infected seeds [235].

Nevertheless, the major strength of non-chemical methods lies in the prevention of seed borne infection originating from surface, or near surface attached pathogens. Currently, there are 213 annotated seeds borne pathogens according to the ISTA Pest list [11], encompassing fungi, bacteria and viruses. Commercially relevant examples are *Fusarium* causing a number of diseases in various plants (head blight in barley and panama disease of banana), *Tilletia* causing common bunt, dwarf bunt and stinking smut of cereals, *Ustilago tritici* causing common and loose smut in barley and rye, *Phoma* causing stem rot in rapeseed, *Typhula incarnata* causing snow mold in rye, and *Pyrenophora graminea/Drechslera tritici-repentis,* causing yellow leaf spot in wheat and barley.

Apart from the localisation of the pathogen on or inside the seed, the complex lifestyle with sexual and asexual cycles of especially fungi makes seed treatment more complicated. Reproduction of fungi is primarily by means of spores which can be produced sexually or asexually. The sexual reproduction cycle (teleomorphic phase) of fungi forms different types of spores via meiosis such as oospores, zygospores, ascospores and basidiospores. In the asexual cycle (anamorphic phase) oidia (formed by fragmentation of hyphae into individual cells), conidia (borne on tips or sides of specialized branches of hyphae) and sporangiospores (a nonmotile spore born in a sporangim or case) are produced by mitosis [2]. The disease cycle of monocyclic fungi usually starts with a primary infection, which involves colonization, growth, and reproduction as well as overseasoning in the absence of the host. Polycyclic fungi on the other hand, produce asexual spores (secondary inoculum) at each infection site that can cause new (secondary) infections to produce more asexual spores for more infections.

In vegetables and herbs, bacterial and viral pathogens are of special concern. Bacterial wilt in tomato caused by *Clavibacter michiganensis, Xanthomonas* causing citrus canker, bacterial leaf spot in many plant species, black rot of crucifers and bacterial blight of rice, *Pseudomonas syringae* causing wilt and spot diseases in many vegetables and legumes are frequently reappearing. Viral pathogens in vegetables and herbs encompass the mosaic virus (TMV, ToMV), Asparagus Virus (AV-2), tobacco ringspot virus (TRSV) and pea early browning virus (PEBV): The efficiency of cold plasma in reducing seed associated pathogens will be discussed in the following Sects. 6.2.2, 6.2.3 and 6.2.4. Not considered in this chapter are losses in seed quality and health caused by nematodes, insects, herbivory, nor post-harvest disease, which will be handled in Sect. 6.4. Notably, almost every study on CAP inactivation of pathogens is unique because they either use a specific plasma source, often build in-house, with specific configurations (e.g., input power, working gas, treatment time), they deal with treatment of different matrices (e.g., suspensions in water, other solutions, seeds from different plant families) and different type of pathogens (fungi of different life cycle stages, bacteria in sporulated or vegetative form, viruses) are used. This diversity makes it difficult to compare results from different studies directly and to define any universal inactivation parameters.

6.2.2 *Effect of Cold Plasma Treatment on Fungi*

Because of their relevance for losses in crop yield, fungal pathogens have been subjected to a number of studies, 39 are listed in Table 6.1. Inactivation of is highly dependent on the treatment properties, and the optimal parameters need to be chosen on a case-by-case basis.

Studies, which can serve as a general proof of concept for inactivation of fungal pathogens, are using spore suspensions as a test object. In these cases, the complex matrix of seed surfaces including topography, texture and chemical composition are absent. Moreover, information on the effect of the individual CAP treatment on seed germination is lacking, which makes it difficult to transfer gained knowledge to actual occurring crop diseases. However, 12 studies demonstrated efficient inactivation of *Alternaria, Ascochyta Aspergillus, Chaetomium, Cladosporium, Colletotrichum, Fusarium (Gibberella Penicillium,), Phomopsis* and *Rhizoctonia.* Reduction was in the range 44% to complete inactivation [13, 127, 143, 146, 193, 232, 243, 250, 252, 254, 316, 320, 360]. A variety of plasma sources was used including DBD in three cases, jets in two, corona, and arc discharge in three cases, microwave induced, and radiofrequency CAP was applied in two cases each.

Several authors used artificial inoculated seeds to investigate the inactivation efficiency of CAP. Important for pre-harvest application is an unimpaired seed germination, making it necessary to at least monitor maximum germination for the respective plasma treatment. Unfortunately, this was not always the case, but

Table 6.1 Efficiency of non-thermal plasma treatment for inactivation of fungal pathogens in pre-harvest

Target	Pathogen	Plasma type and source	Gas type and exposure time	Effect	References
Fungal sporesuspension on agar-coated microscope slides	*Ascochyta pinodella, fusarium culmorum*	Atmospheric pressure dielectric barrier discharge (DBD)	Air, 60, 180, 360 s	After 180 s delay in growth, complete inactivation after 360 s	[13]
Fungal sporesuspension	*Penicillium digitatum*	High-density nonequilibrium atmospheric pressure plasma	Oxygen/argon, 1–6 min	At 0.6% flow rate ratio = in each case1 log unit decrease from 1 to 4 min and 3 log units decrease at 5 min	[127]
Fungal sporesuspension on VM (Vogel's Minimal) agar plate	*Fusarium graminearum, fusarium oxysporum*	Microwave plasmajet	Mixtures of argon with oxygen (5:1), with nitrogen(5:1), or with air (5:1), and nitrogen gas, 3 slm, 10–180 s	After 1 min plasma treatment, hyphal growth in *N. crassa* and *F. oxysporum* no growth in *F. graminearum*	[232]
Fungal sporesuspension inserted by spraying	*Aspergillus niger, penicillun citrinum*	Atmospheric pressure plasma	Air, 1–5 h	99.9% inactivation after 5 h	[243]
Fungal sporesuspension in PBS or saline (0.85% NaCl) solution	*Fusarium oxysporum f.sp. lycopersici*	Atmospheric pressure dielectric barrier discharge DBD	Air or argon, 1, 5, 10 min	Spores in saline solution reduced germination only at 10 min treatment after 3 h post treatment, germination of spores in PBS or air plasma not affected	[250]
Fungal spore suspension	*Colletotrichum gloeosporioide*	Plasma-activated water, atmospheric pressure corona discharge	Artificial air (21% oxygen + 79% nitrogen) or bottle oxygen (99.99%), 2.5 slm, plasma treatment time 5–30 min, PAW exposure time 10 and 30 min	Inactivation rate 10 min exposure = 44% and 85% for Air and O_2 plasma 30 min exposure 4–45% for air and O_2 plasma	[360]

(continued)

Table 6.1 (continued)

Target	Pathogen	Plasma type and source	Gas type and exposure time	Effect	References
Fungal spore suspension in sterile physiological saline solution	*Penicillium crustosum* (DBM 4159) *Aspergillus oryzae* (DBM 4002) *Cladosporium sphaerospermum* (DBM 4282) *Alternaria* sp. (DBM 4004)	Corona discharge, negative and positive discharge	Air, 5–30 min	*Penicillium*: reduction up to 2 log units for 5–20 min treatment time, after 25 min complete inactivation (positive discharge), no reduction for negative discharge *Aspergillus*: up to 20 min no reduction, 30 min 3 log units reduction for positive discharge, reduction of 2 log units for 20 min treatment time for negative discharge *Cladosporium*: positive and negative discharge reduction by 4–5 log units for the first 20 min, complete inactivation >20 min *Alternaria*: complete reduction form 15 min treatment time on for negative and positive discharge	[320]
Fungal spore suspension in water, Spores on rice (*Oryza sativa* L. cv. Hopyeong) seeds submerged in water	Inoculated with *Fusarium fujikuroi*	Arc discharge Plasma	Air, Suspension 1, 5, 10 min, Seeds 5, 10, 20, and 30 min	Spore suspension: maximum reduction by 3.4 log units at 10 min reduction in infected seeds by 20% up to 10 min treatment time, 80% for >10 min treatment time No seed germination determined	[146]
Fungal spore suspension on potato dextrose agar (PDA) Spores on rice (*Oryza sativa* L. cv. Cocodrie) seeds	Inoculated with *Gibberella* (i.e. *Fusarium*) *fujikuroi* isolate, FGSC #8381	Atmospheric pressure dielectric barrier discharge DBD	Air, 0.5, 1, 2, or 3 min	Linear increase in clearing zone form 10–180 s Reduction of spores on seeds by 80% after 1 min No decrease in seedling emergence	[143]

(continued)

Table 6.1 (continued)

Target	Pathogen	Plasma type and source	Gas type and exposure time	Effect	References
Fungal spores on filter paper	*Cladosporium sladosporioides, Penicillium citrinum Chaetomium sp.*	Microwave-induced argon plasma	Argon 100 L per min, 1–30 s	Complete inactivation at 1 s treatment time	[254]
Fungal spores on filter paper or slide glasses coated with poly-L-lysine	*Aspergillus niger Penicillium citrinum*	Microwave-induced argon plasma	Argon 100L per min, 1–30 s	Complete inactivation at 1 s treatment time	[252]
Fungal spores suspension dried on glass plate Spores on tomato seeds (*Solanum lycopersicum* L.)	Inoculated with *Cladosporium fulvum*	Atmospheric pressure plasma jet	Oxygen/argon(volume ratio: 1/99) feed gas, 10–300 s	Inactivation of spores efficient from 11 kV input voltage onwards and 50 s treatment time From 20 to 60 s treatment time decrease in rotting rate of tomato plants by 15% No seed germination determined	[193]
Seeds and plants of winter wheat (*Triticum aestivum* L.), maize (*Zea mays* L.), narrow-leaved lupine (*Lupinus angustifolius* L.)	Inoculated with *Fusarium culmorum* on winter wheat, naturally occurring *Alternaria* sp., *Cladosporium* sp., *Colletotrichum gloeosporioides*, Kabatiella caulivora,	Low-pressure radio frequency plasma	Air at a pressure of 200 Pa., 2, 5, 7 min, vacuum control	Maize: reduction in infection with *Penicillium* spp. by 21%, no reduction in *Fusarium* Lupine: no *Colletotrichum* and *Kabatiella* detected after treatment, reduction in infection level by *Alternaria* by 30% and *Cladosporium* by 60% Winter wheat: *Fusarium* no significant reduction, *Alternaria* infection reduced by 25–35% No decrease in seed germination	[84]
Seeds of barley (*Hordeum vulgare* L. cultivar Tokak 157/37), corn (*Zea mays everta* L.)	Natural fungal load	Low pressure ambient air plasma, 15 Pa	Air, 2–20 min	No effect from 2 to 5 min, decrease by max. 25% at 20 min treatment time in barley, reductions by 40% at 10 min treatment time in corn seeds No decrease in germination of seeds	[37]

(continued)

Table 6.1 (continued)

Target	Pathogen	Plasma type and source	Gas type and exposure time	Effect	References
Seeds of barley, (*Hordeum vulgare* L.) wheat (*Triticum aestivum* L.)	Natural fungal load, barley seeds inoculated with *Penicillium verrucosum*	Atmospheric pressure dielectric barrier discharge DBD plasma	Air, 5, 20 min direct and indirect treatment	Barley: 20 min treatment time: natural load reduction by 2.1 and 1.5 log units for direct and indirect treatment, *P. verrucosum* reduction by 3.6 and 2.7 log units for direct and indirect treatment respectively, seed germination: not determined Wheat: 20 min treatment time: reduction of natural load by 2.5 and 1.7 log units for direct and indirect treatment, seed germination: significant decrease	[189]
Seeds of broccoli (*Brassica oleracea* var. kialica plen)	Natural fungal load (moulds, yeasts)	Corona dischargeplasma jet	Air, 1, 2, 3 min	Reduction of moulds and yeasts by ~1 log unit and 1.5 log units at 2 and 3 min treatment time respectively Decrease in seed germination at 3 min treatment time by 32%	[157]
Seeds of buckwheat CB (*Fagopyrum esculentum* Moench) and TB (*Fagopyrum tataricum* Gaertn.)	Natural fungal load	Radiofrequency (RF) plasma system 1 Pa	Oxygen gas with 202 standard $cm^3\ min^{-1}$, 15, 30, 45, 60, 90, and 120 s	Identification using next generation sequencing, *Alternaria, Didymella (Phoma), Epiccocum, Rhodotorula, Hannaella* Reduction in fungal cover of petri dish by 13% for 15–30 s treatment time and by 20–40% for treatment times >30 s in CB No reduction up o 90 s in TB, reduction by 30% at 120 s treatment time Decrease in seed germination by ~10% from 15 to 45 s and by 50% at treatment times >45 s	[228]

(continued)

Table 6.1 (continued)

Target	Pathogen	Plasma type and source	Gas type and exposure time	Effect	References
Seeds of chinese cabbage (*Brassica campestris* var. amplexicaulis)	Inoculated with *Rhizoctonia solani*	Atmospheric- and low-pressure plasma	Argon 0.5 L/min, 2–40 min	Survival rate decreased from 100 to 3% from 5 min treatment time onwards Strong delay in seed germination by 64%	[241]
Seeds of common bean (*Phaseolus vulgaris* L.) cultivar SCS204	Natural fungal load	Atmospheric pressure dieletric barrier discharge DBD	Gas not defined, 10–30 min exposure	Natural community identified as *Aspergillus* sp. and *Penicillium* sp. by selective plating and visual determination Complete inactivation after 10–30 min treatment time Seed germination as visual radicle formation, no radical formation at 20, 30 min, inferior development at 10 min	[286]
Seeds of cucumber (*Cucumis sativus* L.), pepper (*Capsicum annuum* L.)	Inoculated with *Didymella bryoniae, cladosporium cucumerinum, Didymella licopersici*	Diffuse coplanar surface barrier discharge (DCSBD)	Air, Cucumber 40 s, Pepper 4 s	Cucumber: complete reduction of *C. cucumerinum*, reduction of *D. bryoniae* spores 60–80%, Pepper: reduction of *D. licopersici* spores from 50 to 80% No decrease in seed germination	[325]
Seeds of ginseng (*Panax ginseng*)	Natural fungal load	Atmospheric pressure dielectric barrier discharge- planar-type DBD	Argon and Argon/oxygen mixture (80:20); 10 min each day, 3 days in a row	Identification using next generation sequencing: *Coniochaeta, pyrenochaeta, humicola, clonostachys, fusarium, mortierella* No reduction in *Humicola, Clonostachys*, reduction below 20% in *Fusarium* and *Mortierella*, reduction by >80% in *Coniochaeta* and *Pyrenochaeta*, Survival of fungi decreased to 40% in Argon/oxygen mixture No decrease in seed germination	[174]

(continued)

Table 6.1 (continued)

Target	Pathogen	Plasma type and source	Gas type and exposure time	Effect	References
Seeds of hazelnut (*Corylus avellane*), peanut (*Arachis hypogaea*), pistachio nut (*Pistacia vera*)	Inoculated with *Aspergillus parasiticus*	Low pressure cold plasma (500 m Torr)	Air gas and Sulfur hexafluoride; 1–20 min	Air plasma 1-log reduction SF6 plasma ~5-log decrease (in 5 min) No seed germination determined	[25]
Seeds of hazelnuts (*Corylus avellane*)	Inoculated with *Aspergillus. flavus* (ATCC 327) and *Aspergillus parasiticus* (ATCC 1041)	Atmospheric pressure fluidized bed plasma	Air and nitrogen, 2–20 min	Maximum reductions of 4.5 log for *A. flavus* and 4.2 log units for *A. parasiticus* after 5 min treatments in nitrogen plasma No seed treatment determined	[57]
Seeds of hazelnuts (*Corylus avellane*)	Natural fungal load and inoculated with *Aspergillus flavus* and *A. parasiticus*	Atmospheric pressure fluidized bed plasma	Air and nitrogen; 1–5 min	Reduction in *A. flavus* by (4.2 log) and *A. parasiticus* (4.1 log) after 5 min treatment, reduction in natural load by 3.45 log units at 2 min, no seed germination determined	[58]
Seeds of lentil (*Lens culinaris*)	Inoculated with *Aspergillus niger, Penicillium decumbens*	Diffuse Coplanar Surface Barrier Discharge	Air, 3, 5, 10 min	*P. decumbens.* maximum reduction of 3.1 log CFU/g after 10 min treatment, *A. niger* 1.6 log CFU/g after 10 min, seed germination: 120 s treatment time 90%, 180 s treatment time 42%, 240 s treatment time 5% germination	[356]
Seeds of maize (*Zea mays* L.)	Natural fungal load	Gliding Arc, spatial post-discharge mode	Air, 300, 900 s	Reduction in infection degree by ~48% at 300 s treatment time No decrease in seed germination	[145]
Seeds of maize (*Zea mays* L.)	Natural fungal load and inoculated with *Aspergillus flavus, Alternaria alter-nata, Fusarium culmorum*	Diffuse Coplanar surface barrier discharge AP DCSBD CP	Air; 60–300 s	Reduction by 3.79 log (CFU/g) in *F. culmorum* after a 60 s plasma treatment, 4.21 log (CFU/g) in *A. flavus* and 3.22 log (CFU/g) in *A. alternata* after a 300 s plasma Decrease in seed germination from 180 s onwards	[380]

(continued)

Table 6.1 (continued)

Target	Pathogen	Plasma type and source	Gas type and exposure time	Effect	References
Seeds of narrow-leaved lupine (*Lupinus angustifolius* L.)	Natural infected seeds with *Fusarium* sp., *Stemphilium* sp., *Colletotrichum/Glomerella*, *Didymella pinodes* (syn. *Mycosphaerella pinodes*)	Radiofrequency (RF) capacitively coupled discharge	Air, 5, 10, 15 and 20 min	Maximal reduction of fungal infection in lupine at 15 min treatment time 16% for *Fusarium* sp. 14% for *Didymella pinodes*, 10% for *Stemphilium* sp., no reduction in *Colletotrichum/Glomerella* No decrease in field emergence at 15 min treatment time, decrease by ~7% at 20 min	[81]
Seeds of narrow-leaved lupine (*Lupinus angustifolius*), field pea (*Pisum arvense*)	Natural fungal load	Radiofrequency (RF) capacitively coupled discharge	Air, 5, 10, 15 and 20 min	Fungi were identified via morphological and cultural characteristics = *blue* lupin infected mainly with *Fusarium* and *Alternaria*, seeds of field pea with *Fusarium*, *Alternaria* and *Stemphilium* Field pea: maximum reduction of fungal infection at 10 min treatment time by 4%, 24 and 3% for *Fusarium*, *Alternaria* and *Stemphylium* respectively Lupine: maximum reduction at 15 min treatment time by ~9% and 1% for *Fusarium* and *Alternaria* respectively Seed germination in lab field pea no decrease till 15 min, lupine decrease by 1% at 15 min	[82]
Seeds of pea (*Pisum sativum* L.) Zuccini (*Cucurbita pepo* L.)	Natural fungal load	Fluorinated silicate glass (FSG) plasma	Air (5 L/min), 30, 60 s	No identification of natural load Reduction by 1 and 3 log units for both species after 30 and 60 s treatment time, respectively No decrease in seed germination	[151]

(continued)

Table 6.1 (continued)

Target	Pathogen	Plasma type and source	Gas type and exposure time	Effect	References
Seeds of pine (*Pinus radiata*)	Inoculated with *Fusarium circi-natum* (pine pest)	Diffuse coplanar surface barrier discharge AP DCSBD CP	Air; 5–300 s	Reduction of seed-borne pathogens (14–100%) Decrease in seed germination by 9–10% for 5 and 10 s treatment time No seed germination for treatment times > 60 s	[307]
Seeds of rapeseed (*Brasica napus*)	Natural fungal load (molds and yeasts)	Corona discharge plasma jet	Air, 1, 2, 3 min	Reduction by 2 log units at 3 min treatment Decrease in seed germination by ~ 30% at 3 min treatment time	[270]
Seeds of rice (*Oryza sativa* var. Indica cv. KDML105)	Natural fungal load	Atmospheric hybrid micro corona discharge plasma	Argon ~2.5 l/min, 5,10,15,20 min	Qualitative assessment reduced number of seeds with fungal infection, no quantification nor identification, visually, no fungal growth detected after treatment No decrease in seed germination	[150]
Seeds of soybean (*Glycine max* L.) variety DM 53i53 IPRO	Naturally infected seeds with *Diaporthe/Phomopsis* complex	Atmospheric pressure dielectric barrier discharge DBD	Oxygen/nitrogen mixture 6 NL/min, 60–180 s	Reduction in D/P infected seeds from 15% to minimum of 4% at treatment TN1 No decrease in seed germination	[264]
Seeds of sweet basil (*Ocimum basilicum*)	Natural fungal load	Atmospheric pressure surface dielectric barrier discharge (SDBD)	Humid air 7 slm; 10 to 600 s	Decrease of contaminated seeds by up to 35% at 300 s treatment time No adverse effect on seed germination	[7]
Seeds of wheat (*Triticum aestivum*)	Natural fungal load and inoculated with *Fusarium nivale, Trichothecium roseum, Aspergillus. flavus, Trichothecium roseum*	Diffuse coplanar surface barrier discharge AP DCSBD CP	Air; 10–600 s for NO microflora 1–300 s for AI fungi-infected seeds	Complete inactivation of naturally occurring filamentous fungi after 120 s treatment, complete inactivation of *Fusarium* after 90 s treatment time, *Aspergillus* after 240 s and *Trichothecium* after 180 s Decrease in seed germination from 70 s treatment time onwards	[379]

(continued)

Table 6.1 (continued)

Target	Pathogen	Plasma type and source	Gas type and exposure time	Effect	References
Seeds of wheat (*Triticum durum*), bean (*Phaseolus vulgaris*), chick pea (*Cicer arietinum* L.), soy bean (*Glycine max* cv), barley (*Hordeum vulgare* L. cv.), oat (*Avena sativa*), rye (*Secale cereale*), lentil (*Lens culinaris*), corn (*Zea Mays*)	Inoculated with *Penicillum* sp MS1982 on wheat, *Aspergillus* sp. 798 on all	Low pressure cold plasma, 500 m Torr	Air and Sulfur hexafluoride, 30 s to 30 min	Log reduction of *Penicillum* sp on wheat maximum 3 and 2 log units at 20 min treatment for air and SF6 gas, respectively Log reduction of *Aspergillus* sp. for SF6: Wheat 2.6, Balrey 1.5, Oat 1.2, Lentil 0.8, Rye 0.6, Corn 0.5, Chickpea 0.5 No significant difference in germination treated with 5, 10, 15 min of air and SF6 plasma	[301]
Seeds of wheat (*Triticum vulgare* L.; cv. "Forhand"), barley seeds (*Hordeum vulgare* L.; cv. "Maltz")	Inoculated with *Fusarium culmorum* (CCM F-163)	Diffuse coplanar surface barrier discharge (DCSBD)	Air, 15–300 s	Wheat and barley: reduction at 30 s = 50%, 100% reduction after 120/180 s treatment time Seed germination: 60 s treatment decrease in max. germination by 30%, from 120 s plasma treatment onwards inhibition of germination by up to 54%., strong decrease in seedling vigor by up to 80%	[117]
Seeds of winter wheat (*Triticum aestivum*)	Natural fungal load	Packed bed reactor	Air, 3, 10, 30 s	Reduction of fungal colonies on wheat grains by 77% at the optimum exposure 10 s No decrease in seed germination	[161]
Spore suspension in sterile water	*Aspergillus flavus* hyphae	Diffuse coplanar surface barrier discharge AP DCSBD CP	Air, 5–180 s	Almost complete loss of cell viability from 30 s treatment onwards	[316]

needs to be addressed in future studies. Fungal pathogens investigated belonged to the genus *Alternaria, Aspergillus, Cladosporium, Colletotrichum* (syn. *Glomerella), Didymella* (syn. *Mycosphaerella), Kabatiella, Penicillium Rhizoctonia, Stemphilium, Trichothecium.* Studies with no simultaneous determination of seed published inactivation of pathogens from 1 to 5 log units for *Aspergillus* [25, 56, 58], by ~3 log units for *Penicillium* on barley seeds [189], a reduction in *Fusarium* infected seeds by 20–80% [146], or a significant decreases in *Cladosporium* diseased plants [193]. No decrease in seed germination was detected in four studies using artificial inoculated pathogens and subsequent CAP treatment. Efficiency of inactivation was depended upon the pathogen used with no effect for *Fusarium* inoculated on maize seeds [84], 0.5–3 log units for *Penicillum* and *Aspergillus* [301], 80% reduction of viable spores of *Fusarium* on rice seeds [143], up to complete inactivation of *Cladosporium* on cucumber seeds [325].

On the other hand, significant decrease in seed germination after CAP treatment was reported by seven studies. Efficiency of CAP treatment again depended on the pathogen used, along with plasma source and applied parameter including treatment time. Effects ranged from no effect for *Colletotrichum/Glomerella* along with rather small reduction in seed germination <10% [81]. A reduction by 3 to 4 log units for *Aspergillus, Alternaria* and *Fusarium* on wheat seeds up to complete inactivation using the same plasma source for *Fusarium, Trichothecium* and *Aspergillus* on maize seeds which was accompanied by severe decrease in seed germination for longer treatment times in both cases [379, 380]. A delay in seed germination >50% for Chinese cabbage accompanied by nearly complete inactivation of *Rhizoctonia* [241]. Seed germination of wheat and barley was decreased by ~54% for treatment times >120 s, while at the same time DBD treatment led to a nearly complete reduction of *Fusarium* [117]. Lentil seed germination decreased by 95% after treatment time of 240 s with a coplanar DBD, at the same time reducing viability of *Penicillium* and *Aspergillus* by 3 and 1.6 log units/g seeds, respectively [356]. Additionally, a complete loss of pine seed germination for coplanar DBD treatment >60 s was detected, while a complete inactivation of *Fusarium* was observed [307].

However, 18 studies dealt with natural fungal communities on seeds (often accompanied by artificial inoculation with specific fungi), which includes also non-pathogenic fungi. Two studies specifically focused on natural occurring pathogens like *Diaporthe/Phomopsis* complex on seeds of soybean [264] and *Fusarium sp., Stemphilium sp., Colletotrichum/Glomerella, Didymella pinode* on seeds of narrow-leaved lupine [81]. Pérez Pizá and colleges [264] published reduction in *Diaporthe/Phomopsis* infected soybean seeds from 15% to minimum of 4% after DBD treatment with no decrease in seed germination. Moreover, Filatova and colleges [81] reported the efficacy of at 15 min treatment of lupine seeds using a radiofrequency (RF) capacitively coupled discharge with maximal reduction of 16% for *Fusarium*, 14% for *Didymella*, 10% for *Stemphilium* and no reduction of *Colletotrichum/Glomerella*. At the same time, CAP treatment did not decrease field emergence at 15 min treatment time, while 20 min treatment resulted in a decrease by ~7%. Four publications dealing with natural fungal communities present a detailed identification applying selective plating and visual determination methods or next generation sequencing. The first one by Filatova and colleges [82] identified fungi on lupine and pea seeds using morphological and cultural characteristics. Fungi on lupine consisted mainly of *Fusarium* and *Alternaria*, while on seeds of field pea *Fusarium*, *Alternaria* and *Stemphilium* were identified, using cultring techniques, which are selective and don't include the whole community. Moreover, inactivation using 10 min treatment of a radiofrequency capacitively coupled discharge displayed a maximum reduction of 4%, 24% and 3% for *Fusarium*, *Alternaria* and *Stemphylium* on pea seeds respectively. On lupine seeds, maximum reduction occurred at 15 min treatment time resulting in ~9% and 1% for *Fusarium* and *Alternaria* respectively. Seed germination in the laboratory resulted in no decrease until 15 min CAP treatment for field pea and a decrease by 1% at 15 min for lupine seeds. The second study using selective plating identified mainly *Aspergillus* and *Penicillium* on seeds of common bean [286]. Treatment applying DBD for 10–30 min revealed complete inactivation of both genera detected. However, seed germination presented as visual radicle formation resulted in a complete loss of radical formation at 20 and 30 min CAP treatment and in an inferior radicle development at 10 min treatment time. Two further studies implemented next generation sequencing to disentangle the fugal community. Lee and colleges [174] focused on ginseng seeds, detecting the following genus *Coniochaeta, Pyrenochaeta, Humicola, Clonostachys, Fusarium, Mortierella.* Treatment using DBD for 10 min three days in a row showed no reduction in *Humicola* and *Clonostachys*, a reduction below 20% in *Fusarium* and *Mortierella* and a reduction by >80% in *Coniochaeta* and *Pyrenochaeta* in an Argon/oxygen mixture. Additionally, no decrease in seed germination was observed. Likewise, next generation sequencing, as well as plating and visual identification were applied by Mravlje et al. [229] on the fungal community of buckwheat seeds. *Alternaria, Didymella (Phoma), Epiccocum, Rhodotorula* and *Hannaella* were identified. A radiofrequency plasma system operated at low pressure of 1 Pa was implemented and treatment times were in the range of seconds. After 120 s treatment, filamentous fungi of the genus *Alternaria* predominated, while other genus was detected in lower quantities. Alongside, seed germination decreased by ~10% from 15 to 45 s and by 50% at treatment times >45 s. No in-depth identification of the natural fungal load was presented in 12 other studies, displaying inactivation efficacy on barley, broccoli, sweet basil, hazelnut, maize, pea, rapseed, rice, soybean, wheat as a bulk parameter. Inactivation was in the range of 10% to 3 log units (99,99%) inactivation [7, 37, 58, 145, 150, 151, 157, 161, 189, 271, 379, 380]. Differences in the susceptibility of fungi to CAP compared to bacteria were previously reported with fungi being more resistant to CAP exposure [174, 189, 270, 380]. Nevertheless, the proposed mechanism of inactivation of filamentous fungi by CAP likely resemble the ones described in bacteria (see Sect. 6.2.3).

The plasma-treated fungal spores often show severe morphological degeneration including damage of cell envelope structures [252] also related to lipoperoxidation of cell macromolecules [316] and seem to undergo necrotic death [250]. Panngom and colleges [250] argued that elevated levels of peroxynitrite and nitrite originating from the CAP treatment of the saline solution might have been responsible for the observed fungal spore death. Furthermore, when direct CAP treatment is applied inactivation can occur via different other mechanisms e.g. DNA fragmentation or destruction by UV irradiation, erosion through intrinsic photodesorption or erosion through etching to form volatile compounds as a result of slow combustion using oxygen atoms or radicals emanating from the plasma (reviewed by [165, 217, 218, 228]). As noted before, CAP produces different reactive species (RONS, e.g. atomic oxygen (O), metastable oxygen (O_2^*), superoxide $(\cdot O_2^-)$, ozone (O_3), hydroxyl radical ($\cdot OH$), hydrogen peroxide (H_2O_2), nitride oxide (NO), and nitride dioxide ($\cdot NO_2$)), which play a crucial role in the inactivation of any microbial target by oxidation of cytoplasmic membrane, protein and DNA [229]. Most likely, O and ·OH induces the largest number of hydrogen abstraction reactions [55, 364], while the activity of HO_2 and H_2O_2 is lower, however, the number of main chain and branched chain fractures of cell wall glucan structures appears to be bigger. Consequently, the destructive effect of H_2O and H_2O_2 is more efficient [55]. Fungi might exhibit possible protection from CAP damage when carotene pigments are present protecting spores from oxidative damage by plasma [232].

6.2.3 *Effect of Plasma Treatment on Bacteria*

Inactivation of bacterial pathogens, like fungal ones, is highly dependent on plasma source, configuration, and the treatment properties. The majority of studies applied DBDs for direct treatment or gliding arc for indirect treatment by producing plasma treated water or gas (Table 6.2). There are several proof of concept studies using spore suspension of phytopathogenic bacteria (e.g., *Xanthomonas campestris, Erwinia* sp., *Clavibacter michiganensis, Pectobacterium carotovorum)* showing a successful reduction in the number of viable bacteria from 1.5 log units to complete inactivation in a time-dependent manner [223, 224, 227, 230, 344].

There is an almost equal part of studies dealing with pathogens artificially inoculated on seeds or growth solution and naturally load on seeds with the majority applying DBD plasma sources or jets. Artificial inoculation of hydroponic growth solution for tomato cultivation with the pathogenic bacteria *Ralstonia solanacearum* and subsequent treatment of this solution using a gas-liquid phase discharge plasma reactor displayed a reduction by 5 log units in the solution and a decrease in disease severity of tomato seedlings by 80% after 10 days of growth [247]. Treating tomato seeds with a capacitively coupled plasma (CCP) generated by a radiofrequency discharge at 150 Pa led to an increased resistance of the 30 days old plants to *Ralstonia solanacearum* by 25% [140]. Treating seeds which were artificial inoculating with either non-plant-pathogenic bacteria *Bacillus atrophaeus* and *Escherichia coli* as a model or with actual pathogenic bacteria, e.g. *Xanthomonas*, *Burkholderia plantarii* and *Geobacillus stearothermophilus* often resulted in an efficient reduction of viable bacteria from 2.4 to 6 log units, but simultaneously reduced seed germination in one case [189]. For two studied no information on seed germination after plasma treatment was presented for the same study [42, 242]. Altogether, vegetative cells of *Bacillus atrophaeus* and *Escherichia coli* seemed to be easier to inactivate than spores of *Bacillus atrophaeus* [189]. Disease severity was monitored in one study using

Table 6.2 Efficiency of non-thermal plasma treatment for inactivation of bacterial pathogens in pre-harvest

Target	Pathogen	Plasma type and source	Gas type and exposure time	Effect	References
Bacterial suspension spotted onto aluminium plates	*Xanthomonas campestris pv. campestris* (Xcc)	Roller conveyer plasma device, atmospheric pressure dielectric barrier discharge (APDBP)	Air, 1.8 to 30 min	Inactivation from 1.5 log units after 2 min to complete inactivation from 3.7 min onwards	[344]
Bacterial suspensions in distilled water	*Erwinia amylovora, Clavibacter michiganensis* subsp. *michiganensis*	Gliding arc discharge plasma	Air 12 min	Lag phase during growth expanded, maximum growth delayed	[230]
Bacterial suspensions in 0.85% NaCl	*Clavibacter michiganensis* subsp. *sepedonicus, Dickeya solani, Xanthomonas campestris* pv., *Pectobacterium atrosepticum* (Pba), *Pectobacterium carotovorum* subsp. *carotovorum* (Pcc)	Flow through atmospheric pressure glowdischarge (dc-APGD)	Air, 1 min	Cms, Dsol, and Xcc complete inactivation, Pcc and Pba maximum 3.7 log units	[227]
Bacterial suspensions in sterile LB broth	*Erwinia carotovora carotovora, Erwinia carotovora atroseptica, Erwinia chrysanthemi*	Gliding arc discharge	Mixture of nitrogen and oxygen, 1 to 12 min	Lag phase with complete survival up to 4/5 min, rapid decrease in survival up to 7 min, complete inactivation after 8 min	[224]
Bacterial suspensions in sterile LB broth	*Erwinia carotovora*	Gliding arc discharge	Mixture of nitrogen and oxygen, 1–10 min	Rapid loss of survival up to 90% of the initial bacterial population within 2.5 min, second step with slower kinetics, leading to a complete loss survival within 5 min	[223]

(continued)

Table 6.2 (continued)

Target	Pathogen	Plasma type and source	Gas type and exposure time	Effect	References
Seeds of barley (*Hordeum vulgare* L.), wheat (*Triticum aestivum* L.)	Natural bacterial load, Barley: artificially inoculated with *Escherichia coli*, *Bacillus atrophaeus* (vegetative cells and endospores)	Atmospheric pressure dielectric barrier discharge DBD	Air, 5, 20 min direct and indirect treatment	Barley 20 min treatment time: Natural load reduction by 2.4 and 1.7 log units for direct and indirect treatment, *B. atrophaeus* spores reduction by 2.4 and 1.3 log units for direct and indirect treatment respectively *B. atrophaeus* vegetative cells reduction by 3.2 and 2.7 log units for direct and indirect treatment, respectively *E.coli:* 3.5 and 3.3 log units for direct and indirect treatment Seed germination: strong delay for 20 min direct treatment: Wheat: reduction of natural load by 1.5 and 1.2 log units at 20 min direct and indirect treatment Seed germination: strong delay for 20 min direct treatment	[189]
Seeds of chickpea (*Cicer arietinum*)	Natural bacterial load	Surface microdischarge plasma FlatPlaSter 2.0	Air, 30–300 s	1 log unit reduction till 3 min treatment time, 2 log units reduction from 3–5 min Seed germination reduced by 10–60% for 2–5 min treatment	[216]
Seeds of chinese cabbage (*Brassica campestris* var *amplexicaulis*)	*Xanthomonas campestris*	Low-pressure plasma	Argon 0.5–1.0 L/min, 5–40 min	Reduction of 3–6 log units, no seed germination determined	[242]
Seeds of ginseng (*Panax ginseng*)	Natural bacterial load	Atmospheric pressure dielectric barrier discharge DBD	Argon and argon/oxygen mixture (80:20); 10 min each day, 3 days in a row	Identification using next generation sequencing *Kocuria, Variovorax, Pseudomonas, Duganella, Rahnella, Flavobacterium, Azospirillum, Chryseobacterium* Reduction <20% *Pseudomonas Duganella* Reduction 30–65% for all other No decrease in seed germination	[174]
Seeds of maize (*Zea mays* L.)	Natural bacterial load	Diffuse Coplanar surface barrier discharge AP DCSBD CP	Air; 60–300 s	Complete reduction of bacteria from 60 s treatment onwards Decrease in seed germination from 180 s onwards	[380]

(continued)

Table 6.2 (continued)

Target	Pathogen	Plasma type and source	Gas type and exposure time	Effect	References
Seeds of rapeseed (*Brasica napus L.*)	Natural bacterial load	Corona discharge plasma jet	Air, 1, 2, 3 min	Identification using general and selective growth media *B. cereus, E. coli, Salmonella* spp. Reduction of total aerobic bacteria by 2.2 log units after 3 min treatment Decrease in seed germination by ~30% at 3 min treatment time	[270]
Seeds of rice (*Oryza sativa* 'Haenuki')	*Burkholderia plantarii*	Atmospheric pressure plasma jet	Air, 10 min	Disease severity decreased by 40%, no reduction in seed germination	[246]
Seeds of sweet basil (*Ocimum basili-cum L.*)	Natural bacterial load	Atmospheric pressure surface dielectric barrier discharge SDBD	Humid air; 10–600 s	Decrease in contaminated seeds by maximum 30% at 300 s treatment time No decrease in seed germination	[7]
Seeds of wheat (*Triticum aestivum* L.)	*Geobacillus stearothermophilus*	Atmospheric pressure dielectric barrier discharge DBD	Argon (2.8 nlm), 1–60 min	Up to 15 min reduction by 1 log unit, 3 log units after 60 min No information on seed germination	[42]
Tomato (Solanum lycopersicum L., Rinka 409) seedlings, inoculated via hydroponic nutrient solution	*Ralstonia solanacearum*	Gas-liquid phase discharge plasma reactor	Air, 100 min	Inactivation of pathogen in liquid solution by 5 log units Disease severity of tomato seedlings decreased by 80% after 10 days of growth	[247]
Tomato seeds treated with CAP	*Ralstonia solanacearum* suspension poured onto the soil near the roots of the tomato seedlings at 30 days of growth	Capacitively coupled plasma (CCP) was generated by radiofrequency discharge, 150 Pa	Helium	Increased tomato resistance to *R. solanacearum* with an efficacy of 25%	[140]

artificially inoculated *Burkholderia plantarii* on seeds of rice which were subjected to atmospheric pressure plasma jet, subsequently [246]. Results indicated a reduction in disease severity of seedling blight by 40% along with no reduction in seed germination.

Six studies presented information on the effect of CAP on natural bacterial load on seeds, which of course also encompasses non-pathogenic bacteria. The natural community was identified in two of the six studies [174, 270]. A very recent study by Lee and colleges [174] applied next generation sequencing to elucidate the community on seeds of ginseng identifying the genus *Kocuria, Variovorax, Pseudomonas, Duganella, Rahnella, Flavobacterium, Azospirillum* and *Chryseobacterium.* CAP treatment for thre times 10 min using a DBD releveled a reduction by less than 20% for *Pseudomonas* and *Duganella,* as well as a reduction by 30–65% for all other, while a t the same time no negative effects on seed germination were detected. Puligundla and colleges [270] actually focuses on post-harvest relevant bacteria and therefore used general and selective growth media to quantify *B. cereus, E. coli, Salmonella* spp. on rapeseed. For all detected microorganisms the reduction after treatment with a corona discharge plasma jet for 3 min. was in the range of 1.2–2.2 log CFU/g. However, 3 min CAP treatment also provoked a decrease in seed germination by ~30%. The four remaining studies dealt with an unknown community of natural bacterial load on seed surfaces of sweet basil, barley and wheat as well as chickpea and maize showing inactivation by 1.2–3 log units, up to complete inactivation [380] or a reduction by 30% in contaminated seeds [7, 189, 216]. In two of the latter studies, seed germination was severely negatively affected with a decrease by up to 60% [189, 216].

Altogether, seed decontamination/ inactivation often was accompanied by a reduction in seed germination when applying identical CAP treatment times, impeding a possible application of CAP in pre-harvest seed treatment. It has to kept in mind, that plasma can induce a sub-lethal state of bacteria leaving them viable but nonculturable after CAP treatment (VBNC) state [367]. Further investigations on this effect are needed. Previous studies investigating the effect of CAP treatment on bacteria as well as fungi on seed surfaces demonstrated that bacteria especially in vegetative state are more prone to CAP exposure compared to fungi[174, 189, 270, 380].

Some authors investigated and proposed inactivation mechanisms, which resemble some of those found for fungal inactivation. Previous knowledge originating form plasma medicine and/or food science can be transferred regarding some general patterns and concepts for inactivation. Different effects of CAP treatment were observed for Gram-positive and Gram-negative bacteria [147, 167, 198] that differ in cell envelope structures. Gram-negative bacteria, which possess a cell wall composed of an outer membrane and thin peptidoglycan (murein), displayed substantial damage to the membrane resulting in the cytoplasm leakage. Gram-positive bacteria on the other hand, display cells with a thick cell wall and did not show the significant morphological modifications and decontamination was most probably appeared here due to interactions of reactive compounds with the intracellular components [147, 238]. Bacteria morphometry might also be responsible for differences in inactivation patterns with more resistant spherical cells (cocci) than rod-shaped cells (bacilli) [167, 331]. As pointed out before, CAP produces many reactive oxygen and nitrogen species (RONS), which can oxidize proteins, lipids, and nucleic acids and lead to pathogen destruction [180]. Moreover, inactivation mechanisms might include erosion the surface of microbial cells through etching [218], oxidative damage of intracellular macromolecules, such as membrane lipids, proteins, and DNA, and a reduction in intracellular pH from diffusion into the microbial cells disrupting pH homeostasis [166]. Furthermore, sub-lethal damages can induce viable but non-culturable (VBNC) states in fungi as well as bacteria which is defined as an inactive form of life that is induced by stressful conditions [51] and undergoes recovery under suitable conditions [277]. These state transitions have been reported after CAP treatment [53, 72, 197, 298] and need to be taken into account in future studies.

6.2.4 *Effect of Plasma Treatment on Viruses*

Plant virology is a very dynamic research area with new plant viruses being detected more rapidly. Moreover, awareness of their pathological impact and severity of economic loss caused by reduction in yield by up to 100% [222, 312] or quality of crops has led to efforts for new detection as well as plant treatment methods. Plant pathogenic viruses are mainly transmitted horizontally by biological vectors, usually insects, but can also be transmitted via seeds, tubers, rhizomes and bulbs [294]. Increasing evidence suggest that transmission can also occur via contaminated process water [205].

The majority of studies dealing with the effect of CAP on plant viruses applied DBDs in various configurations, next to jets and torches for direct treatment and indirect underwater treatment (Table 6.3) of viruses in suspension and inoculated onto plant leaves. Only one study dealt with actual seeds, cucumber and pepper, which were naturally infected with cucumber mosaic virus, zucchini yellow mosaic virus and watermelon mosaic virus [325]. Štěpánová and collegues used only one treatment time per plant species (20 s for cucmber and 4 s for pepper) and detected no decrease in viral load after treatment of seeds with a diffuse coplanar surface barrier discharge plasma (DCSBD). Seed germination on the other hand, was not decreased after plasma treatment. Milusheva and colleagues [212] investigated the effect of a surface-wave-sustained argon plasma torch and an underwater diaphragm discharge on plum tree microplants, which were naturally co-infected by M and D strains of Plum pox virus (PPV). Microplant's nodal segments or leaflets were subjected directly to a CAP torch, as well as to electrical discharges in water media. Treating nodal segments without leaves in gas medium using the torch tip tuned out to be most effective with no detection of viable D strains of Plum pox virus along with a decrease in symptomatic plants by 80%. Plant leaves inoculated with specific viruses were the focus of two studies using Tulane virus for Romain lettuce and tobacco mosaic virus for tobacco [104, 213]. Reduction of Tulane viral load by 1.3 ± 0.2 log PFU/g Romanian lettuce and no necrotic lesions cause by tobacco

Table 6.3 Efficiency of non-thermal plasma treatment for inactivation of viral pathogens in pre-harvest

Target	Pathogen	Plasma type and source	Gas type and exposure time	Effect	References
Bacteriophage suspensions	T4, Φ174 and MS2 viral strains	Air surface micro-discharge (SMD)	Argon and artificial air (79% nitrogen plus 21% oxygen) to generate plasma-activated water = 10 to 100 s	inactivation below detection limit after 120 s of treatment Aggregation of bacteriophages and damage to nucleic acid and proteins	[101]
Irrigation water	Potato virus Y	Single electrode cold atmospheric plasma jet	Argon (~99%) and oxygen (~1%), 5, 15, 30, 45, 60 min	RNA degradation in the virus material induced complete viral inactivation One infection out of two for 5 min treatment time in plant infectivity assay using *Nicotiana tabacum*, cv. 'White Burley', no infection from 5 min onwards	[86]
Leaves of Romain lettuce (*Lactuca sativa* L. var. *longifolia*) Post-harvest	Tulane virus	Atmospheric pressure dielectric barrier discharge DBD	Air inside the package, 5 min	Reduction of viral load by 1.3 ± 0.2 log PFU/g lettuce	[213]
Plum tree (*Prunus domestica* L., cv 'Kyustendilska sinya') microplants	naturally co-infected by M and D strains of Plum pox virus (PPV)	Surface-wave-sustained argon plasma torch and an underwater diaphragm discharge	Argon 2 l/min, 5 s	PCR and visual analysis, D strains of Plum pox virus (PPV) inactivated, only M strain detectable with PCR Decrease in number of symptomatic plants by 80%	[212]

(continued)

Table 6.3 (continued)

Target	Pathogen	Plasma type and source	Gas type and exposure time	Effect	References
Seeds of cucumber (*Cucumis sativus* L.) and pepper (*Capsicum annuum* L.)	Naturally infected CMV Cucumber mosaic virus ZYMV Zucchini yellow mosaic virus WMV Watermelon mosaic virus	Diffuse Coplanar Surface Barrier Discharge (DCSBD)	Air, cucumber 20 s Pepper 4 s	No reduction in CMV, ZYMV, and WMV diseases No reduction in seed germination	[325]
Tobacco (*Nicotiana*) leaves	Inoculated with Tobacco mosaic virus solution	Atmospheric pressure dielectric barrier discharge DBD	Air, 0.5–10 min	No necrotic lesions on TMV irradiated leaves for 5 and 10 min irradiation	[104]

mosaic virus detectable in plant leaves after treatment. Filipić and colleges [86] investigated irrigation water inoculated with Potato virus Y wich was treated using a single electrode cold atmospheric plasma jet. No infection was detected in the plant infectivity assay using *Nicotiana tabacum* cv. 'White Burley' from 15 min treatment time of suspension onwards.

Possible mechanisms for viral inactivation were presented in a study by Guo and colleges [100], who did not investigate plant viruses but bacteriophages T4, Φ174 and MS2, which can serve as a general proof of concept for inactivation. Bacteriophage suspension was treated with PTW produced by an air surface microdischarge, showing an inactivation below the detection limit after 120 s treatment. Although not being pathogenic to plants, the proposed model of inactivation is likely to be adaptable from bacteriophages. The proposed model of inactivation includes plasma-generated reactive species, especially singlet oxygen, which efficiently inactivated different kinds of bacteriophages in water, including double-stranded DNA, single-stranded DNA, and RNA bacteriophages by damaging both nucleic acid and proteins and leading to excessive aggregation of the bacteriophages. In addition, knowledge can be transferred form studies dealing with other types of viruses, e.g., animal viruses. Work on Newcastle disease (ND), an infectious viral disease of avian species, reported complete inactivation after PTW treatment resulting most likely from singlet oxygen, which quickly reacts with cysteine, resulting in the formation of cystine (R-cys-S-S-cys-R) with disulfides; thus creating products which lead to aggregation of bacteriophages [362]. Furthermore, enzyme activity can be impaired by hydroperoxides which is formed by the interaction of amino acids, including tyrosine, tryptophan, and histidine, which selectively interact with singlet oxygen [63].

Altogether, studies dealing with the efficiency of CAP to inactivate plant pathogenic viruses on seeds and plants are scarce and efforts should be taken to fill the gaps of knowledge. Unknown up to now is the effect of CAP itself on insects as transmission vectors, which should be examined in the future. In addition, in natural environments mixed-infections with two or more plant viruses are frequent, with viruses being able to interact in multiple and intricate ways. These interactions can be synergistic, antagonistic, or neutral and will likely have an impact on the efficiency of CAP application for phytosanitary purposes.

6.3 Application of Non-thermal Atmospheric Pressure Plasma to Seed Germination and Plant Growth

Major seed dressing methods are aiming to prevent pathogenic attack and outbreak by using e.g. fungicides. Inoculation of seeds with fertilizers, chemical stimulants or plant growth promoting bacteria (PGPB) support seed germination performance to promote proper seedling establishment, further plant growth and stress resilience to finally secure or increase yield. Furthermore, different kinds of chemical and physical seed treatment methods have been studied aiming to stimulate and synchronize germination of seed population and to prime plants against various stresses [9, 78, 251]. Numerous studies have shown that plasma as a physical treatment method can improve seed germination performance and plant growth (Tables 6.4 and 6.5). Recent studies investigated the potential of plasma to prime seeds against biotic and abiotic stressors as well [14, 17, 80, 84, 99, 178, 219, 264].

Unlike in plasma medicine, there is a much greater variability of plasma sources and a higher number of plant species to be treated. In contrast to human or animal tissue surfaces, the surface of seeds consists of dead cellular material and water-repellent polymer layers to protect the plant embryo from physical and chemical influences [26]. Another difference is that, in contrast to animal organs, the entire seed is treated, not single specific parts of it. In addition, seeds are not treated as a single individual, but usually in a batch with a large number of seeds at the same time. Therefore, there is a need to develop devices for treatments on a larger scale, which will be necessary for future agricultural application. Thus, the requirements for plasma source dimensions to treat plant seeds along with a greater flexibility of plasma processes and operation conditions need to be addressed. Section 6.3 focuses on gaseous plasma treatment of seeds under atmospheric and low-pressure conditions comprising the plasma effects on physicochemical alterations of the seed and on germination and developmental processes.

A wide range of options exists to generate non-thermal plasma. This refers to configuration of electrodes, applied pressure, feed gas composition and flow rates, and electrical parameters (voltage, type of electrical current, frequency, power) used to ignite plasma, as well as treatment times and the mode of treatment with respect to direct or indirect plasma exposure of the plant target, as can be seen in Tables 6.4 and 6.5. In general, dielectric barrier discharges (DBD) in different configurations such as surface DBD (planar DBD) or diffuse coaxial DBD (DCSBD), gliding arc discharges, jets, corona discharges, microwave discharges as well as different kinds of radio-frequency (RF) discharges exist and has been applied. For treatment of seeds under atmospheric pressure, dielectric discharges using AC, DC or even RF were most frequently studied so far (Table 6.4). Regarding low-pressure conditions, RF plasmas were mostly investigated (Table 6.5).

Proper seed germination and seedling establishment on the field is the fundamental requirement for resilient plant growth, which ultimately determines the yield. Here, plasma has relevance for potential future application in agriculture as many studies have proven the beneficial effects of non-thermal plasma on seed germination performance. Important agricultural relevant plant species with different usages ranging from food and feed production to pharmaceutical and plant-based industry have been investigated so far (Tables 6.4 and 6.5). Wheat (e.g. [38, 99, 207]), maize (e.g. [381]), rice (e.g. [150, 373]) and barley (e.g. [38, 267]) produce seed-like fruits (botanical term "caryopsis"; caryopses are propagation units and the term "seeds" will be used within this chapter for simplification) containing a starchy endosperm important for feed and food production. Legume seeds such as soybean (e.g. [175]), pea (e.g. [151, 330]), chickpea (e.g. [216]), common and mung bean (e.g. [35, 281,

Table 6.4 Effects of atmospheric pressure plasma on plant seeds

Plant species	Plasma parameters	Observed effects	References
Family: Violaceae papaconha *Hybanthus calceolaria* (L.) Schulze-Menz	DBD driven plasma jet at 8.1 kV and 720 Hz FG: He GF: 2 L min^{-1} TT: 1, 5, 10 min	• Enhanced seed germination speed and max. germination for 1 min plasma TT • Decreased WCA • Elevated imbibition by ~30% for all plasma TTs • No change in electrical conductivity	[60]
Family: Cucurbitaceae cucumber *Cucumis sativus* L. var. Regina F1 and Family: Solanaceae pepper *Capsicum annuum* L. var	DCSBD at AC with 15 kHz and 20 kV peak-to-peak FG: air Power density: 100 W cm^{-3} TT: 10–50 s (cucumber), 4–15 s (pepper)	• SEM pictures showed no evidence of structural damage for shorter TTs (cucumber: 12 s, pepper: 4 s) • XPS revealed a relative increase in oxygen containing groups which were still observable for plasma-treated seeds (cucumber: 12 s, pepper: 4 s) after seed storage of 9 days • Germination percentage monitored at two observation times was increased for both plant species at TTs of 10–40 s for cucumber and at TTs >12 s for pepper	[325]
Family: Acanthaceae green chiretta *Andrographis paniculata* (Burm.f.) Nees	Planar DBD at different voltages 3.4, 4.25, 5.1, and 5.95 kV FG: air TT: 10 and 20 s	• SEM analysis revealed erosion of seed surface that were more pronounced for seeds treated with plasma at 5.95 kV for 10 s • Permeability of seeds was significantly accelerated for seeds treated with 5.95 kV at 10 s TT and significantly decreased for seeds treated with 3.4 kV at 20 s TT • Positive trends on seed germination parameters and seedling emergence could be observed which were not consistently significant upon plasma treatment parameters • SOD activity and MDA content in seedling leaves was lower for almost all applied plasma treatment parameters of seeds, while CAT activity and tended to increase	[343]

(continued)

Table 6.4 (continued)

Plant species	Plasma parameters	Observed effects	References
Family: Alliaceae onion *Allium cepa* L. cv. Wolska	Plasma jet at RF and 1 kV FG: He/O_2, He/air ratio of 3:2 GF: 710 L h^{-1} Power: 45 W TT: 2, 5, 10, 15 s	• SEM analysis revealed no changes in surface structure • Significant increase in germination speed for He/O_2 (TT $\leq$ 10 s) and for He/air (TT $\geq$ 10 s) • Significant increase in max. germination for 10 s plasma-treated seeds with both gas mixtures • Seedling lengths of He/O_2 and He/air plasma-treated seeds were unchanged during the first 7 days of growth, but were significantly increased during further growth for 5 days	[340]
Family: Amaranthaceae spinach *Spinacia oleracea* L.	High voltage nanosecond pulsed plasma at 6 kV and 0.7 kA FG: air Discharge energy: 0.3 J per one shot pulse, TT: 1, 5, 10 shots	• SEM analysis revealed morphological alterations of seed surfaces • Germination and dry weight of seedlings increased after high voltage short pulse shots (1 and 5 shots) • Longer TT (10 shots) had negative effects on seed germination rate and seedling growth • Levels of GA3 hormone and pullulanase mRNA was elevated in 1 day germinated seeds except for 10 shots treated seeds	[136]
Family: Amaranthaceae spinach *Spinacia oleracea* L.	Micro DBD at 6 kV, 14 mA and 22 kHzFG: air, N_2 GF: 1.5 L min^{-1} TT: 0.5, 1, 3, 5 min	• SEM analysis revealed no alteration of seed surfaces • Air DBD plasma exhibited slightly higher germination and seedling growth than those treated with N_2 plasma • Levels of pullulanase mRNA was unchanged in one day germinated seeds • Increased chlorophyll content in 5-weeks old spinach seedlings from air DBD plasma-treated seeds with TTs $\leq$ 3 min • Increase in total polyphenol content in 5-weeks old spinach seedlings from N_2 DBD plasma-treated seeds with TTs $\geq$ 3 min	[136]

(continued)

Table 6.4 (continued)

Plant species	Plasma parameters	Observed effects	References
Family: Amaranthaceae spinach *Spinacia oleracea* L.	High voltage nanosecond pulsed plasma at 6 kV and 0.7 kA FG: air Discharge energy: 0.3 J per one shot pulse TT: 1, 5, 10 shots	• WCA was significantly decreased after 1 shot plasma treatment, while WU was unaffected • Raman and FTIR spectroscopy analyses revealed no chemical modification of the seed coat • Germination percentage of 7-days germinated seeds was significantly higher for 1 and 5 shots plasma treatment, while germination of 10 shots plasma-treated seeds was unchanged • GA hormone level and starch hydrolysis (hydrolytic activity and pullulanase transcript level) was significantly elevated within 24 h in germinating seeds plasma-treated by 1 shot	[138]
Family: Apiaceae cumin *Cuminum cyminum* L.	DBD at 10 kVand 15 kHz FG: Ar GF: 9 L min^{-1} TT: 5 and 10 min	• SEM analysis of seed surfaces revealed damaging effects of plasma for both plasma TTs • WU was significantly accelerated for both plasma TTs • Germination percentage at day 7 of germination time was significantly higher for both plasma TTs • Biomass (dry weight, shoot and root lengths, root area and volume) and biochemical (chlorophyll and carotenoid content) parameters of 4 weeks old seedlings were significantly increased for 5 min plasma-treated seeds while 10 min TT led to reduced seedling growth • Leaves of seedlings from 5 min plasma -treated seeds had higher contents of N, P, K, Mg, Ca and Fe elements • Significant higher ROS levels in leaves were detected for both TTs and more pronounced for 10 min plasma-treated seeds • Proline content in seedling leaves was markedly increased for 5 min plasma-treated seeds while MDA content was strongly increased in seedling leaves of 10 min plasma-treated seeds • SOD, CAT, APX and GR activities were significantly increased in seedling leaves of 10 min plasma-treated seeds, while 5 min plasma treatment resulted significant decrease in APX and GR activity while SOD and CAT activities were unchanged	[280]

(continued)

Table 6.4 (continued)

Plant species	Plasma parameters	Observed effects	References
Family: Apiaceae coriander *Coriandrum sativum* L.	Micro DBD at 15.4 kHz and ~0.6 kV FG: Ar, N_2, air GF: 1 L min^{-1} Energy: 0.94 (Ar), 0.92 (air), 1.04 (N_2) J s^{-1} TT: 30, 60, 180 s (4-times each with 24 h interval)	• SEM analysis revealed structural degeneration of the seed surface with increasing TT of N_2 plasma • Significant positive effects on germination performance was observed for N_2 plasma at TTs $\leq$ 60 s, while germination of plasma-treated seeds at 180 s was unchanged • Positive trends on seed germination for air plasma-treated seeds were overlapped by high standard deviations • Mean values of germination percentage of 60 and 180 s Ar plasma-treated seeds were lower but without significance • Polyphenol content was elevated in 2 weeks old seedlings from 60 s N_2 plasma-treated seeds and dropped down after 4 weeks of growth, while elevated polyphenol content of seedlings from 180 s N_2 plasma-treated seeds were recorded in 4 weeks old seedlings only	[135]
Family: Apiaceae coriander *Coriandrum sativum* L.	Microwave plasma torch at 2.45 GHz FG: N_2/O_2 mixture GF: 10 L min^{-1} for N_2 with 50, 100, 200, 300 sccm for O_2 Power: 400 W TT: 5 and 10 min	• Germination percentage at day 5, max. germination and lengths of 3 weeks old seedlings were significantly increased for plasma treatment parameters at O_2-GF $\geq$ 200 sccm and both TTs • The observed positive effects were attributed to the increasing occurrence of NO with increasing O_2 flow rate in the N_2/O_2 gas mixture	[135]
Family: Araliaceae dehisced ginseng *Panax ginseng* C.A.Mey	Planar DBD at ~8,4 kV and 60 Hz FG: Ar, Ar/O_2 (80%/20%) GF: 1 L min^{-1} TT: 10 min (3 times with 24 h interval)	• Germination was significantly elevated for Ar/O_2 plasma-treated seeds during 10-days oberservation time, while for Ar plasma-treated seeds increased values for germination percentage were only observed after 7 days of germination • Ar/O_2 plasma treatment of seeds resulted in increased root lengths of 10 days old seedlings, while seedlings of Ar plasma-treated were unaltered	[174]

(continued)

Table 6.4 (continued)

Plant species	Plasma parameters	Observed effects	References
Family: Asteraceae sunflower *Helianthus annuus* L. var. Nykrségi fekete	DBD at 7 kV and 14.4 kHz FG: air Power density: 3.05 W cm^{-2} TT: 5, 7, 9, 11 min	• Dry seeds displayed changes in phytohormone levels (ABA, gibberellin, auxin, cytokinins, salicylic acid) after 7 or 11 min plasma treatment • Content of ABA was significantly decreased for both TTs, but gibberellin species GA7 was significantly increased in 7 min plasma-treated seeds while gibberellin species GA3 and GA7 were significantly decreased in 11 min plasma-treated seeds • No significant positive effects on germination parameters (max. germination, median germination time, uniformity) could be observed under laboratory germination conditions • Germination in substrate of 7 and 11 min plasma-treated seeds resulted in negative effects on germination kinetics but max. germination was not affected • Biomass parameters (weight and length) of 9 day old seedlings were not changed, except that 11 min plasma-treated seeds had significantly reduced seedling length • Biomass parameters (plant and leaf weight, root length) and photosynthetic performance of 30 days old plants grown in substrate from 7 min plasma-treated seeds were significantly increased while these parameters were significantly decreased for 11 min plasma-treated seeds	[390]

(continued)

Table 6.4 (continued)

Plant species	Plasma parameters	Observed effects	References
Family: Asteraceae sunflower *Helianthus annuus* L.	Pin-shaped multi-corona discharge plasma at 8, 10, 12 and 14 kV FG: Ar/O_2 mixture GF: Ar: O_2 flow rates of 4: 2, 3: 3, 2: 4 L min^{-1} TT: 1, 3, 5 min	• Most of the applied plasma parameter combinations with different GF ratios, voltages and TTs, resulted in positive effects on seedling growth (dry weight and shoot lengths) • Positive effects on seedling shoot growth at 5, 6 and 7 days of cultivation time observed for all applied plasma parameters (Ar/O_2 mixture with 3 L min^{-1} each at 8. 10, 12 and 14 kV for 1 min), • Plasma treatment (Ar/O_2 mixture with 3 L min^{-1} each, at 8 kV) of seeds for 1, 3 and 5 min resulted in higher shoot lengths of 7 days old seedlings • All applied voltages and TTs for Ar/O_2 mixtures with 3 L min^{-1} each led to higher values for dry weight and shoot lengths of 7 days old seedlings	[203]
Family: Asteraceae sunflower *Helianthus annuus* L.	Planar DBD FG: air Power: 90 W TT: 30, 60, 90, 120 s	• WCA significantly decreased with increasing TT • WU values did not correlate with TT • Slightly increased WU values for 60 and 120 s TTs and slightly decreased WU values for 30 and 90 s TTs were observed • Germination percentage at day 5 of germination time was higher for all plasma TTs • Seedling lengths were slightly increased for TTs > 60 s, while 30 s plasma treatment resulted in slightly lower values during seedling growth for 7 days • Seedling fresh and dry weights were slightly higher for 60 and 90 s plasma-treated seeds	[371]
Family: Brassicaceae broccoli *Brassica oleracea* L. var. kialica plen	Corona discharge plasma jet at 20 kV and 58 kHz FG: air TT: 1, 2, 3 min	• Elevated germination after 1 min plasma TT and strongly impaired germination after 3 min plasma treatment • Seedling weight and length was elevated for 1 and 2 min TTs, but decreased at 3 min TT • No change in reducing sugars content, total phenolic content and radical scavenging activity was observed	[157]

(continued)

Table 6.4 (continued)

Plant species	Plasma parameters	Observed effects	References
Family: Brassicaceae broccoli *Brassica oleracea* L.	Multi plasma jet at 76 kHz FG: Ar GF: 4.2 lpm Power: 412 W TT: 20 – 80 s	• SEM analysis revealed slight smoothening of seed surface which was most pronounced in seeds treated with plasma for 80 s • WCA decreased strongly with the increase in TT • Germination tests and seedling growth with 30 and 60 s plasma-treated seeds revealed higher max. germination but no change in germination rate for both TTs • Shoot lengths of 7 days old seedlings were unchanged but shoot weight was slightly increased for 30 s treated seeds and decreased for 60 s plasma-treated seeds	[321]
Family: Brassicaceae radish *Raphanus sativus* L.	DBD at 7 kV and 14.4 kHz FG: air Power density: 3.05 Wcm^{-2} TT: 1, 3, 5, 7, 9, 15 min	• All applied plasma TTs resulted in significant positive effects on seed germination parameters (max. germination, median germination time) when freshly harvested dormant seeds were treated with plasma • Effects on germination correlated with decreased ABA and increased GA levels in seeds, auxin levels were unchanged but cytokinin levels were significantly elevated, too • Plasma treatment for 5 min of less dormant seeds (5 month storage) did not lead to improved germination parameters but decreased ABA and increased GA levels were observed as well	[67]
Family: Brassicaceae radish *Raphanus sativus* L. var. caudutrus	Multicorona air Plasma at 30 kV and 13.5 kHz FG: air TT: 2 and 4 min	• Slight acceleration of germination was observed for 2 min plasma-treated seeds with increased dry weight in 7 days old seedlings • Significantly increased shoot length without change in dry weight in 7 days old seedlings from 4 min plasma-treated seeds • Moisture content was elevated in 7 days old seedlings for both TTs	[337]

(continued)

Table 6.4 (continued)

Plant species	Plasma parameters	Observed effects	References
Family: Brassicaceae thale cress *Arabidopsis thaliana* (L.) Heynh ecotype Columbia	DBD at 10 kV and 9.7 kHz FG: air TT: 15 min plasma jet FG: He GF: 3 L min^{-1} TT: 15 min	• SEM analysis of air DBD plasma treated seeds revealed strong roughening of seed surface • Germination defined as testa rupture was significantly elevated after He jet and air DBD plasma-treatment and the effect was stable during observation time of 40 h, as monitored for air plasma-treated seeds • Decrease in seed coat permeability (tetrazolium test) of air DBD plasma-treated was observed	[16]
Family: Brassicaceae thale cress *Arabidopsis thaliana* (L.) Heynh	DBD at 10 kV and 10 kHz FG: air TT: 15 min	• Germination of wild type seeds was elevated under saline conditions up to 65 h of germination time • Germination of mucilage-deficient mutant seeds was significantly accelerated und under normal and saline germination conditions, while germination of cuticle-deficient mutant seeds was strongly impaired after plasma treatment under both germination conditions • No change in max. germination was observed for all seed types and germination conditions • A decrease in seed permeability (tetrazolium red) after plasma treatment was observed for all seed types • Hydrophobic compounds on seed surface were changed for all seed types	[17]

(continued)

Table 6.4 (continued)

Plant species	Plasma parameters	Observed effects	References
Family: Brassicaceae thale cress *Arabidopsis thaliana* (L.) Heynh and Family: Brassicaceae camelina *Camelina sativa* (L.) Crantz	DBD at 10 kV and 10 kHz FG: air TT: 1 and 15 min	• SEM analysis revealed strong roughening of seed surface with increasing TTs for both plant species • WCA was strongly decreased with increasing TTs for both plant species • WU was slightly improved for both plant species, while permeability of seeds was strongly decreased • Lipid analysis of Arabidopsis plasma-treated seeds revealed a decrease in saturated and an increase in unsaturated fatty acids, while in camelina both fatty acid types were reduced • Arabidopsis germination rate was almost unchanged while max. germination was increased especially for 15 min TT • Camelina germination rate and max. germination was positively affected after 1 min, while 15 min plasma treatment had negative effects on germination • The area of camelina cotyledons of 1 min treated seeds were significantly increased while the cotyledons of 15 min plasma-treated seeds had similar size as controls at day 5 of germination	[18]
Family: Brassicaceae thale cress *Arabidopsis thaliana* (L.) Heynh	DBD at 8.47 kV and 7.95 kHz FG: air Power: 2.5 W TT: 0.5, 1, 3, 5, 10 min	• SEM-EDX analysis revealed a plasma dose-dependent etching effect on the seed surface with increased surface oxidation • Short plasma TTs $\leq$ 3 min led to improved seed germination and seedling growth (fresh weight, root length), while plasma TTs < 5 min had inhibitory effects • MDA, ROS and RNS levels in seeds directly after plasma treatment were enhanced • H_2O_2 levels in 7 days old seedlings were higher for $\geq$3 min plasma-treated seeds • CAT, SOD, and POD activities as well as proline level in short-time plasma-treated seedlings were apparently higher	[54]

(continued)

Table 6.4 (continued)

Plant species	Plasma parameters	Observed effects	References
Family: Brassicaceae thale cress *Arabidopsis thaliana* (L.) Heynh	Planar DBD FG: air Power: 2.17 W Power density: 1.49 W cm^{-2} TT: 3 min	• Plasma treatment led to earlier maturity (2,5 days earlier), earlier harvest day (6 days earlier), elevated total seed weight of harvest and elevated number of seeds produced	[160]
Family: Brassicaceae thale cress *Arabidopsis thaliana* (L.) Heynh and Family: Asteraceae zinnia *Zinnia Peruviana* L.	DBD FG: air Power density: 1.49 W cm^{-2} TT: 3 min	*Arabidopsis* • Long term effect: in the third generation, the leaf area is 2 times larger than that without plasma irradiation and the stem length is 1.5 times longer than that without plasma *Zinnia* • In the second generation, the stem length is 2 times longer than that without plasma	[292]
Family: Brassicaceae thale cress *Arabidopsis thaliana* (L.) Heynh	Plasma (not specified) at 7.2 kHz FG: air Power density: 0.042 W cm^{-2} TT: 0.5, 1, 2, 3, 4, 5 min	• Plasma treatment for 1 min significantly promoted seedling growth • ABA level in seedlings increased and peaked 48 h after treatment, but were lower than in the control after 96 h • Transcript levels of ABA signalling genes markedly enhanced at 48 h, but significantly downregulated after 96 h • Plasma treatment reduced stomatal aperture after 24 h, accelerated ROS accumulation in guard cells • Ca^{2+} level in the treatment group higher than in the control at 24 and 96 h	[352]
Family: Brassicaceae radish *Raphanus sativus* L.	Plasma torch FG: O_2/ air GF: 1 slm Power: 5 W TT: 10, 20, 30, 60 min	• Increased germination speed after 24 h • Increased max. germination after 70 h • Elevated sprout length by 5 cm after 95 h of growth for O_2 plasma	[108]
Family: Brassicaceae radish *Raphanus sativus* L.	Surface discharge plasma at 50 Hz, and 15 kV FG: air GF: 1 min^{-1} Power: 2.7 W TT: 20 min	• Little effect on the germination rate, but influenced the early growth of seeds • Sprouts and roots of plasma-treated seeds were longer and heavier than those of control seeds • Best results were obtained for 20 min TT, where an increase of the length of roots and sprouts	[208]

(continued)

Table 6.4 (continued)

Plant species	Plasma parameters	Observed effects	References
Family: Brassicaceae radish *Raphanus sativus* L.	Corona discharge plasma jet at 20 kV 1.5 A and 58 kHz FG: air GF: 2.5 m s^{-1} TT: 1, 2, 3 min	• The plasma seed treatment for up to 2 min showed beneficial effects on seed germination rate and growth of seedlings • Compared to untreated controls, sprouts grown from plasma-treated seeds no significant changes in the levels of moisture reducing sugars total phenolic content and DPPH radical scavenging activity	[271]
Family: Brassicaceae radish *Raphanus sativus* L.	DBD FG: air Power density: 1.49 W cm^{-2} TT: 3 min	• No information on germination • Seedling length elevated after Plasma treatment • Response of plants to the plasma irradiation becomes gradually weak with time, ratio of plant length with plasma irradiation to control decreases from 3.7 at the first day to 1.3 at 7 day	[291]
Family: Brassicaceae radish *Raphanus sativus* L.	DBD at 9.2 kV and 0.2 A FG: air, O_2, NO, He, Ar, N_2 GF: 6 NL min^{-1} Power density: 1.49 W cm^{-2} TT: 60–180 s	• No information on germination • Enhanced plant growth for O_2, air and NO (10%) + N_2 feeding gases plasma • No significant growth enhancement for He, N_2, and Ar gases plasma • Humid air plasma irradiation was more effective in growth enhancement than dry one, >2.3 times faster growth was observed by 3 min air plasma irradiation	[293]
Family: Brassicaceae rapeseed *Brassica napus* L.	Corona discharge plasma jet at 20 kV, 1.5 A and 58 kHz FG: air GF: 2.5 m s^{-1} at inter-electrode gap TT: 1, 2, 3 min	• Plasma treatment of seeds up to 2 min showed positive effects on their germination rate and seedling growth • Physicochemical and sensory characteristics of rape sprouts unaffected due to plasma treatment of their respective seeds	[270]
Family: Cannabaceae hemp *Cannabis sativa* L.	Gliding arc at 50 Hz FG: air GF: 10 L min^{-1} TT: 180, 300, 600 s	• Differences in response among seeds of three hemp cultivars ('Finola', 'Bialobrzeskie', 'Carmagnola') • Positive/neutral effect was observed in all measured characteristics after gliding arc plasma pre-treatment • Gliding arc pre-treatment increased the length of seedlings, seedling accretion and weight of seedling in both cv. 'Finola' and cv. 'Bialobrzeskie' hemp	[306]

(continued)

Table 6.4 (continued)

Plant species	Plasma parameters	Observed effects	References
Family: Cucurbitaceae watermelon *Citrullus lanatus* (Thunb.) Matsum. & Nakai var. Niagara	Atmospheric plasma jet at power supply output 10 kV, 30 mA and 20 kHz FG: N_2 TT: 2, 4, 6, 8, 10 min	• WU was significantly increased with strongest increase at 2 min TT, but no difference to control after 11 h of soaking • Germination after 4 days: significant increase in germination for all TTs, most pronounced at 4 min • Max. germination after 10 day: no difference between control an 2 min, significant increase for all other TTs • Seedling (shoot and root) length after 10 days: elevated for all TTs • Seedling weight: no difference between control and 2 min	[191]
Family: Cucurbitaceae bitter melon *Momordica charantia* L.	DBD FG: Ar GF: 2 L min^{-1} Power density: 0.84 W cm^{-2} TT: 60 and 120 s	• Seed priming with plasma and/or multi-walled carbon nanotubes (MWCNT) led to the dramatic increase in growth-related traits, like root and shoot lengths, fresh and dry mass, vigor index, and leaf length • Simultaneous treatments with MWCNT and plasma amplified their individual effects • Uptake and transportations of MWCNTs from the root to leaves were manifested using an electron microscopy • The ultra-structural study revealed that plasma enhanced MWCNT uptake and accumulation • Modifications in organogenesis and differentiation patterns of tissues, especially vascular system, were provoked by the MWCNT and/or plasma treatments, • Plasma and MWCNT reinforce conducting xylem tissue • Long-term effects of MWCNT and/or plasma treatments on reproductive stage: increase in the number of produced flowers, and the decrease in the time of fructifying, caused by the MWCNT and/or plasma treatments	[299]
Family: Cucurbitaceae pumpkin *Cucurbita pepo* L. cv. Cinderella and *Cucurbita maxima* L. cv. Jarrahdale, cv. Warty Goblin	Plasma jet at 10 and 6 kHz FG: Ar/He GF: 2 L min^{-1} TT: 1 and 25 min	• Plasma jets accelerated the germination of pumpkin seeds • Changes in WCA and hydrophilicity	[349]

(continued)

Table 6.4 (continued)

Plant species	Plasma parameters	Observed effects	References
Family: Fabaceae fenugreek *Trigonella foenum-graecum* L.	Plasma jet at 16 kV and 24 kHz FG: Ar GF: 2 slpm TT: (1st) 30 s, 1, 5, 10, 15, 20 min; (2nd) 10 s, 30 s, 1, 5, 10, 15 min	• Enhanced seed germination rate for 30 s and 15 min, no change in max. germination • Increase in shoot fresh weight was observed, especially at 10 min TT, • Root:shoot ratio was lower in plasma-treated seedlings, compared with the control plants • Accelerating grounded (AG) electrode: O-radical emission line (777.4 nm) enhanced 5 times	[77]
Family: Fabaceae chickpea *Cicer arietinum* L.	Surface microdischarge (SMD) FG: air Power density: 10 mW cm^{-2} TT: 0.5–5 min	• Germination speed elevated for 0.5–4 min TT • Max. germination enhanced for 0.5–2 min • Seedling length and dry weight: increased for 0.5–2 min TT	[216]
Family: Fabaceae white leadtree *Leucaena leucocephala* (Lam.) de Wit	DBD at 17.5 kV and 990 Hz with pulsed DC mode FG: air TT: 3, 9, 15 min	• WCA was significantly decreased for all plasma TTs • Germination increase after plasma treatment by 3% (15 min) after 11 days of counting	[5]
Family: Fabaceae mung bean *Vigna radiata* (L.) R. Wilczek	Plasma jet FG: Ar GF: 13 slm Power: 80, 140, 200 W TT:1 min	• Induce significantly more water absorption and lead to a higher speed of germination, max. germination did not differ • WCA decrease after plasma treatment with 200 W only • Seedling morphology changed to short radicle and longer hypocotyls with a larger diameter • GABA in plasma-treated beans was approximately 3 times higher than the untreated group at 80 W treatment	[49]
Family: Fabaceae lentil *Lens culinaris* Medik	Seed-packed dielectric barrier at 6 kV in amplitude at 600 Hz device (SP-DBD) FG: He/N_2 GF: 2 slm/ 150 sccm TT: 20 min	• No impact is evidenced on germination rates • Median germination time decreasing from • Addition of molecular oxygen to the helium discharge does not promote seeds' vigor • Increased water absorption	[74]

(continued)

Table 6.4 (continued)

Plant species	Plasma parameters	Observed effects	References
Family: Fabaceae pea *Pisum sativum* L. and Family: Cucurbitaceae zucchini *Cucurbita pepo* L.	FSG plasma device at applied voltage was 15 kV FG: air GF: 5 L min^{-1} TT: 30 and 60 s	Pea: • Max. germination elevated for 30, 60 s • Germination speed slower for 30 s • Seedling length and dry weight increased for 30, 60 s, Zucchini: • Max. germination elevated for 30, 60 s • Germination speed slower for 30 s • Seedling length and dry weight increased for 30, 60 s • Drought resistance and germination of seedlings increased after plasma was applied to seeds at 30 s, while seeds treated whiten	[151]
Family: Fabaceae soybean *Glycine max* (L.) Merr	Planar needle array DBD at 50 Hz FG: O_2 and N_2 GF: 6 slm Power: 65 W or 85 W TT: 1, 2, and 3 min	• No significant effect on seed germination • Reductions in CAT activity and increments in glutathione content after plasma treatment, reversing the oxidative damage caused by pathogenic fungi	[264]
Family: Fabaceae lentil *Lens culinaris* Medik	SCDBD at 9.25 kV, 13.33 kH FG: air Power: 858.5 W TT: 0–10 min	• Seed quality: number of dead seeds increased for TTs > 180 s • Accelerated growth of seedlings for short plasma treatment (<120 s) as seen in percentage of hypocotyl and leaf emergence at 3 days • WCA revealed increased seed wettability	[357]
Family: Fabaceae common bean *Phaseolus vulgaris* L.	Plasma jet at 9 kV, 16 mA, 6 kHz FG: He GF: 2.0 L min^{-1} TT: 12, 25, 40, 45, 50 min	• WCA: decrease at 50 min • Germination: increase with increasing cap exposure time • Seedling length: elevated for 1 and 15 min plasma treatment	[348]
Family: Fabaceae lentil *Lens culinaris* Medik	Stacked volume DBD at 6 kV and 500 Hz FG: He/N_2 GF: 1 slm/150 sccm TT: 2–20 min	• Increased speed in germination • Accelerated for different sequences of stirring or randomizing for repeated treatment cycles instead of one long time period • Decrease in WCA greatest for stirring and repeated treatment	[73]

(continued)

Table 6.4 (continued)

Plant species	Plasma parameters	Observed effects	References
Family: Fabaceae mimosa *Mimosa caesalpiniafolia* Benth	DBD at 990 Hz (pulsed) and 35 kV FG: air Power density: 0.18 W cm^{-2} TT: 3, 9, and 15 min	• Wettability (WCA) and imbibition were found to be directly related to the treatment duration, but saturation of the imbibition was found for treatment durations greater than 9 min • Plasma treatment strongly improved germination, especially for short TT	[59]
Family: Fabaceae mulungu *Erythrina velutina* Willd	DBD FG: He GF: 0.03 L s^{-1} Power: 150 W TT: 60 s	• Only slightly better germination performance with 5% higher germination percentage after 25 days • Plasma treatment changed the wettability (WCA) of the hilum more effectively than it changed the micropyle	[6]
Family: Fabaceae pea *Pisum sativum* L. var. Salamanca	Surface DBD at 6 to 12 kV and 3.0 kHz FG: air TT: 1, 2, 3, 5, 10 min	• Significant increase in germination after 24 h for 1, 2, 3 min, after 48 h elevated for 3 and 5 min but not significant • Shoot length of 16-days old seedlings was reduced and only a very minor effects on the dry weight were observed • Seedlings from 5 min plasma-treated seeds had lower max. photochemical efficiency of photosystem II • Plasma treatment had effects on flavonoid glycosides in seedlings	[40]
Family: Fabaceae pea *Pisum sativum* L.	Planar DBD FG: air Power: 9–35 W TT: 1, 3, 5, 7, and 10 min	• Elevated seedling length, chlorophyll a, dry weight, with max. change at 15 W • Increased WU most pronounced for 15 W	[89]
Family: Fabaceae pea *Pisum sativum* L. var. Prophet	DBD at 10 kV, 14 kHz FG: air Power density: 2.3 W cm^{-2} TT: 60–600 s	• Increased in germination percentage and growth parameters root and shoot length • Increased biosynthesis of auxin and cytokinins as well as their catabolites and conjugates • 600 s plasma treatment changed seed surface morphology with abrasion and loosening of testal areas • WU most noticeable increased within the first 2 h but later saturation and comparable to controls	[330]

(continued)

Table 6.4 (continued)

Plant species	Plasma parameters	Observed effects	References
Family: Fabaceae pea *Pisum sativum* L. cv. Prophet	DCSBD at 14 kHz sinusoidal high voltage with an amplitude up to 20 kV peak-to-peak FG: air, O_2, N_2 GF: 3 L min^{-1} Power density: 80 W cm^{-3} TT: 60, 180, 300 s	• Significant positive effect of air and N_2 plasma treatment TT of 60 s on germination and growth parameters • Increased levels of radicals in young 3-day old seedlings and activation of antioxidant enzymes	[334]
Family: Fabaceae pea *Pisum sativum* L.	DBD at 14 kHz, 20 kV FG: ambient air, N_2, O_2, mixtures (O_2:N_2 = 20:80; 40:60; 60:40; 80:20) GF: 3 L min^{-1} Power: input 400 W TT: 60, 180, 300 s	• Ambient air plasma appears to be the most advantageous for the plasma treatment due to no significant DNA damage • More DNA damage than in non-treated samples; an ambient air plasma had the least damaging effects on seed DNA, compared to plasma treatment with different mixtures of O_2 and N_2	[342]
Family: Fabaceae soybean *Glycine max* (L.) Merr	Needle to plane DBD at 0–25 kV operating at 50 Hz FG: N_2 and O_2 GF: 6 NL min^{-1} TT: 60 – 180 s	• Plasma treatment increased germination, plant growth (root, shoot length, foliar area, fresh weight), promoted a normal and healthy physiological performance and incremented the yield of plants • No change in photosynthetic activity • Plants grown from infected seeds did not trigger oxidative stress due to the reduction of pathogen incidence in seeds treated with cold plasma • Vegetative growth revealed a similar pattern for plants grown from treated seeds than that found for the healthy control • Infected control, by contrast, showed clear signs of damage	[264]
Family: Fabaceae soybean *Glycine max* (L.) Merr	Needle to plane DBD at 0–25 kV operating at 50 Hz FG: N_2 and O_2 GF: 6 NL min^{-1} TT: 2 and 3 min	• Complement to [264] • Biometrical parameters of 40-d-old plants grown from plasma-treated seeds • Significant increase in shoot length of seedlings from O_2 plasma-treated seeds was observed, while both applied FGs led to an increase in root length and total leaf area • Elevated nitrogenase activity in nodules growing on plants of plasma-treated seeds	[265]

(continued)

Table 6.4 (continued)

Plant species	Plasma parameters	Observed effects	References
Family: Fabaceae soybean *Glycine max* (L.) Merr	Needle to plane DBD at 0–25 kV and at 50 Hz FG: N_2 and O_2 GF: 6 NL min^{-1} Power: 16 W TT: 2 min for N_2 and 3 min for O_2	• Complement to [264, 265] • Changes in phenotypes but no significance • Plasma treatment with both FGs induced DNA methylation changes with respect to the Control plants, with higher differentiation at 20 days after sowing than at 6 days after sowing • Epigenetic variability and the phenotypic variability correlated only at 20 days after sowing	[266]
Family: Fabaceae soybean *Glycine max* (L.) Merr	Needle array DBD FG: O_2 and N_2 gas GF: 6 NL min^{-1} Power: 65 W and 85 TT: 60–180 s	• All plasma treatments had a significant stimulatory effect on seed germination and vigor • CAT, SOD and guaiacol peroxidase: high CAT activity and GSH content exhibited by infected seeds (IC) suggested that they coped with the oxidative damage • Reduction in WCA after treatment, positive correlation between seed hydrophilicity and lipid peroxidation of seed coats • ABA levels were decreased, and auxin level was increased	[264]
Family: Fabaceae soybean *Glycine max* L. cv. Nížina	DCSBD at 20 kV and 14 kHz FG: air, O_2, N_2 GF: 3 L min^{-1} TT: 30, 60, 90, 120 s	• Positive effect on seed germination for O_2 60, 90 s, no significant increase for air and N_2 • Seedling length was elevated for N_2 plasma-treated seeds at 30 s and for air plasma-treated seeds at 60 and 90 s • Longer N_2 plasma TTs significantly inhibited succinate dehydrogenase activity, but stimulated lactate and alcohol dehydrogenase activities, suggesting anoxygenic metabolism • Significant DNA damage was found for N_2 plasma • Higher level of DNA damage was also detected in the negative control (untreated seeds), which might be associated with the age of seeds followed by their lower germination capacity	[335]

(continued)

Table 6.4 (continued)

Plant species	Plasma parameters	Observed effects	References
Family: Fabaceae soybean *Glycine max* (L.) Merr	Needle to plane DBD at voltage from 10.8 to 22.1 kV FG: Ar GF: 2 L min^{-1} Power: 3.4 to 15.6 W TT: 12, 24, 48, 60, 120, 180 s	• 2 min of exposure duration, max. soybean sprout growth was obtained at 17.3 kV, but decreased with further increase in plasma potential till growth is eventually inhibited at >21.2 kV • Germination and production rates of soybean sprouts exposed to plasma at 22.1 kV for 1 min showed the highest increases compared with those of the control group, but longer plasma exposure duration of 2 min had significant contrary effects • A slightly higher germination rate and enhanced root and shoot growth; changes in DNA methylation level, an increased SOD, POD and CAT enzyme activity	[382]
Family: Lamiaceae sweet basil *Ocimum basilicum* L.	Surface DBD FG: air Power density: 80 mW cm^{-2} TT: 10–600 s	• SEM only "etched" at microplyar region, no alterations on other surface regions • SEM-EDX, increase in relative O-content • Swelling behavior, seed size increase during imbibition • Improved seed germination, growth parameters of seedlings (total length, weight, leaf extension) considerably increased compared to the controls."	[7]
Family: Lamiaceae sweet basil *Ocimum basilicum* L.	Volume DBD FG: humid air GF: 7 slm Energy density: 151 kJ m^{-3} for seed-filled system TT: 10, 20, 30 s, 1, 3 min	• Elevated germination at all treatments parameters • Significant increase in seedling weight at 30 s and 3 min	[8]
Family: Lamiaceae woodland sage *Salvia nemorosa* L.	Planar DBD FG: Ar GF: 1–2 slm Power density: 0.84 W cm^{-2} TT: 80 and 100 s	• Significant increase in shoot, root fresh weight, root length for 80 s more pronounced than 100 s	[91]
Family: Lamiaceae lemon balm *Melissa officinalis* L.	DBD at 13 kHz and 10 kV FG: Ar GF: 2 L min^{-1} Power density: 0.84 W cm^{-2} TT: 50, 90, 120 s	• 50 s treatment significantly improved the stem length • Plasma significantly improved root length at 50 and 90 s TT • Toxicity of zinc oxide and selenium was reversed after plasma treatment • Improved the total plant fresh mass for 50 and 90 s TT	[14]

(continued)

Table 6.4 (continued)

Plant species	Plasma parameters	Observed effects	References
Family: Malvaceae cotton *Gossypium hirsutum* L. var. Sicot 74BRF	Needle-shaped DBD at 1 kHz sine wave with 38 kVpp for air and 11 kVpp for argon FG: Ar, air GF: 1 L min^{-1} TT: air 3, 27 min. dry air, Ar 81 min	• 27 min plasma treatment air significantly increase water absorption of seed, improves warm germination, metabolic chill test germination and chilling tolerance in cotton • Effect stable for 4 months	[66]
Family: Malvaceae garden tree-mallow *Lavatera thuringiaca* (L.) Vis (L.) Vis	DBD plasma jet at 17 kHz FG: N_2, He GF: 1.6 dm^3 min^{-1} of helium with 0.03 dm^3 min^{-1} of nitrogen Power: mean power of 6 W TT1, 2, 5, 10 and 15 min	• Increase in germination speed and max. germination after plasma treatment especially for 1, 5 min • No significant decrease in WCA	[260]
Family: Malvaceae garden tree-mallow *Lavatera thuringiaca* (L.) Vis var. Uleko	Gliding arc at 50 Hz, applied RMS voltage amounted to 680 V (max. 3.8 kV), RMS current and mean power from the mains were 33 mA FG: N_2 GF: 8 L min^{-1} Power: 40 W TT: 1, 2, 5, 10, 15 min	• Highest germination parameters were obtained for seeds stimulated with plasma for the exposure times of 2 and 5 min • Decrease in germination • WCA no significant decrease • Longer exposure of seeds to plasma resulted in affecting the deeper zone of cuticle and damage or fracture of some parts of the cuticle	[259]
Family: Pinaceae common pine *Pinus sylvestris* L. black pine *Pinus nigra*, J.F. Arnold mountain pine *Pinus mugo* Turra	DBD FG: air Power: density 75 W cm^{-3} TT: 1, 3, 5, 10, 30, 60 s	• Trends but non- significant elevated germination speed and max. germination, as well as early growth • Decrease in mean seedling length for black pine for 60 s plasma TT	[308]
Family: Pinaceae black pine *Pinus nigra* J.F. Arnold	DCSBD at 14 kHz sinusoidal high voltage, amplitude of up to 20 kV peak-to-peak FG: air TT: 1, 3, 5, 10, 30, 60 s	• WCA: significant decrease • Significant elevated germination for 3 s TT after two days • 60 s plasma treatment lowered germination • Early growth: 30 and 60 s treatments had significantly smaller dry seedling weight than the other samples • 5 s samples had the heaviest seedlings	[309]

(continued)

Table 6.4 (continued)

Plant species	Plasma parameters	Observed effects	References
Family: Piperaceae pepper *Capsicum annuum* L. and Family: Cucurbitaceae melon *Cucumis melo* L.	Volume DBD at 20.7 kV peak-to-peak, fAC = 5 kHz) FG: air GF: 1 slm TT: 2–300 s	• Decrease in germination for pepper seeds from 15 s onwards, for melon from 100 s onwards, strong decrease at 300 s • Decrease in WCA for pepper and melon, • WU for 60 s treatment small differences for the first 60 min of imbibition • Element ratio: pepper seeds, O:C ratio gradually increased with increasing the plasma TT, the O:C ratio of melon seeds increased only for short TTs < 5 s, and plasma treatment for >5 s led to a profound decrease in the O:C ratio	[116]
Family: Poaceae barley *Hordeum vulgare* L. cv. Maltz	DCSBD FG: air, O_2 or N_2 GF: 3 Lpm Power density: surface (1–3 W cm^{-2}) and volume (~80 W cm^{-3}) Power: 400 W TT: 10, 20, 30, 60, 180, 300 s	• Stimulating effects on germination for 10 s plasma TT with all feed gases but significant decreased germination for >20 s plasma, • Slight positive effects of plasma treatment on seedling growth (lengths and weight of shoot and root) were observed for all applied FGs and TTs $\leq$ 20 s, • Significant increase in DNA double-stand breaks for 60 s air plasma treatment revealed genotoxic potential of plasma treatment	[267]
Family: Poaceae rice *Oryza sativa* L. var. Zhu Liang You 06, var. Qian You No. 1	Non defined non-thermal plasma generator at 10 kV, 2 kHz FG: N_2 (79%) and O_2 (21%) GF: 3 L min^{-1} TT: 1 min	• Seed germination and seedling growth enhanced • Photosynthetic pigments, photosynthetic gas exchange, and chlorophyll fluorescence improved by cold plasma treatment	[311]
Family: Poaceae rice *Oryza sativa* L. var. NSCI RC298	Plasma jet at AC, 232 V and 1.99A input FG: air GF: 25 cfm Power: 450 W TT: 1, 2, 3 s	• No to small change in germination speed nor max. germination • Seedling length decease for 1 s, increase for 2 and 3 s	[262]
Family: Poaceae wheat *Triticum aestivum* L. var. Buryatskaya Ostistaya, var. Buryat BRIA	Atmospheric pressure glow discharge (APGD) FG: Ar Power: 5 W TT: 30 s	• WCA strongly decreased after plasma treatment • Elevated germination, no quantification presented	[22]

(continued)

Table 6.4 (continued)

Plant species	Plasma parameters	Observed effects	References
Family: Poaceae wheat *Triticum aestivum* L. var. Julius Family: Poaceae barley *Hordeum vulgare* L. var. Kosmos	Volume DBD FG: Ar GF: 4 slm Power density: 53.5 mW cm^{-3} TT: 10–60 s	• Cumulative germination was enhanced after 27 h of germination for wheat seeds treated from 10 to 60 s, • Barley seeds respond to the short plasma TTs with increased germination of after 24 h of germination • Decrease in the water contact angle • WU by seeds was moderately increased after 2 h of imbibition • Increase in hydrophilic functional groups being detected by x-ray photoelectron spectroscopy	[38]
Family: Poaceae barley *Hordeum vulgare* L. var. Saechal (waxy hull-less barley) and var. Saessal (non-waxy hullless)	DBD FG: N_2, air GF: (3 lpm) mixed with bubbled air (0.1 lpm, O_2 0.65% containing) as feed gas Power: 400 W TT: 10, 20, 40 and 80 s	• Growth of plasma-treated barley was increased • GABA content of plasma-treated Saechal barley was slightly increased under no germination process, while DPPH activity was decreased at the same condition • Denser and longer roots and higher shoots in plasma-exposed seedlings	[256]
Family: Poaceae brown rice *Oryza sativa* L. six cultivars	DBD driven by RF FG: Ar GF: between 18, 20 24 ml min^{-1} Power: 100, 135, 170, 200 W TT: 25, 50, 75, 100, 150, 200, 300 s	• SEM analysis revealed eroded seed surfaces with softer and smoother structures after plasma treatment • Germination percentage and seedling biomass parameters (root length and seedling length) were increased for applied plasma powers $\leq$170 W and all gas flow rates with max. values for seeds treated with plasma at 135 W and gas flow rates of 20–24 ml min^{-1} • Applied power of 200 W and all applied gas flow rates resulted in inhibitory effects on seed germination and biomass parameters • Exposure times up to 150 s at undefined plasma power and gas flow rates, resulted in increased values for germination and seedling lengths • Similar positive effects on germination and seedling growth were observed for 5 further brown rice cultivars, • Within the first 48 h after germination, the total content of phenolic compounds, of vitamin E and of γ-oryzanol significantly increased in all six rice cultivars, while the antioxidant activity was not significantly altered	[373]

(continued)

Table 6.4 (continued)

Plant species	Plasma parameters	Observed effects	References
Family: Poaceae maize Zea *mays* L.	DCSBD FG: air Power: 100 Wcm^{-3} TT: 60 and 120 s	• 6-day old seedlings obtained from 60 s plasma treatment showed >10% increase in root length, root fresh and dry weight • 120 s plasma treatment had inhibitory effects on seedling growth with shorter root lengths and lower weight • CAT, G-POX, SOD and DHO in seedlings of plasma-treated seeds were differently affected	[110]
Family: Poaceae maize Zea *mays* L. cv. Ronaldinio	DCSBD FG: O_2, N_2, air GF: 3 L min^{-1} Power: 400 W input TT: 30, 60, 90, 120, 180 and 300 s	• No change in germination for 30/60 s • Prolonged treatment >90 s = negative impact on the germination, growth, and production indexes • No change in imbibition • WCA: significant reduction form 30 s onwards	[115]
Family: Poaceae maize *Zea mays* L. cv. Ronaldinio	DCSBD at 14 kHz sinusoidal high voltage with an amplitude up to 20 kV peak-to-peak, input power of 400 W FG: air Power density: 80 W cm^{-3} TT: 15, 30, 60, 120, 180, 240, 300 s	• WCA strong decrease • Increase in wettability, resulting in a better WU and in an enhancement of growth parameters • WU for 60 and 300 s treated seeds: the most significant difference was visible after the first hour. Samples plasma-treated for 60 and 300 s absorbed more water than control samples • Shorter TT (about 60 s) had beneficial effects on the growth parameters of seedlings • No significant response of changes of germination found at TT of 60 and 120 s in comparison to the control • Greater reduction in germination was observed at the exposure time of 240 s • After 300 s nearly complete loss of germination • TT of 60 s stimulated root length by 12% and shoot length increased • Plasma treatment of maize seeds affected germination and growth of primary roots (length, fresh and dry weight production) but does not significantly affect root anatomy and morphology	[380]

(continued)

Table 6.4 (continued)

Plant species	Plasma parameters	Observed effects	References
Family: Poaceae wheat *Triticum aestivum* L.	Surface DBD FG: air GF: 1 L min^{-1} Power: 2.7 W TT: 15 min	• WCA: strong decrease • WU: significantly accelerated • Little effect on the germination rate while a substantial positive impact on growth parameters (roots and shoot lengths, dry weight of roots) of 4 days old seedlings • Increased root-shoot-ratio	[71]
Family: Poaceae wheat *Triticum aestivum* L.	Planar DBD at 13.0 kV, 50 Hz FG: air GF: 1.5 slm TT: 4 min	• Germination potential and germination rate increased, and the root length and shoot length of the wheat seedlings also increased • Increased proline and soluble sugar levels in normal water conditions and in seedlings exposed to drought conditions • Decrease in MDA content in seeds under drought stress • Increase in SOD, POD and CAT enzyme activity	[99]
Family: Poaceae wheat *Triticum aestivum* L. cv. Xiaoyan 22	Planar DBD at 50 Hz, 9–17 kV FG: air GF: 1.5 slm TT: 4 min	• Germination rate, germination potential, root length, and shoot length of the wheat seedlings increased after DBD treatment at 11.0 kV	[101]
Family: Poaceae wheat *Triticum aestivum* L.	Packed-bed at 100 Hz and 83 kHz voltage at 8 kV FG: air TT: 3, 10, 30 s	• Elevated germination for all TTs • No effect on leaf and root length, dry matter	[161]
Family: Poaceae wheat *Triticum aestivum* L. cv. Xiaoyan 22	Planar DBD FG: air GF: 1.5 slm Power: 1.5 W TT: 1 – 13 min	• Time of germination accelerated • Growth accelerated • For 4 min plasma TT, activity of antioxidant enzyme increased • Increased root and shoot growth • Increase in proline and soluble sugar levels	[179]
Family: Poaceae wheat *Triticum aestivum* L.	DBD at 80 kV and 50 kHz FG: air TT: 5 and 20 min	• WCA was strongly reduced for direct plasma and both TTs, but not for indirect treatment mode • No effect on germination for 5 min TT, but reduction in germination for 20 min TT	[189]

(continued)

Table 6.4 (continued)

Plant species	Plasma parameters	Observed effects	References
Family: Poaceae wheat *Triticum aestivum* L.	DBD at 80 kV and 50 kHz FG: air TT: 30, 60, or 180 s	• WCA was strongly reduced • Treatments of 30–60 s significantly enhanced germination rate and showed positive effects on seedling growth • Changes in seed pH and total titratable acidity, as well as nitrites, nitrates, and MDA contents • No significant increase in WU for both direct and indirect plasma treatment • Retention time 24 h before usage: significant elevated WU 60, 180 s	[190]
Family: Poaceae wheat *Triticum aestivum* L.	Plasma jet at 2.6 kV FG: N_2 GF: 14 L min^{-1} TT: 2, 4, 6, 8, 10 min	• After 2 h elevated WU after plasma treatment • No difference in WU after 12 h of soaking • Significant increase in max. germination • Significant increase in shoot, root length, fresh and dry weight after 10 days of growth	[192]
Family: Poaceae wheat *Triticum aestivum* L.	DBD at 13 kV and 50 Hz FG: O_2, air, Ar, N_2 GF: 1.5 slm TT: 4 min	• SEM analysis revealed etching effects on the seed coat for air, N_2 and Ar plasma-treated seeds, which affected the hygroscopicity and permeability of the wheat seed • Germination potential significantly increased by after 4 min of the air, N_2 and Ar plasma treatments • Shoot and root length was increased but not after O_2 plasma treatment	[207]
Family: Poaceae wheat *Triticum aestivum* L.	Planar DBD at 16 kHz and 20 kV FG: He GF: 5 slm Power: 30 W TT: 10–900 s	• Decrease in WCA • Surface analysis revealed a remarkably enhanced wettability of plasma-treated seeds due to the insertion of oxygen containing functionalities on their surface • Short plasma exposures were shown to enhance WU and accelerate seed germination, especially under water-scarcity conditions at TT < 200 s • WU accelerated after plasma treatment under low water availability for the first 4 h of imbibition • Long plasma exposures damaged the outermost layers of the pericarp due to a pronounced oxidative etching effect	[220]

(continued)

Table 6.4 (continued)

Plant species	Plasma parameters	Observed effects	References
Family: Poaceae wheat *Triticum aestivum* L.	Coaxial DBD at 10 Hz FG: Ar/He GF: 2 slm Power: 1-10 W TT: 3 to 30 min	• Faster germination acceleration and a lower WCA for Ar plasma treated seeds	[240]
Family: Poaceae wheat *Triticum aestivum* L.	DBD at 22.5 kHz and 10 kV FG: air Power: 30 W TT: 2, 5, 10 s RF Jet at 13.56 MHz and 10 kV FG: Ar GF: 4 L min^{-1} Power: 30 W TT: 15–300 s	DBD: • Enhancement of hydrophilicity of the seed surface and reduction of WCA • Ability of water imbibition by wheat seeds increased • Significant decrease in mean germination time for all plasma treatments • No change in root, shoot length Jet: • Decrease in WCA • Significant decrease in mean germination time for 2 s plasma treatment • Significant elevated mean germination time for >50 s • Elevated root length, dry weight >50 s TT	[347]
Family: Poaceae wheat *Triticum aestivum* L. cv. Eva	DBD at 14 kHz and at 20 kV peak to peak FG: air Power density: 70–100 W cm^{-3} TT: 10–80 s	• WU slightly tended to increase with increasing TT • Germination percentage 10 days after germination was significantly higher for TTs of 20–40 s • Significant positive effects on dry weight and vigor of 10 days old seedlings were observed for 30 and 40 s plasma-treated seeds • Plasma TTs of 70 and 80 s had significantly negative effects on seed germination and seedling growth	[379]
Family: Solanaceae pepper *Capsicum annuum* L. var. 'Superhot'	Plasma jet at 25 kHz FG: Ar GF: 3 slm Power: 0.41–0.61 W TT: no treatment time indicated	• Significant increase in germination speed for all plasma treatment parameters after 1 week • No change in max. germination (after 2 weeks) • No significant difference for lengths of seedlings and canopy and number of leaves	[341]

(continued)

Table 6.4 (continued)

Plant species	Plasma parameters	Observed effects	References
Family: Solanaceae pepper *Capsicum annuum* L.	Planar DBD at 11 kV FG: Ar GF: 1–2 slm Power density: 0.84 W cm^{-2} TT: 60 and 120 s	• Plasma treatment of 1 min had an improving effect on the shoot and root lengths as well as total leaf area • Plasma treatment of 2 min had an adverse effect. significantly impaired growth and hence reduced the total biomass • Alterations in stem diameter and differences in tissue patterns (especially in the vascular system) occurred, and were mainly dependent on the plasma exposure time and/or the presence of hormones	[288]
Family: Solanaceae pepper *Capsicum annuum* L.	DBD at 11 kV and 23 kHz FG: Ar Power density: 0.84 W cm^{-2} TT: 60 and 120 s	• Plasma-treated seeds seemed to gernate earlier according to visual observation • 35 days old plants of 120 s plasma-treated seeds displayed slight but significant increased biomass parameters (plant fresh weight, total leaf area) and biochemical parameters (chlorophyll and carotenoid content) • For both plasma TTs, a significant increase in POD and PAL activities and increased phenolic contents in leaves and roots were found in 35 days old seedlings	[125]
Family: Solanaceae tomato *Lycopersicon esculentum* Mill hybrid Belle F1	Coaxial DBD at 13–17 kV and 50 Hz FG: air GF: 15 slm Power: 0.55–1.43 W TT: 5, 7, 13, 30, 45 min	• Germination percentage after 5 days of germination was higher for all applied plasma TTs (5, 30, 45 min) at 17 kV and 1.43 W • 5 days old seedlings from 5 and 30 min plasma-treated seeds displayed significantly higher root and shoot lengths and higher plant dry weight while seedlings from 45 min plasma-treated seeds were unaffected, all plasma treatments at different applied voltages and TTs (13 kV and 13 min, 15 and 7 min, 17 kV and 5 min) resulted in faster germination, increased biomass parameters (fresh and dry weight) and higher root length with higher branches of seedlings monitored up to 20 days of growth • Shoot lengths of 12 days old seedlings were only increased for 17 kV and 5 min plasma-treated seeds while root-to-shoot ratio increased for all applied plasma treatment parameters	[196]

(continued)

Table 6.4 (continued)

Plant species	Plasma parameters	Observed effects	References
Family: Tropaeolaceae nasturtium *Tropaeolum majus* L	DBD at 20 kV and 16 kHz FG: He with air impurity GF: 5 L min^{-1} Power: 30 W TT: 10, 30, 120, −300 s	• SEM analysis revealed no changes in seed surface morphology after 300 s plasma treatment • Analysis of seed coat by XPS revealed a relative increase of O, N and K elements and a relative decrease of carbon for all plasma TTs • FTIR-ATR analysis of the seed coat revealed no change in chemical composition and a decrease in OH band intensity in plasma treated seeds was attributed to water loss from the seed coat during plasma exposure • Weight loss of seeds increased with increasing plasma TTs due to loss of water and rising temperatures during plasma exposure • WCA of the seed surface was significantly decreased as observed for 30 s plasma-treated seeds • WU was accelerated for 30 and 300 s TT and was unchanged for 10 s plasma-treated seeds • Germination speed and Germination of plasma-treated seeds was lower during the 12 h observation under sufficient water supply conditions, while seeds treated with plasma for 10 and 30 s displayed better germination under conditions of low water supply • Germination of 300 s plasma-treated seeds was impaired under all water supply conditions	[219]

ABA—abscisic acid; APGD—atmospheric pressure glow discharge; APX—ascorbate peroxidase; CAT—catalase; cv.—cultivar; DBD—dielectric barrier discharge; DCSBD—diffuse coplanar surface barrier discharge; DPPH—2;2-diphenyl-1-picrylhydrazyl; FG—feed gas; GA—gibberellin; GA3—gibberellin A3; GA7—gibberellin A7; GABA—γ-aminobutyric acid; GF—gas flow rate; GR—glutathione reductase; GSH—glutathione (γ-glutamyl-cysteinyl-glycine); MDA—malondialdehyde; nZnO—zincoxide; PAL—phenylalanine ammonia lyase; POD—peroxidase; PPO—polyphenol oxidase ROS-reactive oxygen species; SEM-EDX—scanning electron microscope-energy dispersive X-ray spectroscopy; SOD—superoxide dismutase; TT—treatment time; WCA—water contact angle; var.—variety; WU—water uptake; XPS—X-ray photoelectron spectroscopy

Table 6.5 Effects of low-pressure plasma on plant seeds

Plant species	Plasma parameters	Observed effects	References
Family: Amaranthaceae lamb's quarters *Chenopodium album* agg	Microwave discharge P: 40 mbar FG: Ar/O_2 and Ar/N_2 Power: 100 W TT: 6–48 min	• Germination tests revealed no changes in seed viability or medium germination time even after longest plasma TT	[304]
Family: Amaranthaceae Quinoa *Chenopodium quinoa*, Willd. var. Atlas	DBD at 1 kHz and 8.2 kV P: 500 mbar FG: dry air Power: 6.4 W TT: 10, 30, 60, 180 and 900 s	• XPS: decrease in C % and increase in O % and N %, • WU slightly accelerated • SEM images revealed etching and damage of seed surface for longest TT • Significant improved germination for all treatment times longer than 10 s	[96]
Family: Amaranthaceae Quinoa *Chenopodium quinoa*, Willd. var. Atlas	RF at 13.56 MHz P: 0.1 mbar, FG: dry air Power: 15 W TT: 10, 30, 60, 180 s	• XPS: decrease in C % and increase in O % and N % • WU slightly accelerated • SEM images revealed etching and damage of seed surface for longest TT • Shortest TT of 10 s resulted in most stimulating effect on germination while 180 s treatment slowed down germination	[96]
Family: Amaranthaceae spinach *Spinacia oleracea* L. var. Beiouzhizun	Magnetized arc plasma P: 10–20 Pa FG: air Power: 400 W TT: 0.45 s	• Plasma treatment resulted in higher germination rates with increased germination percentage after 7 and 21 days of observation time	[310]
Family: Apiaceae Sprague Ajwain *Trachyspermum ammi* (L.)	RF at 13.56 MHz P: 1 Pa FG: air Power: 50, 80, 100 W TT: 2 min	• WCA significantly decreased after plasma exposure at 50 W • 50 W plasma treatment resulted in significantly higher cumulative germination values for 7 days germination time and improved germination index while 100 W plasma treatment resulted in germination performance similar to untreated seeds • Root length of seedlings significantly increased for 50 W plasma treated seeds	[92]

(continued)

Table 6.5 (continued)

Plant species	Plasma parameters	Observed effects	References
Family: Asparagaceae wild asparagus *Asparagus acutifolius* L.	RF at 13.56 MHz P: 800 mTorr FG: O_2, N_2 and O_2/N_2 mixture (20%/80%) Power: 50 W TT: 1, 15, 30 min	• 1 min O_2/N_2 and N_2 plasma and 15 min N_2 plasma had positive effects on germinability with reduced T_{50} value • Oxygen plasma led to negative effects on seed germination parameters (germinability, mean germination time, 50% time of germination) • All other plasma treatments resulted in higher T_{50} values and mean germination times	[188]
Family: Asteraceae artichoke *Cynara scolymus* var. *scolymus* L.	RF at 13.56 MHz P: 1.8 Pa FG: N_2 Power: 10 W TT: 3, 10, 15 min	• SEM analysis revealed occurrence of cracks and holes even after 3 min TT • WCA drastically reduced even after 3 min TT and complete wettability was achieved after 15 min plasma treatment • WU significantly accelerated • Germination parameters (rate, vigor index) and seedling growth (root and shoot length) positively affected for all TTs • 10 and 15 min TT led to significant root and shoot dry weight • POD and CAT activities slightly increased after the 15 min plasma treatment	[119]
Family: Asteraceae purple coneflower *Echinacea purpurea* (L.) Moench	RF at 5.28 MHz P: 60 Pa FG: air Power density: 0.35 W cm^{-3} TT: Power density: 0.35 W cm^{-3}	• All treatments resulted in faster germination without change in final germination percentage • Plasma treatment resulted in increased plant height, leaf number, and root weight • Content of vitamin C and phenolic acids as well as radical scavenging activity leaf extract significantly increased	[210]

(continued)

Table 6.5 (continued)

Plant species	Plasma parameters	Observed effects	References
Family: Asteraceae safflower *Carthamus tinctorium* L.	RF at 13.56 MHz P: 1.6 and 16 Pa FG: Ar Power: 20 W TT: 30 and 130 min	• SEM analysis displayed surface modification at the hilum • Increased germination rate after 10 days for plasma treatment at 20 W for 30 min and at 16 Pa for 130 min	[69]
Family: Asteraceae sunflower *Helianthus annuus* (L.)	RF at 5.28 MHz P: 200 Pa FG: air Power density: 0.35 W cm^{-3} TT: 2, 5, 7 min	• No significant trend in changes of germination characteristics (final germination percentage, median germination time) with increasing TTs could be observed • Plasma treated seeds stored for 4 days displayed higher GA3 levels • 7 min plasma treated seeds had increased levels in IAA while ABA levels were unchanged • An increase of ABA in 2 min plasma treated seeds and a decrease in 5 min plasma treated seeds could be observed • Biomass parameters (lengths and weight of roots and shoots, leaf weight) of seedlings 2 weeks after sowing were not improved for different TTs • Leaf protein expression data from 2 weeks old seedlings revealed long-term effects of plasma treatment on photosynthesis related metabolism and regulation	[209]
Family: Brassicaceae radish *Raphanus sativus* L.	RF at 13.56 MHz P: 40 Pa FG: air Power: 50 W TT: 10–90 min	• All TTs led to increased seedling lengths observed 3 days of germination • 20 min treatment displayed the highest values and germination was significantly accelerated	[109]

(continued)

Table 6.5 (continued)

Plant species	Plasma parameters	Observed effects	References
Family: Brassicaceae radish *Raphanus sativus* L.	RF at 13.56 MHz P: 20, 40, 80 Pa FG: Ar and O_2 Power: 60 W TT: 5, 15, 30, 60 min	• Radish seedling lengths were increased for all TTs at 80 Pa • 30 min plasma treatment at 40 or 80 Pa resulted in increased radish seedling lengths as well, while seedling length from 20 Pa plasma treated seeds were unchanged	[233]
Family: Brassicaceae radish *Raphanus sativus* L.	RF at 13.56 MHz P: 100 Pa FG: O_2 and N_2 Power: 50 W TT: 30, 60, 90, 120 min	• SEM analysis revealed no alteration of surface structures even after 120 min treatment time for O_2 plasma treated seeds • FTIR-ATR displayed no chemical modifications after O_2 plasma treatment • Accelerated seedling growth of O_2 plasma treated seeds monitored after 7 days • No effects on seedling growth after N_2 plasma treatment of seeds for 30 min • No effects were observed for seed germination	[158]
Family: Brassicaceae rapeseed *Brassica napus* L. cv. Zhongshuang 7 and11	RF at 13.56 MHz P: 150 Pa FG: He Power: 100 W TT: 15 s	• WCA was significantly decreased after plasma exposure of seeds • Under drought stress conditions, germination parameters (germination time, rate, germination and vigor index) were improved while maximum germination and median germination were unaffected • Biomass parameters (shoot and root lengths, dry weights and lateral root number) of 7 days old seedlings were significantly increased for both cultivation conditions (drought stress and non-stress) • Under drought stress conditions, protein and soluble sugar contents and activity of antioxidant enzymes (CAT, SOD) were significantly increased in seedlings, while MDA contents were decreased	[178]

(continued)

Table 6.5 (continued)

Plant species	Plasma parameters	Observed effects	References
Family: Brassicaceae rapeseed *Brassica napus* L.	DBD at 3 kV and 4.5 kHz P: 10 Torr FG: Air/O_2 (Air: 60%, O_2: 40%) and Ar/Air (Ar: 60%, Air: 40%) Power: 30 W TT: 90 s	• Germination of plasma treated seeds accelerated after 3 days of germination time with higher α-amylase activity in germinating seeds • Activities of SOD and CAT accelerated for Ar/Air plasma treated seeds, while APX activity was unaltered • Shoot dry weights increased and shoot lengths were significantly higher in 5 days old seedlings of plasma treated seeds • Biomass parameters of roots unaltered • No changes in H_2O_2 content detected in germinating seeds or shoots and roots of seedlings • Seedlings from Air/O_2 plasma treated seeds exhibited significant higher chlorophyll and protein contents in shoots and tremendous higher APX activity and mRNA level in roots • SOD and CAT activities in roots and shoots and APX activity in shoots not affected in seedlings of plasma treated seeds	[129]
Family: Brassicaceae rapeseed *Brassica napus* L.	DBD at 5.7 kHz and 8.7 kVpp P: 300 mbar FG: Ar TT: 15 s	• Higher germination percentage observed after 24 and 48 h	[297]
Family: Brassicaceae rapeseed *Brassica napus* L. cv. Zhongshuang 9	RF at 13.56 MHz P: 150 Pa FG: He Power: 60, 80, 100, 120 W TT: 15 s	• WCA decreased with increasing applied power • WU significantly accelerated after plasma treatment • Germination characteristics (germination rate, index, vigor) improved for all applied powers • Increased dry weight more prominent for roots accompanied by increased root length for seedlings • Field experiments showed that plasma treated seeds resulted in plants with higher pod numbers per plant and 1000 seed weights • 100 W plasma treatment resulted in best results for studied germination and field parameters	[176]

(continued)

Table 6.5 (continued)

Plant species	Plasma parameters	Observed effects	References
Family: Cannabaceae hemp *Cannabis sativa* L. cv. Finola cv. Bialobrzeskie cv. Carmagnola	Microwave plasma discharge at 2.45 MHz P: 140 Pa FG: Ar, O_2 TT: 3, 5, 10 min	• Effects of plasma treatments had different effects on germination depending on cultivar and observation times of 3, 4 and 5 days • No effects on germination were observed for Finola, while negative effects were observed for Bialobrzeskie and Carmagnola • All cultivars respond negatively to microwave plasma treatment resulting in reduced seedling growth with smaller seedling weights	[306]
Family: Cannabaceae industrial hemp *Cannabis sativa* L. cv. Futura 75	RF at 5.28 MHz P: 200 Pa FG: air Power: 8.4 W TT: 5 min	• Maximum germination and median germination time was significantly improved while uniformity of germination was unchanged under laboratory conditions while no stimulation of germination under field conditions could be observed • Biomass parameters of 4 months grown plants were either negatively affected for female plants (shoot weight) or positively affected for male plants (shoot lengths and weight) • Number of inflorescences per female plant was unaltered • Cannabidiolic acid levels in leaves and inflorescences of female plants were markedly decreased	[131]
Family: Fabaceae alfalfa *Medicago sativa* L. cv. Zhongmu 6	RF at unknown frequency P: 130–160 Pa FG: Air/He mixture Power: 20–280 W TT: 15 s	• Effect of different plasma treatments on the germination simulated drought stress conditions was investigated. PEG 6000 with 5, 10, and 15% (w/w) • Plasma effects on germination parameters (potential, rate, index) and seedling parameters (root and shoot length, vigor index) under non-stressed and under different stress levels depended on applied power, with positive effects up to 140 W and partially negative effects at higher power values • 40 W was denoted as the most effective treatment power for improved germination and seedling growth under different germination conditions simulating drought stress	[80]

(continued)

Table 6.5 (continued)

Plant species	Plasma parameters	Observed effects	References
Family: Fabaceae black gram *Vigna mungo* (L.) Hepper cv. Barimash 3	DBD at 4.5 kHz and 5 kV P: 400 Torr FG: air Power: 45 W TT: 20, 40, 60, 90, 120, 180 s	• SEM analysis revealed smoothening of seed surface structure with increased TT • WU values significantly increased for TTs with max. values at 180 s • Germination at day 3 significantly higher for all TTs • Biomass parameters (shoot and root lengths and dry weight) tended to increase at all TTs, with significant higher values for shoot dry weight • Chlorophyll content of seedlings from 40–189 s plasma treated seeds had significant higher values • H_2O_2 content in leaves and roots were significantly increased for 90–180 s plasma treated seeds • NO content in leaves and roots tended to increase • Total soluble protein content in roots and leaves significantly increased for TTs > 60 s • Total soluble sugar content increased in leaves of plasma treated seeds >40 s • APX activity in leaves and roots (120 s, 180 s TT) and CAT activity in leaves (180 s) and roots (90–180 s) were significantly accelerated • SOD activity was unchanged in leaves and roots of plasma treated seeds	[31]
Family: Fabaceae common bean *Phaseolus vulgaris* L.	RF at 13.56 MHz P: 5, 15, 20 Pa FG: O_2 Power: 400 W TT: 0.5, 3, 10 s	• Increased surface roughness (SEM analysis), • WCA: drop down from 85° to 0°, but hydrophobic recovery observed up to 42° after 30 h storage for 0.5 s and up to 15° • XPS: decrease in C % and increase in O % • WU accelerated • Depending on TT and applied power germination performance was affected positively and negatively • Despite this, radicle lengths of seedlings were increased with increased TT and power values	[281]

(continued)

Table 6.5 (continued)

Plant species	Plasma parameters	Observed effects	References
Family: Fabaceae common bean *Phaseolus vulgaris* L.	RF at 10 MHz P: 0.067 FG: air Power: 20 W TT: 2 min	• Weight of seeds clearly decreased with increasing TT up to 5 min • WCA of seed surface significantly decreased after plasma treatment, while WCA of surfaces from underlying tissue layers (mesotesta, cotylodones) were unaffected • Exposure of uncovered mesotesta and cotylodones to plasma resulted in a decrease in WCA • WU significantly accelerated for seeds with and without sealed micropyle • WU visualized by bromophenol blue dye revealed proved the plasma effect and the role of the micropyle structure during imbibition • Plasma treatment slightly accelerated germination but did not change final germination percentage	[35]
Family: Fabaceae lentil *Lens culinaris* Medik		• SEM analysis revealed no alterations of seed surfaces after 15 s plasma treatment of lentil, bean and wheat seeds • TOF-SIMS MS analysis of seeds revealed increase in nitrogen and oxygen containing groups of seed surface	
Family: Fabaceae common bean *Phaseolus vulgaris* L.	RF at 10 MHz P: 0.067 Pa FG: air Power: 20 W TT: 15 s – 2 min	• WCA was severely decreased for all plant species treated for 15 s, but WU was only slightly increased • Longer TTs did not increase wettability any further • WCA still decreased for one month stored seeds	[34]
Family: Poaceae wheat *Triticum* spec. L.		• Seed germination clearly accelerated for all plant species during different observation time (6 d for beans, 1 d for lentils and 12 h for wheat)	

(continued)

Table 6.5 (continued)

Plant species	Plasma parameters	Observed effects	References
Family: Fabaceae mung bean *Vigna radiate* (L.) R.Wilczek	RF at 13.56 MHz P: 20 Pa FG: air Power: 40 and 60 W TT: 10, 15, 20 min	• WCA was significantly decreased after plasma treatment • WU values increased significantly • All plasma treatment parameters resulted in faster germination, and increased shoot lengths of seedlings • Higher reserve mobilization was accompanied with increased soluble protein and sugar content and higher amylase and phytase activities • Trypsin inhibition activity and phytic acid decreased in plasma treated seeds	[287]
Family: Fabaceae peanut *Arachis hypogaea* L. cv. Eyou 7	RF at 13.56 MHz P: 150 Pa FG: He Power: 60, 80, 100, 120, 140 W TT: 15 s	• WCA decreased significantly with plasma power >80 W • Maximum germination values were not changed for plasma treated seeds • Germination potential was significantly increased for 80–120 W plasma treated seeds, while germination rate was only significantly increased for 120 W plasma treated seeds • Median germination time and uniformity of germination were slightly improved for 80–120 W plasma treated seeds as well • Shoot dry weight of 7 days old seedlings tended to increase with applied power, but only 120 W plasma treatment resulted in significant changes • Shoot dry weight of 7 days old seedlings was significantly increased for 80–120 W plasma treated seeds • Field trials with 120 W plasma treated seeds revealed improved biomass parameters of shoots (height, diameter, branch number, dry weight, leaf area and thickness) and roots (dry weight) of growing plants • Nitrogen content and SPAD value was increased in leaves of field grown plants from plasma treated seeds • Yield parameters (pod number and weight, yield in kg ha^{-1}) were markedly improved	[177]

(continued)

Table 6.5 (continued)

Plant species	Plasma parameters	Observed effects	References
Family: Fabaceae red clover *Trifolium pratense* L. cv. Vyčiai cv. Sadūnai	RF at 5.28 MHz P: 200 Pa FG: air power density: 0.025 W cm^{-3} TT: 5 and 7 min	• 7 min treatment had positive effects on germination speed of both cultivars without effecting the maximum germination value • For both TTs lengths of 7 days old seedlings were decreased in cv. Sadūnai but were increased in cv. Vyčiai • Both cultivars tended to have higher seedling weight for TTs applied • Seed population of cv. Vyčiai harvested on different years respond differently to plasma treatment and showed that maximum germination values cannot be increased • Plants grown on the field experiment for 5 months displayed higher shoot biomass, but number of inflorescences was almost not altered • Protein content in leaves increased noticeably in both cultivars while other nutritional values were only slightly improved for cv. Vyčiai • Changes of major leaf isoflavone levels biochanin A and formononetin were dependent on vegetation stage • 5 min plasma treatment resulted in a significant increased biochanin A/formononetin ratios in both cultivars in vegetative stage, but ratios were decreased in flowering stage	[211]

(continued)

Table 6.5 (continued)

Plant species	Plasma parameters	Observed effects	References
Family: Fabaceae red clover *Trifolium pretense* L. cv. Arimaiciai	RF at 5.28 MHz P: 200 Pa FG: air Power density: 0.025 W cm^{-3} TT: 5 and 7 min	• Seeds with different coat colors (yellow, brown, and dark purple) displayed distinct responses to plasma exposure • Maximum germination and uniformity of germination was not altered for all seed populations, a significant decrease in median time of germination was observed for both TTs with faster germination for yellow seeds compared to dark purple and brown seeds • Changes in phytohormone levels (ABA, gibberellins, auxins, cytokinins, salicylic acid) of dry seeds after plasma treatment were dependent on seed color and did not correlate with germination kinetics • Seedling biomass parameters (lengths and dry weight) of 7 days old seedlings were positively affected for yellow seeds, while darker colored seeds did not respond significantly to plasma treatment • Root lengths of 5 weeks old plants from plasma treated yellow and dark purple seeds were positively affected while root weight was only significantly improved for dark purple plasma treated seeds accompanied with markedly increase in nodule number	[132]

(continued)

Table 6.5 (continued)

Plant species	Plasma parameters	Observed effects	References
Family: Fabaceae soybean *Glycine max* (L.) Merr cv. Zhongdou 40	RF at 13.56 MHz P: 150 Pa FG: He Power: 0, 60, 80, 100 and 120 W TT: 15 s	• WCA revealed increased seed surface wettability after plasma treatment • WU values tended to be higher for plasma treated seeds with maximum values observed for 80 W plasma treated seeds during 24 h of imbibition • Germination (potential, rate, germination and vigor index) parameters were either unaffected or moderately improved in case of 80 W plasma treated seeds • Biomass parameters (shoot and roots lengths and dry weight) of 7 days old seedlings were increased up to 100 W plasma treated seeds while 120 W treated seeds were unaffected • Plasma treated germinating seeds displayed higher seed reserve utilization and higher soluble sugar and protein contents	[175]
Family: Lamiaceae sweet basil *Ocimum basilicum* L.	RF at 13.56 MHz P: 0.40 mbar FG: mixture of O_2 (80%) and Ar (20%) Power: 30–270 W TT: 10 min	• 150 W with highest stimulatory effect on seed germination and seedling vigor • 150 W treatment with highest values for carbohydrate and protein content and catalase activity of 7 days old seedlings	[317]
Family: Moringaceae Moringa *Moringa oleifera* Lam	RF at 13.56 MHz P: 2 Torr FG: Ar Power: 100 W TT: 1, 5, 10, and 15 min	• SEM images reveled etching effects for all TT • Highest germination parameters (percentage, potential) were obtained for seeds treated for 1 min • Plant length and weight 1 and 5 min TT was larger than of untreated seeds • TT above 5 min have damaging effects on seeds viability	[64]

(continued)

Table 6.5 (continued)

Plant species	Plasma parameters	Observed effects	References
Family: Paulowniaceae Empress tree *Paulownia tomentosa* (Thunb.) Steud	RF at 13.56 MHz P: 200 mTorr FG: air Power: 50, 100, 200 W TT: 1–40 min	• Depending on applied plasma power and TT times light-induced seed germination were either positively or negatively affected • A linear correlation for power intensity and exposure time on effects of germination could be observed • All TTs times at 50 W up to 30 min at 50 W resulted in an increase in germination percentage 10 days after imbibition • TTs < 10 min at 100 W had stimulatory effects on germination while TTs of 15 or 30 min inhibited germination • 1 min plasma treatment at 200 W stimulated seed germination while germination was impaired at longer TTs • UV shielding of seeds (glass cover of seeds) during plasma treatment at 100 W resulted in improved germination for 15 min TT, while 30 min plasma treatment still impaired germination	[389]
Family: Paulowniaceae Empress tree *Paulownia tomentosa* (Thunb.) Steud	RF at 13.56 MHz P: 100, 200, 400 mTorr FG: Ar, air Power: 100, W TT: 1–40 min	• At 100 W and 100 mTorr, air plasma TTs < 15 min had stimulatory effects on seed germination while plasma treatments at 100 W and 400 mTorr resulted in inhibition of germination when seeds were treated longer than 1 min • Compared to air plasma, Ar plasma inhibited germination by TTs > 1 min at 100 W and 200 mTorr • A clear markedly reduction in pH of water from imbibed plasma treated seeds was observed • The decrease in pH with a difference value of 3 was more pronounced for air plasma compared to Ar plasma • The nitrogen content of seeds was increased after air plasma treatment	[269]

(continued)

Table 6.5 (continued)

Plant species	Plasma parameters	Observed effects	References
Family: Pinaceae Norway spruce *Picea abies* (L.) H. Karst	RF at 5.28 MHz P: 60 Pa FG: air Power density: 0.35 W cm^{-3} TT: 2, 5, 7 min	• Seed germination characteristics (final germination percentage, median germination time, dispersion of germination time) differed upon germination conditions in petri dishes and in peat substrate • Negative effects of all TTs on germination observed when seeds germinated in petri dishes for 17 days, while a positive effect on maximum germination of 7 min plasma treatment on germination was observed when seeds were germinated in peat substrate for 33 days • H_2O_2 content of seeds lowered for all TTs observed within 48 h after imbibition • Long-term effects of plasma treatment on biomass production revealed significant shoot lengths for 5 and 7 s plasma treated seeds 17 months after sowing • Depending on observation time during 17 months of growth after sowing, number of needles and number of branches tended to increase	[258]
Family: Poaceae brown rice *Oryza sativa* L. acc. Tai Keng 9, japonica cultivar	Glow discharge P: 800 Pa FG: air DC voltage: 1, 2, 3 kV TT: 10 min	• For all applied voltages, significant increased germination percentage, seedling lengths and WU values were recorded during 24 h observation time • Among all applied voltages, 3 kV treatment resulted in highest a-amylase activity, highest gamma-aminobutyric acid (GABA) levels, highest total phenol content and highest radical scavenging activity during several time points within 24 h of germination	[45]

(continued)

Table 6.5 (continued)

Plant species	Plasma parameters	Observed effects	References
Family: Poaceae Maize *Zea mays* L. Family: Fabaceae lupine *Lupinus angustifolius* L. Family: Poaceae winter wheat *Triticum aestivum* L.	RF at 5.28 MHz P: 200 Pa FG: air Power density: 0.025 W cm^{-3} TT: 2, 4, 5, 7 min	• Germination percentage at day 10 of germination time and shoot lengths of 7 days old seedlings tended to increase for maize after plasma or vacuum treatment • Proline and phenolic contents were markedly increased in maize seedling roots and reduced in shoots after plasma treatment • Anthocyanin contents were increased in both shoots and roots of maize seedlings after plasma treatment • Yield of maize grown in the field tended to increase by 1.7% as a result of the plasma treatment • Germination percentage at day 10 of germination time were unaffected for all treatment conditions, while biomass parameters (shoot and root lengths) tended to increase after plasma or vacuum treatment of seeds • Wheat plants grown on field displayed higher phenolic contents during different growth stages • Yield of wheat grown in the field tended to increase by 2.3% as a result of the plasma treatment • Lupine shoot biomass of 7 days old seedlings was unaffected under all treatment conditions • Field trials with 4 min plasma treated lupine seeds resulted in plants having 26.8% higher yield with higher thousand seeds values	[84]

(continued)

Table 6.5 (continued)

Plant species	Plasma parameters	Observed effects	References
Family: Poaceae oat *Avena sativa* L. Family: Poaceae wheat *Triticum aestivum* L.	Microwave discharge at 2.45 MHz P: 140 Pa FG: air Power: 500 W TT: 3, 5, 10, 20, 40 min	• SEM analysis of 10 min plasma treated oat revealed no changes of seed surface while plasma or vacuum treatment resulted in formation of cracks on wheat seed surface • Plasma treatment slightly improved wheat germination at day 4 of germination time but not at day 8 or 12 • Maximum germination of oat was reached at day 3 and was close to 100% for all TTs • Biomass parameters of seedling roots (lengths and weight) were either unchanged in oat and wheat or tended to have reduced values in wheat • Biomass parameters of seedling shoots (lengths and weight) displayed similar trends except an observed positive response in wheat after 3 min plasma treatment • Analysis of phenolic compounds of wheat radicles and coleoptiles 23 h after germination by reversed phase HPLC revealed the presence of at least 5 species each, which showed an altered pattern after 10 min of plasma treatment • Total phenolic content was unaltered in wheat coleoptiles and reduced in radicle after 10 in plasma treatment • In oat coleoptiles, total phenolic content including vitexin was decreased after 10 min plasma treatment and decrease in radicles was more pronounced	[305]

(continued)

Table 6.5 (continued)

Plant species	Plasma parameters	Observed effects	References
Family: Poaceae rice *Oryza sativa* L. variety Shoniaush	DBD at 4.5 kHz and 5 kV P: 10 Torr FG: Ar/ Air Power: 45 W TT: 2, 4, 6, 8, 10 min	• SEM analysis revealed smoothening of the seed surface with increasing plasma TT • All TTs resulted in higher germination percentage recorded at day 3 of germination • Biomass parameters (shoot and root length, dry weight) and chlorophyll content of 8 days old seedlings were significantly accelerated for all TTs • Soluble protein contents in leaves were significantly higher for all TTs, while no changes in roots were observed • Soluble sugar contents were significantly higher for TTs > 6 min in leaves and for TTs > 8 min in roots • H_2O_2 contents were increased in shoots and roots for TTs > 8 min, while NO levels were unchanged • SOD and APX activities in shoots significantly increased for all TTs, while CAT activities was increased for 2, 6 and 8 min plasma treatments • SOD activity was unchanged in roots and root APX activity was accelerated for 8 and 10 min TTs, while significant increase of CAT activity was observed for TTs > 6 min	[30]
Family: Poaceae wheat *Triticum aestivum* L. variety Darya			
Family: Poaceae maize *Zea mays* L. hybrid Nemo 216 SV	RF at 5.28 MHz P: 40–80 Pa FG: air Power density: 0.34–0.65 W cm^{-3} TT: 2–10 min	• SEM analysis revealed smoothening of wheat and lupine surface structure after 5 min plasma treatment at 66.5 Pa and 0.34 W cm^3 • Seed viability for wheat and maize treated with plasma or vacuum were improved compared to untreated seeds • Seed viability for lupine seeds treated for 10 min with plasma was lowered and all other treatment parameters were unchanged • Shoot lengths of wheat and maize seedlings tended to increase while shoot lengths of lupine seedlings tended to decrease after plasma or vacuum treatment	[83]

(continued)

Table 6.5 (continued)

Plant species	Plasma parameters	Observed effects	References
Family: Fabaceae narrow-leaf lupine *Lupinus angustifolius* L. Variety Pershatsvet			
Family: Poaceae wheat *Triticum* spp. L.	RF at 3×10^9 MHz P: 150 Pa FG: He Power: 60, 80, 100 W TT: 15 s	• Germination percentage at day 3 and 7 significantly higher for 80 W plasma treated seeds, while germination of 60 and 100 W plasma treated seed was unaffected • Field experiments performed with 80 W plasma treated seeds revealed significant higher biomass parameters (shoot and root lengths and fresh weight) of seedlings 70 days after seeding • Biomass parameters (plant height, root length, fresh weight, stem diameter, leaf area and leaf thickness) of growing plants together with chlorophyll and nitrogen content at boot stage (approx. 5 months after seeding) were increased as well • Yield (t ha^{-1}) of plasma treated wheat was increased by 5.89% more than that of the control	[139]
Family: Poaceae wheat *Triticum aestivum* L.		• Germination of wheat was accelerated for applied powers up to 160 W	
Family: Poaceae maize *Zea mays* L.	RF at 13.56 MHz P: 30–200 Pa FG: air and air/He mixture Power: 60–180 W TT: 5 - 90 s	• Plasma treatment improved germination parameters (germinability, germination percentage) of aged seeds from wheat, maize and pumpkin • Field experiments revealed a slight increase in wheat grain number, thousand grain weight and ears per plant for different plasma treatments (60, 100, 140 W)	[381]
Family: Cucurbitaceae pumpkin *Cucurbita* L.		• Field experiments with 160 W plasma treated pepper seeds revealed increased plant height and higher shoot branch number	
Family: Solanaceae pepper *Capsicum annuum* L.			

(continued)

Table 6.5 (continued)

Plant species	Plasma parameters	Observed effects	References
Family: Poaceae wheat *Triticum aestivum* L. cv. BARI Gom 22	DBD at 4.5 kHz and 5 kV P: 10 Torr FG: Ar/O_2 and Ar/air mixture Power: 45 W TT: 90 s	• SEM analysis revealed rougher and chapped seed surface for both plasma treatments • Plasma treatment improved viability of germination without changing median germination time • Shoot biomass parameters of seedlings increased (length, dry weight) while root biomass parameters were decreased • H_2O_2 content in seeds, shoots and roots of seedlings significantly increased • CAT activity significantly accelerated in seeds while SOD and APX activity were unchanged • Plasma treatment of seeds did not lead to membrane damage or cell death in roots and shoots of seedlings • SOD activity and mRNA abundance in roots and APX activities in shoots significantly increased for Ar/O_2 plasma treated seeds, while enzyme activities (CAT, SOD and APX) and mRNA abundance of enzymes were unchanged in the respective organs and treatment modes • Iron content significantly increased in roots while zinc content was dramatically decreased • Total sugar contents in shoots significantly increased	[276]
Family: Poaceae wheat *Triticum aestivum* L. cv. BARI GOM 28	Glow discharge at 3–5 kHz P: 10 Torr FG: air and air/O_2 mixture Power: 60 W TT: 3, 6, 9, 12, 15 min	• WU increased with TT for both FGs applied • Germination parameters (rate, index, vigor, viability, median germination time) positively affected for all TTs and FGs applied • Shoot dry weight of 20 days old plants accelerated for all plasma treatment parameters applied • Increase in chlorophyll content more pronounced for air/O_2 mixture at all TTs • Plants from 3 and 6 min plasma treated seeds displayed higher grain weight and yield	[285]

(continued)

Table 6.5 (continued)

Plant species	Plasma parameters	Observed effects	References
Family: Poaceae wheat *Triticum aestivum* L. cv. Shannong 12	RF at unknown frequency P: 130–160 Pa FG: air/He mixture Power: 80 and 100 W TT: 15 s	• Wheat seed germination and field growth improved for three different generations of seeds • Values of seed germination parameters (rate, potential, index) and values of biomass parameters of seedlings (seedling shoot lengths, root lengths and number of lateral roots) increased after plasma treatment • Field experiments showed that plasma treated seeds displayed better field growth performance resulting in improved tillering, biomass production and finally yield	[123]
Family: Polygonaceae common buckwheat *Fagopyrum esculentum* Moench cv. VB Vokiai cv. VB Nojai	RF at 5.28 MHz P: 200 Pa FG: air Power density: 0.025 W cm^{-3} TT: 5 and 7 min	• Maximum germination and median germination time was not altered in cv. VB Vokiai after plasma treatment, but median germination time was significantly improved for cv. VB Nojai • Field trials revealed a reduced seedling emergence (11–20%) of plasma treated seeds at day 6 after sowing • Biomass parameters (plant height, dry weight, root and leaf weight, leaf number and lateral branches) of 4 weeks old plants from cv. VB Vokiai were significantly improved for 7 min plasma treated seeds • Positive effects on biomass parameters were less pronounced for cv. VB Nojai after 4 weeks of growth but significantly improved after 8 weeks of growth including weight of generative organs and number of shoots per plant • Photosynthetic activity was much more enhanced in cv. VB Nojai • Yield (seed weight, number of seeds per plant) was significantly increased for both cultivars and both TTs • Total phenolic contents were decreased, and the flavonoid quercetin contents were increased in the harvested seeds from plasma treated cv. VB Vokiai plants	[131]

(continued)

Table 6.5 (continued)

Plant species	Plasma parameters	Observed effects	References
Family: Solanaceae bell pepper *Capsicum annuum* L. cv. California wonder	Glow discharge at 500 V and 0.2 A P: 0.2 mbar FG: O_2 Power: 100 W TT: 3, 6, 9, 12, 15 min	• SEM analysis revealed etching of seed surface after plasma treatment at undefined TT • WCA was reduced after plasma treatment at undefined TT • Positive effects on germination parameters (germination percentage, speed) and seedling vigor were observed for all TTs • Positive effects on germination parameters were still detectable after 12 months of seed storage • In combination with osmopriming after plasma treatment, all germination and seedling parameters were further increased even after storage of up to 12 months	[234]
Family: Solanaceae tomato *Solanum lycopersicum* L. cv. Shanghai 906	RF at 13.56 MHz P: 150 Pa FG: He Power: 60, 80, 100 W TT: 15 s	• Germination percentage at 3 and at 7 days of germination was significantly improved for 80 W plasma treated seeds • 36 days old plants of 80 W plasma treated seeds displayed significant biomass parameters (total dry weight, leaf area, root length and dry weight, root surface area and volume) • Nitrogen and phosphorous content in plants 80 W plasma treated seeds were significantly increased • Germination and plants' biomass parameters, nitrogen and phosphorous contents of 60 and 100 W treated seeds displayed a positive but not significant trends	[142]

(continued)

Table 6.5 (continued)

Plant species	Plasma parameters	Observed effects	References
Family: Solanaceae tomato (*Solanum lycopersicum* L.)	RF at 13.56 MHz P: 150 Pa FG: He Power: 80 W TT: 15 s	• Positive effect on germination potential and germination rate • Long-term effects of plasma treatment on biomass, stress and antioxidant related enzymes and pathogen infection were observed • Increased biomass parameters (plant length, leaf thickness, stem diameter, dry weight) of plants grown for 30 days in pots • Plasma treatment resulted in increased levels of H_2O_2 and increased activities of POD, PPO and PAL in leaves • Disease development of plants inoculated with the bacterial wilt pathogen *Ralstonia solanacearum* reduced after plasma treatment	[140]

RF—radio frequency; P—pressure; FG—feed gas; TT—treatment time; WCA—water contact angle; WU—water uptake; SOD—superoxide dismutase; APX—ascorbate peroxidase; CAT—catalase; POD—peroxidase; PPO—polyphenol oxidase; PAL—phenylalanine ammonia lyase

287]), and lentil (e.g. [34]) belong to staple food, while alfalfa [80], blue lupine or clover (e.g. [211]) are used for feed production and are relevant in crop rotation because of the symbiotic activity with nitrogen-fixing bacteria. Increase in seed germination after plasma treatment has been also detected for seeds from rapeseed (e.g. [178, 270]) or sunflower (e.g. [203, 371]) known as oil plants. Moreover, this also applies for seeds from vegetables (e.g. radish [202]); spinach [136, 310]; tomato [196], zucchini [151], spices (e.g. pepper, [325, 341]), herbs (e.g. coriander, [135]; sweet basil [7, 317]), pharmaceutical relevant plants (e.g. ginseng, [174]); ajwain, [92]; hemp, [306]; safflower, [69] or trees (e.g. empress tree, [269, 389]); Norway spruce [258] and black pine [309].

6.3.1 Plasma Effects on Seed Surface Morphology

Depending on the plasma intensity of direct plasma treatment mode, outer seed surfaces can be modified leading to cracks, holes and fissures caused by etching and erosion events. Optical analysis by Scanning Electron Microscopy (SEM) is frequently applied to detect changes on seed surfaces by atmospheric DBD or low-pressure RF plasma treatment.

Several studies using DBDs documented surface modifications of wheat seeds after non-thermal plasma treatment [99] detected cracks on seed surface after 4 min air plasma treatment. [207] showed etching effects on the seed coat, which occurred after the air, nitrogen and argon plasma treatments, causing the change in hygroscopicity and permeability of the wheat seed. Li et al. [179] observed gradual destruction of square mesh structures and occurrence of cracks with elevated treatment time of air plasma. Molina et al. [220] found that the seed pericarp was progressively etched and damaged with increasing helium plasma exposure. Changes started with random nano-grooves on the outer layer at treatment times of 5 min, which extended when the treatment time was further increased to 15 min.

Moreover, other plant species were subjected to DBD treatment like barley, pea, thale cress or quinoa [256] investigated barley seeds and reported that plasma treated seed surface were etched and eroded after nitrogen/air plasma treatment for 40 s. Pea seeds were used by Gao et al. [89] displaying distorted and partially destroyed surfaces and ridges on the seed epidermis which gradually dissolved caused by seed coat erosion via bombardment of seeds with free radicals and ions of air plasma treatment at 15 W for 3 min. In addition [330], applied air DBD treatment to pea seeds and observed an uneven disruption, abrasion or even loosening of original structures in testal areas near the plumule- and radicle apex, especially after 10 min exposure. Effects of plasma on the model plant thale cress has been studied as well [54] presented dose-dependent etching effects of air plasma on seed surface encompassing slight shrinkages at 1 min treatment time up to detached epidermis at 10 min plasma exposure. Similarly, Bafoil et al. [16, 18] found changes on the seed surface after air plasma treatment for 15 min. The authors observed a physicochemical etching of the surface by plasma treatment due to rearrangement of macromolecular structures and exudation of lipid compounds from the seed.

In further studies, no damage (e.g. cracks, holes) of seed surface structure was observed after plasma treatment using a DBD plasma sources, e.g. pea [334], radish [159], spinach [136], sweet basil [7], maize [380], onion [340] or wheat [190].

Low-pressure plasma is also able to modify seed structure (Table 6.5). Flax seeds experienced etching of the cuticle and an accompanied weakening of the underlying mucilage secretory cell (MSC) walls [61]. Although longer RF plasma treatments (15–20 min) induced extensive cracking of the outer integument, the water uptake was not affected [281]. reported a rougher seed surface and an increasing amount of material being removed at 20 Pa with elevated treatment times, which resulted from energetic ions that impinge on the surface. On safflower seeds treated with low-pressure argon RF plasma for 130 min, changes in seed structure and a smoothening relative to the untreated control seeds appeared [69]. Quinoa seeds displayed plasma etching affecting the pericarp after non-thermal plasma treatment [96].

In general, observed modifications of seed surfaces after plasma treatment can be related to following factors: particle bombardment of highly energized species, local heat generation and possibly the individual nature with respect to heterogeneous morphological structures and chemical compositions of outer seed layers. Overall, there are critical methodological aspects for the visual detection of these changes by SEM: 1. SEM analysis provides only a spatially limited section of the entire seed surface. 2. Most seed surfaces are not homogeneous but are highly structured, requiring extensive surface analysis. 3. Eventually, a great number of single SEM pictures have to be analyzed, in addition to different individual seeds to conclude generalizations. 4. SEM analysis of seeds is performed in high or low vacuum and hence plasma effects on seed surfaces could be intensified.

6.3.2 Chemical Modification of the Seed Surface

During direct plasma treatment, bombarded by exited particles such as radicals and ions that can lead to erosion, etching and even chemical modification of the seed surface. This changes the chemical structure and morphology of the surface (e.g., roughness). Interaction of electrons and ions with outer surface layers result in modification and finally higher wettability [34]. Chemical modifications provoked by treatments using gaseous plasma that contain certain proportion of oxygen have been detected in several studies in which oxidation of seed surfaces irrespective of plant origin was observed (e.g. wheat, barley quinoa; [38, 96, 219, 220, 240, 325]) (see also Tables 6.4 and 6.5). Three different methods have been applied to study potential chemical modification: (1) X-ray photoemission spectroscopy (XPS), (2) attenuated Total Reflectance-Fourier Transform Infrared Spectroscopy (ATR-FTIR) and (3) Time of Flight-Secondary Ion Mass Spectrometry (ToF-SIMS). Mainly XPS has been applied to detect chemical changes, because this method can detect elements and their chemical state (e.g. oxidation and binding energy) located at seed surfaces, such as carbon, nitrogen, oxygen, potassium, sodium, magnesium and calcium (e.g. [96, 99, 116, 281, 325]). A correlation between relative oxygen abundance and plasma exposure times from 10 s up to 15 min were noticed on wheat seed surfaces [220]. The increased carbon-oxygen bonds were attributed to air impurities of the DBD device operated using helium at 5 slm. [380] analyzed maize seed surfaces using ATR-FTIR after exposure to DBD treatment for 60 s and found an increased occurrence of polar nitrogen and oxygen containing groups. RONS generated by DBD led to oxidation of lipids located at pea seed surface and subsequent increase in water uptake performance of peas [334]. The decrease of C-H bonds, typical for fatty acids, was more pronounced when air or oxygen was used as feeding gas. However, it was not stated if band intensity typical for oxygen containing groups were increased in FTIR spectra. Bormashenko et al. [34] analyzed the surfaces of wheat and lentil seeds treated with low-pressure air RF plasma using ToF-SIMS and found that a higher proportion of O-containing groups and N-containing groups could be detected on seed surfaces.

6.3.3 Alterations of Seed Surface Hydrophobicity

Physicochemical alterations of seed surfaces can result in changes of surface hydrophobicity. Wettability of seed surfaces can be estimated by measuring the water contact angle (WCA) of a tiny water droplet placed onto the surface. Depending on the contact angle, surfaces are referred as hydrophilic (<90°) or hydrophobic (>90°) [87, 168]. The WCA of plant seed surfaces are usually above 90° but can range from 130° to 76° depending on plant species [315]. Increased wettability of seed surfaces are observed after direct plasma exposure in e.g. wheat [34, 71, 190], soybean [175, 264], rapeseed [178], maize [380], lentil [34, 357], bean [34, 281] or barley [38]. Atmospheric pressure plasmas (e.g. DBD, plasma jets) and RF low-pressure plasmas either operating with air [18, 35, 59, 92, 371], oxygen[234, 281], (Piza et al. 2018), nitrogen[115, 119] or with nobles gases argon [38, 49] or helium [5, 175, 177, 219] displayed effects on seed surface wettability. By using a DCSBD system, different applied feed gases (air, oxygen and nitrogen) resulted in similar strong decreases of WCA values of maize seed surface with increasing plasma treatment times from 30 s to 5 min [115]. Comparable changes in surface wettability of wheat seeds by plasma have been reported for various DBD systems working with feed gases air [71, 190, 347], argon [38, 240] or helium [220].

Indirect plasma treatment does not lead to any significant changes in wettability of seed surfaces from wheat [190], Thuringian mallow [259] and rapeseed, barley or lupine [355].

6.3.4 Alterations of Seed Water Absorbance

The seed coat consists of several layers of dead cells. Seeds from several kinds of plant species such as legumes contain a cuticle as the outer layer that is enriched with phenolic compounds and fatty acid derivates resulting in a hydrophobic seed surface. Naturally occurring cracks on seed surfaces of soybean (*Glycine max* (L.) Merr.) can contribute to water uptake during imbibition [194]). Surfaces of caryopses from wheat or barley contain carbohydrate polymers (e.g., cellulose, hemicelluse) and lignin, which renders the surfaces to hydrophobic state. The mandatory initial step for germination is the uptake of water (imbibition process) to enable physiological processes [244]. Uptake of water does not occur evenly along the seed surface area. In seed science, several methods exist to deduce the route of water entry to the inner parts of the seeds, e.g., using dyes or stable isotopes as tracer. Furthermore, prior to imbibition tests, seed structures can be blocked by water impermeable material or dyes and thus can be used to trace the influx of water [65, 186, 385]. Bafoil et al. [17] measured seed permeability of thale cress by absorbance of tetrazolium red. This test is based on the enzymatic oxidization of tetrazolium red by dehydrogenases in the respiratory chain. Interestingly, the permeability was decreased by plasma treatment which seems to be contradictory to most of published research upon seeds.

Soybean seeds exposed to different gaseous plasma such as DBD displayed increased water absorption after one hour of imbibition [335]. The observed alterations were correlated with treatment time from 30 to 120 s and were more pronounced for nitrogen containing plasma compared to air and oxygen plasma. Similar observation using the same experimental setup was found for one hour imbibed pea seeds treated for 60, 180 and 300 s [334]. Wheat seeds with higher water uptake after plasma treatment simultaneously displayed a decrease in weight due to plasma etching process proved by SEM analysis [220]. Interestingly, water uptake of spinach or wheat seeds was unaffected after plasma treatment even though strong decrease in seed surface hydrophobicity was observed [138, 190].

Future research on plasma treatment of seeds should consider analysis of seeds by nuclear magnetic resonance (NMR) or magnetic resonance imaging (MRI). These techniques give more detailed information about water imbibition process as distribution of water under real time conditions can be monitored [33, 128, 263]. These techniques could provide a better understanding of the plasma-induced effects on whole-seed water permeability.

6.3.5 Plasma Effects on Seed Germination and Plant Growth Parameters

It is assumed that enhanced wettability due to physicochemical seed surface modifications is one of the major factors improving seed germination performance. The stimulating effects on seed germination is frequently discussed with the ability of plasma to break physical dormancy by reducing seed coat hardness and increased water permeability (e.g., [6, 54, 59, 304]). Plasma treated seeds from e.g. artichoke [119], cumin [280], lentil [34], mimosa [59], water melon [191], mung bean [287], pea [330], rapeseed [176], rice [45, 287], wheat [285, 379], or quinoa [96] had a higher water uptake accompanied with faster germination (see also Tables 6.4 and 6.5).

Analysis of germination kinetics to evaluate the effects of plasma on germination performance essentially involves visual monitoring of germination during different time points. The counting of germinated seeds is based on the macroscopic visible emergence of the radicle protruding from the seed [28, 163]. In studies, 1–2 mm minimal radicle length of germinating wheat and barley, [38, 189], half of the length of germinating soybean [175] or approximately 5 mm radicle length of germinating rice and sunflower [150, 322] were defined for germination. The germination value of a defined time point is presented as germination percentage or (cumulative germination percentage) and often referred as 'germination rate' or 'germination potential' in literature. Germination potential and germination rate of control and RF plasma treated wheat seeds were determined at day 3 and at day 7 of germination, respectively [139]. The values for germination rate were slightly higher compared to the germination potential values. Here, 80 W plasma treatment influenced both germination values significantly positive in comparison to unaffected 60 and 100 W plasma treatment. On the other hand, soybean germination potential after 3 days and germination rate after 7 days of control and plasma treated seeds did not vary from each other [175]. This indicate that final germination was almost reached after 3 days of germination and plasma treatment had no effect on the final germination of soybean. Treatment of black gram seeds for 120 s with air DBD plasma under low-pressure improved germination rate recorded at day 3 of up to 10% [31]. Moreover, observation time points and intervals vary among studies. The intervals of observation times can range from hours to several days and observation can last up to 3 weeks or longer. The final germination value and the speed to reach maximum germination depends on several plant related factors such as dormancy state and age of seeds along with plant origin and studied cultivar or variety. Soybean [175] or wheat [38] displayed maximum germination within 3 days with $\geq$80%, irrespective of plasma treatment. In other studies, variation in germination times and maximum germination values can be observed for the same plant species [192, 335], which can be attributed to different applied varieties, cultivars and/or to germination conditions. Molina et al. [220] analyzed wheat germination after plasma treatment using different water supply with 3, 6, and 12 ml. Interestingly, even the longest exposure time of 15 min did not impair seed germination, and seeds from all plasma treatment times displayed the similar maximum germination close to 100% compared to controls after three days. However, plasma treatment times below 2 min resulted in higher germination percentage after 20 and 24 h with 6 and 12 ml water supply. Maximum germination can be affected positively by plasma as shown for e.g., wheat, mimosa or mulungo (Tables 6.4 and 6.5). da Silva et al. [59] found a remarkable increase of final germination for mimosa from 6% for untreated seeds and 50% after 3 min air DBD treatment. Helium DBD treated mulungu seeds had 5% higher maximum germination after 25 days [6]. An increase of more than 10% in final germination recorded after 10 days was observed in wheat after treatment with air DBD plasma for 20 and 30 s [379]. The maximum germination of hemp seeds treated for 5 min with RF air plasma under low pressure was 20% higher compared to untreated seeds [131].

Time-resolved observation of seed germination include several observation times until final germination value is reached, allows more detailed assumptions about velocity and homogeneity of germination. The kinetics of germination can be described with a sigmoidal or logistic function since the rate of germination is not homogeneous over time. The Richard function [105, 283] has been applied to describe plasma effects on maximum germination, median germination time, uniformity and synchrony of germination in hemp [131], lamb's quarter [304], mimosa [59], mulungu [6], rapeseed [176], red clover [130], soybean [175], sunflower [209] and wheat [276, 285].

Next to monitoring of seed germination via observation at several distinct time points, biomass production such as root and shoot fresh and dry weight, lengths of shoot and roots or total seedling lengths are frequently recorded to deduce the effects of plasma. Furthermore, from those parameters different indices can be calculated such as seedling vigour index, and seedling length index (e.g., [309, 330, 335, 379]).

Seedling growth was monitored for thale cress [160], radish [159, 291, 293], sunflower [336, 390], wheat [207], and sweet basil [7]. Soybean seeds were treated with ceramic DBD fed with argon using different voltages and incubation times, and optimum germination was observed in the treatment with 22.1 kV for 12 s [382]. Germination was also higher for up to 1 min treatment and decreased when seeds were treated longer than 2 min [382]. Biomass parameters (shoot and root weight and length) were positively affected in the treatments from 12 s to 1 min and decreased when seeds were treated longer for 2 min [382]. Besides observable positive effects, extensive exposure of seeds to plasma can lead to inhibitory effects on germination and seedling development (Tables 6.4 and 6.5, e.g., [136, 188, 306, 373]). These can be attributed to high levels of radicals and reactive species within plasma such as ozone or nitric oxides (NO_x), next to heat and/or high electrical fields leading to deep entrance of electrons to the inner parts of the seeds.

6.3.6 Plasma Effects on Seed and Plant Physiology

Despite the fact that plasma treatment can accelerate germination speed, the simplest explanation for the frequently observed enhancement in seedling growth would be that plasma treated seeds exhibit a time advantage and therefore, higher biomasses of seedling shoots and roots is achieved. However, this would result in similar level of shoot and root growth compared to untreated plants, and thus, shoot/root ratios (or root/shoot ratios) would not be altered. A clear shift of growth to either shoot or root could be monitored for e.g. tomato seedlings [196] and wheat [305, 347]. Moreover, view studies noted alterations of root morphology [123, 142, 178, 196, 280]. The observed alterations on seedling development can be attributed to further effects of plasma components which are different from only physicochemical modification of the seed surface with accompanied wettability and improved imbibition. Here, reactive oxygen and/or nitrogen species (RONS) derived from plasma are the most versatile candidates that can trigger physiological modification and thus have impact on seed physiology with related development and growth processes as well as stress responses [124, 126].

Reactive oxygen species are known to play pivotal role during plant life cycle and are involved in many responses to biotic and abiotic stress factors [121]. During several steps of germination process reactive oxygen species are formed and play a positive role for dormancy release [244]. Externally applied hydrogen peroxide can stimulate pea seed germination with different effects on phytohormone levels of ABA, auxin, SA, JA and cytokinins [24]. In general, primary metabolism, growth and development related as well as stress relevant factors are frequently analysed in plasma studies.

Soybean seedlings six days after seed treatment with plasma showed an increase in levels of soluble protein, ATP, superoxide dismutase (SOD), peroxidase (POD), catalase (CAT), and adenosine triphosphate (ATP) and a decrease in malondialdehyde (MDA) [382]. Alterations in antioxidant activities in seedlings of plasma treated seeds were found for various plant species and different plasma exposures as well (see Tables 6.4 and 6.5). Expression of chloroplast ATP synthase subunits was accelerated, and methylation level in *ATP a1, ATP b1, TOR, GRF 5,* and *GRF 6* genes decreased [382]. Altogether, the argon plasma used promoted germination and growth by increasing the concentrations of soluble protein and antioxidant enzymes and regulating the demethylation levels of ATP, TOR, and GRF.

Any observed changes in germinating seeds after plasma treatment is likely a result of the time advantage of germination process, which may be associated with a change in levels of phytohormones. Phytohormones such as abscisic acid (ABA) and gibberellins (GA) are involved in regulation of germination initiation and germination process. ABA plays a role in seed dormancy (and stress responses), and gibberellin contributes to the initiation of germination. Two phytohormones, auxin and cytokinin, play a pivotal role during entire life cycle of plants, and ratio(s) of phytohormones is important for seed germination and seedling development. Moreover, these phytohormones are mandatory for development of shoots and roots.

Total cytokinin content increases during the first two days after imbibition in germinating *Tagetes minuta* L. seeds and declined during further seedling growth [326, 327]. Similarly, pea, maize, oat, and alfalfa display species specific dynamics of cytokinin levels during germination process and seedling growth [133, 328]. Comparable to phytohormone changes, levels of amino acids and sugars display different pattern during seedling development and growth [44, 79]. Plasma treated dry seeds showed significant changes in phytohormones content and ratios of gibberellins and ABA as well as auxins and cytokinins [1, 99, 175, 179, 209, 256, 330].

Polyphenols are secondary metabolites and belong to markers for different kinds of stresses such as excessive light, heat, drought, flooding etc. Phenolic contents in upper parts of the plants were assessed in barley [319], shoot and roots of wheat

[305], spinach [136], and pea [40] after plasma treatment. In wheat, levels of some phenols were slightly increased while others were unchanged or even decreased. In barley shoots, 6 min plasma treatment led to a significant increase in level of total phenolic compounds [319]. When barley seeds were indirectly treated by DBD driven air plasma for 6 min, increase in seedling weight and shoot length as well as increased levels in primary and secondary metabolites, like phenols, in leaves were observed [319].

Analysis of plasma effects on seed germination and seedling growth are mostly undertaken under laboratory conditions. However, proof of concept of stimulating effects of plasma on plant performance needs to be evaluated under agricultural relevant cultivation conditions. These include growth in soil and soil-like substrates but also cultivation in greenhouse and on fields are mandatory. Few studies exist so far performing green house or field trials. Field trials were performed with peanut and rape by Li et al. [176, 177], hemp [130], red clover [211], maize and wheat [84], wheat [123, 139] and maize and pepper [381]. Important traits to evaluate the efficiency of plasma treatment are biomass parameters that are correlated to yield which include number of flowers, number of seeds per plant, seed weight and weight of seeds per harvest area.

6.4 Application of Non-thermal Plasma to Food Sanitation

Food sanitation is the most actively explored area in the application of non-thermal plasma in the food industry. The antimicrobial activity of plasma *in vitro* has been demonstrated in numerous studies using food poisoning and spoiling microorganisms in planktonic and biofilm states [290]. Furthermore, experimental data are accumulating on sanitation and inactivation of microorganisms contaminating fresh produce, packaged foods, and processed foods, by plasma [346].

6.4.1 Vegetables and Fruits

Post-harvest fruits and vegetables are most frequently examined for microbiological sanitation using non-thermal plasma. Microbial contamination of fruits and vegetables can originate from pre-harvest infection or contamination during storage. To improve the shelf life and storage period of harvested fruits and vegetables, it is essential to inactivate microorganisms. Non-thermal plasma can efficiently deactivate the inoculated microbes and natural microflora on post-harvest fruits and vegetables, as demonstrated in previous studies (Table 6.6). Therefore, it is considered a potential tool for post-harvest sanitation. In most studies, plasma has been applied to fruits and vegetables after artificial inoculation with microorganisms (Table 6.6). However, there are also studies showing the plasma-mediated deactivation of natural microflora associated with fruits and vegetables [32, 93, 107, 164, 183, 221, 272, 338, 363, 366].

Regardless of whether they were inoculated or naturally contaminated with microorganisms, fruits and vegetables were directly exposed to plasma flame or plasma-generated gas. In relatively few studies, plasma-treated water has used for microbial decontamination [363, 366]. Treatment with dry plasma compared to plasma-treated water may be helpful in preventing the introduction of moisture, which can promote microbial growth. Microbial inactivation by dry plasma or plasma-treated water in post-harvest fruits and vegetables shows a proportional increase in response to the treatment time. Roughly about 0.3–7 log CFU reduction depending on treatment time, plasma sources, and feeding gas was observed in most studies (Table 6.6). The difference in inactivation efficiency between bacteria and fungi was not obvious.

Various non-thermal plasma sources such as plasma jets, DBD plasma, gliding arc plasma, corona discharge plasma, and microwave plasma are used for decontamination (Table 6.6). In most of the studies, plasma was generated mostly under atmospheric pressure. However, a group used plasma generated under low pressure [302]. Fruits and vegetables used to analyze the antimicrobial activity of non-thermal plasma are categorized into three groups: fresh fruits such as grape, banana, lemon, strawberry, blueberry, palm, melon, citrus, cantaloupe, and apple; dry nuts such as almond, hazelnut, and pistachio; and fresh vegetables such as corn salad leaves, lettuce, tomato, carrot, black pepper, red chicory, spinach, perilla, mung bean sprout, and argula leaves (Table 6.6). Additionally, Xu et al. [366] investigated the antimicrobial effects of plasma-treated water on button mushrooms and demonstrated that mushrooms had less microorganisms and could be stored for longer after soaking in the plasma-treated water.

Mycotoxin, a secondary metabolite produced by some fungi, is a food contaminant that threatens human and animal health [201]. Fruits and vegetables infected with mycotoxin-producing fungi have recently become a major concern in food safety [329]. Non-thermal plasma is also used to inactivate toxin-producing fungi and remove mycotoxins. The removal and degradation of mycotoxins by plasma in vitro have already been demonstrated in several studies [113, 253, 339]. Studies have also demonstrated that mycotoxins associated with dry nuts and grains, particularly aflatoxin B1, are efficiently degraded by non-thermal plasma[68, 114, 273, 303]. Additionally, mycotoxin levels have been controlled by inactivating producer fungi on fruits and vegetables using plasma. [248] observed that germination of spores and the levels of aflatoxin B2 and ochratoxin A decreased after date palm fruits inoculated with *Aspergillus niger* were exposed to a plasma jet.

Impact of plasma treatment on the quality of fruits and vegetables as food was analyzed together with antimicrobial activity in most studies (Table 6.6). This analysis is very important to determine whether plasma doses sufficient to kill microorganisms negatively affect the quality of fruits and vegetables as food. The most frequently analyzed quality factors are color, flavor, pH, and antioxidant activity. Studies have demonstrated that maximal antimicrobial efficiency of plasma does not always result in no damage to the food quality of fruits and vegetables. This indicates that there is an optimal plasma treatment condition (mostly treatment time) that produces efficient antimicrobial activity without significant damage to the quality of fruits and vegetables. It may be necessary to identify a proper point balancing between

Table 6.6 Application of non-thermal plasma to foods

Target foods (microorganisms)	Plasma types (feeding gas)	Effects	References
Food microbial safety			
Fruits and vegetables			
Grape and banana (Natural microflora)	High field plasma (ambient air)	No increase in bacterial and mold load on fruit surface during storage	[183]
Almonds (*Salmonella Escherichia coli* O157:H7)	AC plasma jet (air, N_2)	Maximum 1.34 log CFU reduction	[239]
Corn salad leaves (*E. coli*)	RF plasma jet (Ar)	Maximum 2.1–3.6 log CFU reduction No significant change in quality	[21]
Lettuce, tomato, and carrot (*E. coli*)	plasma (Ar)	Maximum 1.6 log CFU reduction No change in color parameters and cell structure	[27]
Rice and lemon (Natural microflora)	Surface discharge plasma (ambient air)	Inactivation of mold spores	[107]
Cherry tomato, strawberry (*E. coli*, *Salmonella* Typhimurium, *L. monocytogenes*)	DBD plasma (ambient air)	3.1–6.7 log CFU reduction	[388]
Black pepper (*Bacillus subtilis* spores, *B. atrophaeus* spores, *S. enterica*)	RF plasma, microwave plasma (Ar)	2.4–4.1 log CFU reduction	[112]
Blueberry (Natural microflora)	AC plasma jet (air)	0.8–2.0 log CFU/g reduction	[164]
Cabbage, lettuce, and dried figs (*Salmonella* Typhimurium)	Microwave plasma (N_2, He + O_2)	0.3–1.8 log CFU/g reduction	[170]
Date palm fruit(*Aspergillus niger*)	AC plasma jet (Ar)	Decrease in spore viability and the level of aflatoxin B_2 and ochratoxin A	[248]
Maize (*Aspergillus falvus Aspergillus parasiticus*)	Plasma jet (air, N_2)	Maximum 5.20–5.48 log CFU/g reduction	[56]
Romaine lettuce (*E. coli* O157:H7, *Salmonella* isolates, *L. monocytogenes*, Tulane virus)	DC DBD plasma (ambient air)	0.4–1.3 log CFU/g reduction	[213]
Grape, strawberry, and cherry tomato (Natural microflora)	Low frequency powered DBD plasma (air)	Lower survival of yeasts and molds during storage	[221]
Red chicory (*E. coli* O157:H7, *L. monocytogenes*)	DBD plasma (air)	1.35 log MPN/cm^2–2.2 log CFU/cm^2 reduction	[257]
Fresh cut melon (Natural microflora)	DC DBD plasma (air)	Delayed growth of microorganisms Weak effects on melon quality	[338]

(continued)

Table 6.6 (continued)

Target foods (microorganisms)	Plasma types (feeding gas)	Effects	References
Button mushrooms (Natural microflora)	Plasma treated water	Maximum 0.5–1.5 log CFU reduction Delayed softening and no significant change in quality	[366]
Citrus (*Penicillium digitatum*)	DBD plasma (ambient air)	Significant deactivation of spores No significant change in citrus surface	[370]
Grape tomato, spinach, and cantaloupe (*E. coli* 0157:H7, *Salmonella* Typhimurium, *L. innocua*)	Plasma activated H_2O_2 aerosol	1.0– 5.0 log CFU/piece reduction No significant change in quality	[141]
Almonds (*Salmonella* Enteritidis PT30)	DC SBD plasma (air, O_2, N_2, CO_2, CO_2 + Ar)	2.0 → 5.0 log CFU reduction Browning of surface color	[111]
Mandarin (*P. italicum*)	Microwave plasma (N_2, He, N_2 + O_2)	Significant decrease in spore viability by N_2 plasma Increase in total phenolic content and antioxidant activity of mandarin peel	[359]
Berries (Natural microflora)	Microwave plasma (Ar)	Decontamination efficiency depending on surface roughness No significant change in antioxidant activity of berries	[32]
Perilla leaves (*Staphylococcus aureus*, *E. coli*)	AC DBD plasma (air)	1.6–4.8 log CFU/ml reduction	[137]
Almonds (*E. coli*)	Gliding arc plasma (air)	Complete inactivation	[149]
Apple (*Salmonella*, *E. coli*)	Corona discharge plasma (air)	0.6–5.5 log CFU/cm^2 reduction	[152]
Black pepper (*B. subtilis*, *E. coli*, *Salmonella* Enteritidis)	SBD plasma (ambient air)	1.0–6.6 log CFU/g reduction No significant change in surface morphology	[226]
Kumquat fruit(Natural microflora)	Corona discharge plasma jet (ambient air)	0.77–1.57 log CFU/g reduction No significant change in taste, flavor, color, texture	[272]
Fresh cut lettuce and fresh mung bean sprouts (*E. coli*, *Listeria innocua*, *Pseudomonas fluorescens*, *P. marginalis*, *Pectobacterium carotovorum*)	Microwave plasma generated gas (ambient air)	Maximum >5 log CFU/ml reduction	[296]
Paprika (*Fusarium oxysporum*)	Plasma jet (air)	50% inhibition of fungal growth No significant change in color and hardness	[94]
Hazelnut (*A. parasiticus*, *A. flavus*)	Atmospheric pressure plasma jet (air, N_2), low pressure plasma (air, N_2, O_2)	4.7–5.6 log CFU/g reduction	[302]

(continued)

Table 6.6 (continued)

Target foods (microorganisms)	Plasma types (feeding gas)	Effects	References
Mun bean sprouts (Natural microflora)	Plasma activated water	2.32–2.84 log CFU/g reduction No significant change in total phenolic and flavonoid content and sensory characteristics	[363]
Arugula leaves (Natural microflora)	SDBD plasma (ambient air)	0.57–1.02 log CFU reduction No significant change in quality	[93]
Black pepper and Allspice berries Juniper berries (*Bacillus subtilis*, *A. niger*)	Microwave plasma (Ar)	Partial inactivation of *A. niger* and less effective against *B. subtilis*	[358]
Pistachio nuts (*A. flavus*, aflatoxin B_1)	DBD plasma (air)	Maximum 4 log CFU reduction Effects on some qualities	[199]
Citrus (*P. venetum*)	DBD plasma (air)	90% reduction in cell viability	[289]
Mycotoxin removal			
Hazelnuts (aflatoxin)	DBD plasma (O_2, N_2)	Over 70% reduction in total aflatoxin and aflatoxin B_1 concentration	[313]
Groundnuts (aflatoxin)	RF plasma (ambient air)	70–90% reduction in aflatoxin B_1	[68]
Hazelnuts (aflatoxin)	Atmospheric pressure plasma jet (air), low pressure RF plasma (air)	70–71% reduction in aflatoxin B_1	[303]
Rice and wheat (aflatoxin)	Corona discharge plasma jet (ambient air)	45–56% decrease in aflatoxin B_1 concentration	[273]
Corn kernels (aflatoxin B_1)	DC SBD plasma (ambient air)	Complete removal minor effects on quality	[114]
Meats, meat products, and fishes			
Bacon (*L. monocytogenes*, *E. coli*, *Salmonella* Typhimurium)	Glow discharge plasma (He, He + O_2)	2.6–4.58 log CFU/g reduction No significant change in quality	[153]
Chicken breast, ham (*L. monocytogenes*)	Plasma jet (He, N_2, He + O_2, N_2 + O_2)	1.37–6.52 log CFU/g reduction	[172]
Chicken skin and breasts (*L. innocua*)	Plasma (He, O_2)	1–3 log CFU/cm^2 reduction	[245]
Chicken breast and thigh (*S. enterica*, *Campylobacter jejuni*)	DBD plasma (air)	1.25–3.11 log CFU reduction	[70]
Porcine (Natural microflora)	Microwave plasma (air)	10^2–10^3 CFU/g reduction No significant change in color, pH, fluorescence, and reflectance	[88]

(continued)

Table 6.6 (continued)

Target foods (microorganisms)	Plasma types (feeding gas)	Effects	References
Pork loin (*L. monocytogenes*, *E. coli*)	DBD plasma (He + O_2)	0.55–0.59 log CFU/g reduction Significant reductions in sensory quality parameters (appearance, color, odor, acceptability)	[154]
Chicken breast (*E. coli*)	Plasma jet (N_2, O_2)	1.85 log CFU/g reduction	[375]
Beef loin, pork shoulder, and chicken breast (murine norovirus, hepatitis A virus)	Plasma jet (N_2)	90–99% reduction of viral titer	[15]
Pork butt and beef loin (*L. monocytogenes*, *E. coli* O157:H7, *Salmonella* Typhimurium)	Flexible thin-layer DBD plasma (air)	1.90–2.68 log CFU/g reduction Minor deterioration of meat quality	[134]
Filefish fillets (*Cladosporium cladosporioides*, *Penicillium citrinum*)	Cold oxygen plasma (O_2; BioZone Scientific International Inc.)	0.91–1.04 log CFU/g reduction No significant deleterious change in physicochemical and sensory qualities	[255]
Pork (Natural microflora)	Pulsed plasma (He, Ar, N_2)	2.7–3.08 log CFU/cm^2 reduction No significant change in color and pH	[345]
Pork slices (*L. monocytogenes*, *E. coli* O157:H7)	Corona discharge plasma jet (air)	1.0– 1.5 log CFU reduction No significant changes in oxidation and sensory qualities Changes in color and appearance	[48]
Chicken breasts (*L. monocytogenes*, *E. coli* O157:H7, *Salmonella* Typhimurium)	Flexible thin-layer DBD plasma (air)	2.14–3.36 log CFU/g reduction Minor changes in sensory properties	[171]
Eggs (*Salmonella* Enteritidis)	Plasma jet (Ar, Ar + O_2; kINPen 09®)	0.22–2.27 log CFU reduction	[225]
Smoked salmon (*L. monocytogenes*, *L. innocua*, *Salmonella* Typhimurium, *Salmonella* Enteritidis, *S. aureus*, *E. coli* O157:H7, *Aeromonas hydrophila*, *Plesiomonas shigelloides*)	Plasma (air)	0.1–1.57 log CFU reduction Minor effects on quality	[50]
Ham (*Salmonella* Typhimurium, *L. monocytogenes*)	Plasma (ambient air)	Maximum 1.02–1.14 log CFU reduction	[182]
Sliced pastirma (*S. aureus*, *L. monocytogenes*)	Plasma (Ar, O_2)	4.88–4.93 log CFU/cm^2 reduction Decrease in moisture content	[95]

(continued)

Table 6.6 (continued)

Target foods (microorganisms)	Plasma types (feeding gas)	Effects	References
Ham (*L. innocua*)	DBD plasma (ambient air)	1.43–1.78 log CFU/cm^2 reduction Some changes in quality	[368]
Oysters (Human norovirus)	Micro DBD plasma (N_2)	Over 90% reduction in copy number with propidium monoazide pre-treatment No significant change in quality	[47]
Salmon sashimi (Human norovirus)	Plasma jet (air, O_2, N_2)	About 20–100% reduction in copy number No significant change in quality	[122]
Chicken drumsticks (Natural microflora)	Plasma activated lactic acid	Maximum >1 log CFU/g reduction and increase in storage time No significant change in color components and decrease in MDA content Improvement of aggregation of myofibrillar proteins	[275]
In-package foods			
Pork butt and beef loin (*Listeria monocytogenes*, *E. coli* O157:H7, *S. typhimurium*)	DBD plasma (O_2 and N_2)	1.90–2.68 log CFU reduction Minor deterioration in meat quality	[134]
Broiler breast fillets (Natural microflora)	DBD plasma (ambient air)	1.53–5.53 log CFU/ml reduction	[162]
Pistachio (*A. flavus*)	DBD plasma (ambient air)	Spore deactivation	[318]
Chicken breast fillets (Natural microflora)	DBD plasma (ambient air, $O_2 + CO_2 + N_2$)	Significant reduction in viability under $O_2 + CO_2 + N_2$ Negative effects on chicken meat appearance	[353]
Beef jerky (*E. coli* O157:H7, *L. monocytogenes*, *Salmonella* Typhimurium, *A. flavus*)	Flexible thin-layer DBD plasma (ambient air)	2–3 log CFU/g reduction Slight changes in the physicochemical quality	[377]
Herring fillets (Natural microflora)	DBD plasma (air)	Significant reduction in microbial load Maintain key quality factors at low voltage	[3]
Chicken breast meat(*Campylobacter jejuni*, *Salmonella* Typhimurium)	AC DBD plasma (air)	>90% reduction in microbial CFU Paler meat color	[386]
Chicken breasts (*Salmonella* Enteritidis, *S.* Montevideo, *S.* Typhimurium, Tulane virus)	DBD plasma (air)	0.7–1.7 log CFU/cube reduction No change in pH, color, volatile nitrogen, lipid oxidation, tenderness, sensory qualities	[169]

(continued)

Table 6.6 (continued)

Target foods (microorganisms)	Plasma types (feeding gas)	Effects	References
Ham (*L. monocytogenes*)	DBD plasma ($O_2 + CO_2 + N_2$, $CO_2 + N_2$, CO_2)	>2 log CFU/cm^2 reduction Significant changes in lipid peroxidation and color	[369]
Processed foods			
Brown cereal bars (*A. flavus*)	RF plasma jet (Ar)	Inhibition of fungal growth for up to 20 days	[332]
Red pepper powder (*A. flavus*, *B. cereus*)	Microwave plasma (N_2 and O_2, He and O_2)	1–2.5 log CFU/g reduction for *A. flavus* 3.4 log CFU/g reduction in the combination with heat treatment for *B. cereus*	[155]
Sliced cheese (*E. coli* 0157:H7, *Salmonella* Typhimurium, *L. monocytogenes*)	DBD plasma (ambient air) flexible thin-layer DBD plasma (ambient air)	1.65–3.10 decimal CFU reduction 2.1–5.8 log CFU/g reduction Effects on cheese quality	[374, 376]
Wheat flour (Natural microflora)	DC plasma (ambient air)	No change in bacterial and fungal counts Modification of flour functionality	[20]
Insect flour ((Natural microflora)	SDBD plasma (air)	4.73–7.1 log CFU/g reduction Effects on quality	[41]
Onion powder (*B. cereus*, *A. brasiliensis*, *E. coli* O157:H7)	Microwave plasma (He)	1.6–2.1 log CFU/cm^2 reduction No significant change in quality	[156]
Orange juice (*S. enterica*)	DBD plasma (ambient air, $O_2 + CO_2 + N_2$)	2.2 → 5 log CFU reduction Effects on quality	[365]
Saffron (Naturally contaminated fungi)	Low pressure RF plasma (O_2)	Eradication of fungi at 60 W and 15 min treatment Minor effects on quality	[118]
Apple juice (*E. coli*)	DBD plasma (ambient air)	3.98–4.34 log CFU/ml reduction Mino effects on quality	[181]
Apple juice (*Zygosaccharomyces rouxii*)	Surface discharge plasma (air)	6.58–6.82 log CFU reduction	[354]
Kimchi cabbage (Natural microflora)	Plasma activated water	0.9–2.2 log CFU/g reduction	[46]
Orange juice (*Penicillium expansum*, *P. buchwaldii*, *P. bialowiezense*)	Plasma jet (N_2)	1–3 log CFU reduction	[97]
Coconut water (*Salmonella* Typhimurium)	DBD plasma (air)	1.30–5 log CFU reduction Minor effects on quality	[195]

(continued)

Table 6.6 (continued)

Target foods (microorganisms)	Plasma types (feeding gas)	Effects	References
Tiger nut milk (Natural microflora)	DBD plasma (air)	>4 log CFU/ml reduction Changes in quality	[231]
Tomato juice (Natural microflora, *Candida albicans*, *Saccharomyces cerevisiae*)	AC gliding arc plasma (N_2)	>3 log CFU/ml reduction Minor effects on quality	[323]
Cheese (*L. innocua*, *E. coli* K12)	DBD plasma (ambient air, $O_2 + CO_2 + N_2$)	1.6–5.0 log CFU/g reduction 1.4–3.5 log CFU/g reduction Minor effects on quality	[350, 351]
Powdered Spirulina algae (*B. subtilis*)	SMD plasma (air, N_2)	2 log CFU reduction Some changes in quality	[29]
Tomato juice (Natural microflora)	Gliding arc plasma (N_2)	$3 \rightarrow 5$ log CFU/g reduction No significant change in quality	[324]
Mixed nuts snack	Plasma (ambient air)	Significant decrease in microbial load Significant increase in the peroxide level and ΔE index	[19]
Rice germ, black pepper powder, and sesame (Natural microflora)	Plasma jet (air)	– 1.4 log CFU/g reduction No change in quality	[173]
Kiwi turbid juice (Natural microflora)	DBD plasma (ambient air)	83.97% bactericidal effect Minor effects on quality	[184]
Milk (Natural microflora)	Low pressure DBD plasma (ambient air)	~95% reduction of coliform microbes	[200]
House cricket powder (Natural microflora)	SMD plasma (ambient air)	1.6–1.9 log CFU reduction No significant change in quality	[268]
Food quality and functional property			
Lamb's lettuce	Plasma jet (Ar)	Changes in phenolic profile	[98]
Fructooligosaccharides (food ingredient)	Plasma (ambient air)	No significant change in quality during food processing	[85]
Alkaline phosphatase (native enzyme in milk)	DBD plasma (ambient air)	Inactivation of enzyme Change in secondary protein structure	[300]
Soybean oil	AC DBD plasma (H_2, $H_2 + N_2$)	Partially hydrogenated soybean oil without the formation of trans-fatty acids	[372]

(continued)

Table 6.6 (continued)

Target foods (microorganisms)	Plasma types (feeding gas)	Effects	References
Cut apple and potato	Microwave plasma (ambient air)	62–89% reduction in polyphenol oxidase and peroxidase activities to inhibit undesirable browning	[39]
Meat batter	DBD plasma ($N_2 + O_2$)	Addition of nitrite into meat	[144]
Peanut oil	AC plasma (Ar)	No deterioration of quality Increase in storage period	[282]
Inulin	DBD plasma(air)	Depolymerization of inulin to fructose Production of fructooligosaccharides	[236]
Lemon verbena	Low pressure microwave plasma (N_2, Ar, O_2)	0.9–1.2% reduction in essential oil content Increase in the content of monoterpene hydrocarbons and oxygenated sesquiterpenes	[75]
Peanut	SDBD plasma (air)	Decrease in unsaturated fatty acid and moisture content Increase in saturated fatty acids, peroxide value, acid value, and total polyphenols	[90]
White grapes	Plasma jet (air)	Faster drying and formation of raisins Improvement of quality	[120]
Tomato-based beverage	DBD plasma (air)	Retaining and improving the content of ascorbic acid, chlorogenic acid, sinapic acid and gallic acid	[206]
Corn starch	Plasma jet (air)	Increase in solubility and paste clarity of starch	[361]
Pork jerky	DBD plasma (ambient air)	Increase in redness, nitrosoheme pigment content, and residual nitrite content Decrease in microbial load	[378]
Maize starch	Plasma jet (air)	Modification of starch properties	[384]
Grape pomace	AC DBD plasma (He)	10.9–22.8% increase in yield of phenolic extracts Improved nutritional quality	[23]
Wheat flour	DBD plasma (ambient air)	Increase in hydration, pasting, viscosity of flour Depolymerization of starch	[43]
Fenugreek	DBD plasma (air)	67–122% increase in galactomannan extraction	[279]

(continued)

Table 6.6 (continued)

Target foods (microorganisms)	Plasma types (feeding gas)	Effects	References
Pearl millet	Plasma (air)	Improvement of hydration and pasting properties	[185]
Flaxseed extracts	Plasma jet (air)	Tailoring techno-functionality of extracts	[237]
Fresh cut endive	Plasma treated water	Improvement of color and texture properties	[295]
Starch	RF or DBD plasma (Ar)	Increase in the ordering and stability of starch molecules by modifying amylose chains	[314]
Wheat gliadin	AC plasma jet (He)	Modification of physiochemical and functional properties	[333]
Soybean protein isolate	DBD plasma (ambient air)	Improved functionality and reduced allergenicity	[383]

antimicrobial activity and food quality control, even though complete microbial decontamination cannot be achieved.

6.4.2 *Meats, Meat Products, and Fishes*

Despite their high nutritional value as a protein source, meat, meat products, and fish can be easily contaminated with microorganisms, causing food poisoning and other foodborne diseases [62]. Non-thermal plasma has been used for the decontamination of meat and meat products such as chicken, pork, beef, ham, bacon, and eggs (Table 6.6). Recently, seafood such as fish filets, oysters, and salmon sashimi have been actively explored in plasma applications [47, 122, 255]. In many studies, the antimicrobial activity of plasma was assessed on microbes inoculated onto meats and fish. Several studies have focused on the contamination of natural microflora [88, 275, 345]. Among microorganisms, food poisoning bacteria such as *E. coli, L. monocytogenes, Salmonella* spp., and *Campylobacter jejuni* were more frequently targeted than fungi (Table 6.6). Park et al. [255] inoculated the *P. citrinum* fungus on filefish filets and decontaminated it with oxygen plasma. Norovirus was also targeted for plasma-mediated decontamination. Norovirus is a pathogenic virus that causes vomiting and acute gastroenteritis; consumption of contaminated foods is one of the routes for disease outbreaks [284]. Several studies have shown the efficient decontamination of beef, pork, chicken, oysters, and salmon sashimi contaminated with norovirus of human or murine origin by plasma [15, 47, 122]. As observed in the decontamination of fruits and vegetables, plasma flame or plasma-generated gas generated from various plasma sources was more frequently applied to meats, meat products, and fish than plasma-treated water or liquids (Table 6.6). Interestingly, Qian et al. [275] found that plasma-activated lactic acid was more effective in decreasing microbial load contaminated on chicken drumsticks than plasma-treated water.

Generally, non-thermal plasma treatment, either gaseous plasma or plasma-treated liquid, can deactivate bacteria inoculated or contaminated on meats, meat products, and fish, with an efficiency of 0.1–6.52 log CFU reduction depending on treatment time and conditions. Regarding viruses, over 90% of murine norovirus and hepatitis A virus copy numbers were reduced after plasma treatment of contaminated beef, pork, and chicken [15]. In oysters contaminated with human norovirus, the virucidal effect of plasma was negligible (<1 log copy number/μL) without propidium monoazide pre-treatment and greater (>1 log copy number/μL) with propidium monoazide pre-treatment [47]. Huang et al. [122] found that N_2 plasma deactivated the human norovirus in salmon sashimi with an efficiency of 20% reduction in copy number, and the level of norovirus was undetectable after treatment with O_2 plasma.

No or minor deterioration in food quality, such as lipid peroxidation, pH, and sensory properties (appearance, color, odor, acceptability), was observed in the majority of studies (Table 6.6). However, Kim et al. [154] demonstrated that sensory quality parameters such as appearance, color, odor, and acceptability were significantly reduced after plasma treatment of bacteria-contaminated pork loin. This indicates that condition tuning or the development of methods for quality control may be necessary for industrial and market applications.

6.4.3 Packaged Foods

Various food products and fresh produce are often distributed as packaged materials in the market and industry. Prevention of microbial contamination during the packaging process can play an important role in ensuring a long shelf life. Thermal treatment is a routine method for sanitation of packaged foods. However, deterioration of food quality has limited the range of applications of thermal sanitation. Fresh produce is more frequently distributed in packaged states in recent markets. The demand for non-thermal sanitation technologies has increased, particularly in the management of packaged foods.

Non-thermal plasma has demonstrated the potential for microbial decontamination of packaged foods over the last decade. A distinguishing point in these studies was that foods were treated with plasma generated inside the package. Recently, Misra et al. [215] reported an excellent review of the application of non-thermal plasma technology to the sanitation of packaged foods. In Table 6.6, studies excluding those mentioned in Misra et al.'s review are indicated. Various designs of plasma systems specialized for in-package treatment have been developed; a package is placed between two electrodes, electrodes are placed on one side of the package, and electrodes are placed inside the package [215]. In most studies, dielectric barrier discharge (DBD) or surface dielectric barrier discharge (SDBD) plasma was used in the treatment of packaged foods (Table 6.2) [215]. Foods inoculated with food poisoning and spoiling bacteria are most frequently targeted for in-package plasma treatment [215], whereas fungi and viruses have rarely been explored [169, 213, 318, 377]. Foods contaminated with natural microflora were also analyzed after in-package plasma treatment [3, 4, 162, 189, 204, 214, 353, 387]. The efficiency of in-package food sanitation using plasma is good; a >1 log reduction in CFU number has been observed in most studies, and complete eradication of microorganisms has been demonstrated in some cases [4, 103, 172, 387, 388].

In-package food quality after plasma treatment is also an important factor to be considered. Most studies have demonstrated that in-package plasma treatment causes minor or no changes in physiological, physical, and sensory properties (Table 6.6) [215]. However, a recent study demonstrated that plasma treatment could result in lipid peroxidation and significant color changes in packaged ham [369].

6.4.4 *Processed Foods*

Plasma has been actively applied to the sanitation of processed foods such as juice, milk, cheese, pepper powder, insect powder, and snacks (Table 6.6). Non-thermal tools such as ultrasonification, UV, ionizing radiation, and electrical fields have been applied to the sanitation of heat-sensitive foods [274]. Non-thermal plasma is also considered a promising technology that can efficiently remove microbial contamination during food processing and packaging. Liquid foods such as fruit juice and milk have been frequent targets for plasma sanitation, and a greater than 1 log reduction in bacterial CFU number was obtained after plasma treatment (Table 6.6). In various studies, the quality of juices and milk was not significantly affected by plasma (Table 6.6). However, Xu et al. [365] found that direct treatment with 90 kV high voltage atmospheric cold plasma reduced vitamin C content by 22% and pectin methylesterase activity by 74–82% in orange juice. Muhammad et al. [231] showed that DBD air plasma caused a significant reduction in pH, protein content, and peroxidase activity in tiger nut milk, whereas no significant changes in soluble solids and fat contents were observed.

The sanitation of dry foods and powders using plasma resulted in an efficient >1 log reduction in CFUs in most cases (Table 6.6). Bacteria inoculated on sliced cheese were efficiently inactivated in encapsulated or flexible thin-layer DBD plasma systems, and some food qualities such as flavor, overall acceptance, and off-color were significantly affected by plasma [374, 376]. Dry powders, such as onion powder, black pepper powder, and insect powder, were efficiently decontaminated with no dramatic changes in food quality [156, 173, 268]. Several studies have demonstrated that plasma treatment can alter protein solubility and the amount of lipids, chlorophyll a, carotenoids, phycobilin, and total phenolic compounds in wheat flour, insect powder, and algae powder [20, 29, 41]. Particularly, Bahrami et al. [20] observed no significant changes in total aerobic bacterial count or total mould count in wheat flour after treatment with 0.19 and 0.43 W/cm^2 air plasma.

6.5 Application of Non-thermal Plasma to Food Quality and Functional Property

Non-thermal plasma has also been used to enhance the quality and functionality of foods and food ingredients (Table 6.6). The quality and nutritional value of fresh produce are investigated together during plasma sanitation to determine whether plasma treatment can affect food quality. Color, texture, pH, proteins, carbohydrates, vitamins, lipids, and antioxidant activity are major properties frequently analyzed in previous studies [249]. These factors are mostly related to the taste, nutritional value, and senescence of the fresh produce. In many studies, plasma treatment did not cause significant damage to the quality of fresh produce. Improvement in antioxidant activity and increase in phenolic content are often observed in lettuce, cut apples, potatoes, peanuts, and grapes [23, 39, 90, 98]. Rinsing with plasma-treated water can improve the color and texture of fresh-cut endives [295]. Plasma can also increase the speed of drying and improve the quality of raisins from fresh grapes [120].

Furthermore, studies have demonstrated that plasma can affect the quality and functionality of food ingredients and processed food products (Table 6.6). The redness of meat can be improved by increasing the amount of nitrite in the meat after plasma treatment [144, 378]. The nutritional value of several herbs, such as fenugreek, pearl millet, and lemon verbena, is also enhanced by plasma. Plasma can facilitate the acquisition of galactomannan from fenugreek, improve the hydration of pearl millet, and elevate the contents of monoterpene hydrocarbons and oxygenated sesquiterpenes in lemon verbena [75, 185, 279]. Moreover, starch structure can be modified by plasma, which can further alter the properties of starch such as solubility, depolymerization, and paste viscosity, making it more suitable for food and non-food industries [43, 314, 361, 384]. Plasma can improve the storage of soybean and peanut oils and the functionality of wheat and soybean proteins [333, 383]. Additionally, plasma can increase depolymerization of inulin for the production of fructooligosaccharides without changing its quality as a food ingredient [85, 236].

6.6 Conclusion and Future Perspectives

Non-thermal atmospheric- and low-pressure plasma are promising tools for several applications, such as microbial decontamination and activation of seed germination and growth, in the food and agriculture industries. However, the mechanisms underlying plasma action, standardization of applied plasma dose, and development of industrial-scale treatments still need for intense further study. The scale addressed in the agriculture and food industries is relatively large compared to that in the medical field, and this should be considered when developing a plasma system. Another future direction in plasma application may be that plasma can be explored to find a potential solution to agricultural and food issues resulted from climatic change. Due to climate change, the current agriculture and food industry is facing a big challenge, and improvement in stress tolerance and storage of fresh produce has received increasing attention as emerging areas wherein plasma can be applied.

Acknowledgements This work was supported by the National Research Foundation of Korea (NRF) (2020R1F1A1070942, 2021R1A6A1A03038785).

References

1. B. Adhikari, M. Adhikari, B. Ghimire, B.C. Adhikari, G. Park, E.H. Choi, Cold plasma seed priming modulates growth, redox homeostasis and stress response by inducing reactive species in tomato (*Solanum lycopersicum*). Free Radic. Biol. Med. **156**, 57–69 (2020)

2. G. Agrios, *Plant Pathology*, 5th edn. (Elsevier Academic Press, Burlington, MA, USA, 2005), pp.79–103
3. I. Albertos, A.B. Martin-Diana, P.J. Cullen, B.K. Tiwari, K.S. Ojha, P. Bourke, D. Rico, Shelf-life extension of herring (*Clupea harengus*) using in-package atmospheric plasma technology. Innov. Food Sci. Emerg. Technol. **53**, 85–91 (2019)
4. I. Albertos, A.B. Martin-Diana, P.J. Cullen, B.K. Tiwari, S.K. Ojha, P. Bourke, C. Alvarez, D. Rico, Effects of dielectric barrier discharge (DBD) generated plasma on microbial reduction and quality parameters of fresh mackerel (*Scomber scombrus*) fillets. Innov. Food Sci. Emerg. Technol. **44**, 117–122 (2017)
5. C. Alves-Junior, D.L.S. da Silva, J.O. Vitoriano, A.P.C.B. Barbalho, R.C. de Sousa, The water path in plasma-treated Leucaena seeds. Seed Sci. Res. **30**, 13–20 (2020)
6. C. Alves Junior, J. de Oliveira Vitoriano, D.L.S. da Silva, M. de Lima Farias, N.B. de Lima Dantas, Water uptake mechanism and germination of *Erythrina velutina* seeds treated with atmospheric plasma. Sci. Rep. **6**, 33722 (2016)
7. P.F. Ambrico, M. Šimek, M. Morano, R.M.D.M. Angelini, A. Minafra, P. Trotti, M. Ambrico, V. Prukner, F. Faretra, Reduction of microbial contamination and improvement of germination of sweet basil (*Ocimum basilicum* L.) seeds via surface dielectric barrier discharge. J. Phys. D: Appl. Phys. **50**, 305401 (2017)
8. P.F. Ambrico, M. Šimek, M. Ambrico, M. Morano, V. Prukner, A. Minafra, I. Allegretta, C. Porfido, G.S. Senesi, R. Terzano, On the air atmospheric pressure plasma treatment effect on the physiology, germination and seedlings of basil seeds. J. Phys. D Appl. Phys. **53**, 104001 (2020)
9. S.D.S. Araújo, S. Paparella, D. Dondi, A. Bentivoglio, D. Carbonera, A. Balestrazzi, Physical methods for seed invigoration: advantages and challenges in seed technology. Front. Plant Sci. **7**, 646 (2016)
10. N.K. Arora, Impact of climate change on agriculture production and its sustainable solutions. Environ. Sustain. **2**, 95–96 (2019)
11. Association IST, *ISTA Reference pest List* (2021). https://www.seedtest.org/en/ista-reference-pest-list-_content---1--3477.html
12. P. Attri, K. Ishikawa, T. Okumura, K. Koga, M. Shiratani, Plasma agriculture from laboratory to farm: a review. Processes **8**, 1002 (2020)
13. G. Avramidis, B. Stüwe, R. Wascher, M. Bellmann, S. Wieneke, A. von Tiedemann, W.J.S. Viöl, Fungicidal effects of an atmospheric pressure gas discharge and degradation mechanisms. Surf. Coat. Technol. **205**, S405–S408 (2010)
14. A. Babajani, A. Iranbakhsh, Z.O. Ardebili, B. Eslami, Seed priming with non-thermal plasma modified plant reactions to selenium or zinc oxide nanoparticles: cold plasma as a novel emerging tool for plant science. Plasma Chem. Plasma Process. **39**, 21–34 (2019)
15. S.C. Bae, S.Y. Park, W. Choe, S.D. Ha, Inactivation of murine norovirus-1 and hepatitis A virus on fresh meats by atmospheric pressure plasma jets. Food Res. Int. **76**, 342–347 (2015)
16. M. Bafoil, A. Jemmat, Y. Martinez, N. Merbahi, O. Eichwald, C. Dunand, M. Yousfi, Effects of low temperature plasmas and plasma activated waters on *Arabidopsis thaliana* germination and growth. PLoS ONE **13**, 16 (2018)
17. M. Bafoil, A. Le Ru, N. Merbahi, O. Eichwald, C. Dunand, M. Yousfi, New insights of low-temperature plasma effects on germination of three genotypes of *Arabidopsis thaliana* seeds under osmotic and saline stresses. Sci. Rep. **9**, 10 (2019)
18. M. Bafoil, M. Yousfi, C. Dunand, N. Merbahi, Effects of dielectric barrier ambient air plasma on two *Brassicaceae* seeds: *Arabidopsis thaliana* and *Camelina sativa*. Int. J. Mol. Sci. **22**, 9923 (2021)
19. H. Bagheri, S. Abbaszadeh, M. Sepandi, Simultaneous effect of cold plasma and MAP on the quality properties of mixed nuts snack during storage. J. Food Process. Preserv. **45**, 12 (2021)
20. N. Bahrami, D. Bayliss, G. Chope, S. Penson, T. Perehinec, I.D. Fisk, Cold plasma: a new technology to modify wheat flour functionality. Food Chem. **202**, 247–253 (2016)
21. M. Baier, J. Foerster, U. Schnabel, D. Knorr, J. Ehlbeck, W.B. Herppich, O. Schluter, Direct non-thermal plasma treatment for the sanitation of fresh corn salad leaves: evaluation of

physical and physiological effects and antimicrobial efficacy. Postharvest Biol. Technol. **84**, 81–87 (2013)
22. B.B. Baldanov, T.V. Ranzhurov, M.N. Sordonova, L.V. Budazhapov, Changes in the properties and surface structure of grain seeds under the influence of a glow discharge at atmospheric pressure. Plasma Phys. Rep. **46**, 110–114 (2020)
23. Y.W. Bao, L. Reddivari, J.Y. Huang, Enhancement of phenolic compounds extraction from grape pomace by high voltage atmospheric cold plasma. LWT-Food Sci. Technol. **133**, 9 (2020)
24. G. Barba-Espin, P. Diaz-Vivancos, M.J. Clemente-Moreno, A. Albacete, L. Faize, M. Faize, F. Pérez-Alfocea, J.A. Hernández, Interaction between hydrogen peroxide and plant hormones during germination and the early growth of pea seedlings. Plant Cell Environ. **33**, 981–994 (2010)
25. P. Basaran, N. Basaran-Akgul, L. Oksuz, Elimination of *Aspergillus parasiticus* from nut surface with low pressure cold plasma (LPCP) treatment. Food Microbiol. **25**, 626–632 (2008)
26. J. Baskin, C. Baskin, Seed germination of gynodioecious species: theoretical considerations and a comparison of females and hermaphrodites. Planta **252**, 73 (2020)
27. D. Bermudez-Aguirre, E. Wemlinger, P. Pedrow, G. Barbosa-Canovas, M. Garcia-Perez, Effect of atmospheric pressure cold plasma (APCP) on the inactivation of *Escherichia coli* in fresh produce. Food Control. **34**, 149–157 (2013)
28. J.D. Bewley, Seed germination and dormancy. Plant Cell **9**, 1055–1066 (1997)
29. M. Beyrer, M.C. Pina-Perez, D. Martinet, W. Andlauer, Cold plasma processing of powdered Spirulina algae for spore inactivation and preservation of bioactive compounds. Food Control. **118**, 8 (2020)
30. M. Billah, S. Karmakar, F.B. Mina, M.N. Haque, M.M. Rashid, M.F. Hasan, U.K. Acharjee, M.R. Talukder, Investigation of mechanisms involved in seed germination enhancement, enzymatic activity and seedling growth of rice (*Oryza sativa* L.) using LPDBD (Ar + Air) plasma. Arch. Biochem. Biophys. **698**, 108726 (2021)
31. M. Billah, S.A. Sajib, N.C. Roy, M.M. Rashid, M.A. Reza, M.M. Hasan, M.R. Talukder, Effects of DBD air plasma treatment on the enhancement of black gram (*Vigna mungo* L.) seed germination and growth. Arch. Biochem. Biophys. **681**, 108253 (2020)
32. T. Bogdanov, I. Tsonev, P. Marinova, E. Benova, K. Rusanov, M. Rusanova, I. Atanassov, Z. Kozáková, F.C. Kršma, Microwave plasma torch generated in argon for small berries surface treatment. Appl. Sci. **8**, 1870 (2018)
33. L. Borisjuk, H. Rolletschek, T. Neuberger, Surveying the plant's world by magnetic resonance imaging. Plant J. **70**, 129–146 (2012)
34. E. Bormashenko, R. Grynyov, Y. Bormashenko, E. Drori, Cold radiofrequency plasma treatment modifies wettability and germination speed of plant seeds. Sci. Rep. **2**, 741 (2012)
35. E. Bormashenko, Y. Shapira, R. Grynyov, G. Whyman, Y. Bormashenko, E. Drori, Interaction of cold radiofrequency plasma with seeds of beans (*Phaseolus vulgaris*). J. Exp. Bot. **66**, 4013–4021 (2015)
36. P. Bourke, D. Ziuzina, D. Boehm, P.J. Cullen, K. Keener, The potential of cold plasma for safe and sustainable food production. Trends Biotechnol. **36**, 615–626 (2018)
37. M. Braşoveanu, M. Nemţanu, C. Surdu-Bob, G. Karaca, I. Erper, Effect of glow discharge plasma on germination and fungal load of some cereal seeds. Roman. Rep. Phys. **67**, 617–624 (2015)
38. H. Brust, T.M.C. Nishime, N. Wannicke, T.S.M. Mui, S. Horn, A. Quade, K.D. Weltmann, A medium-scale volume dielectric barrier discharge system for short-term treatment of cereal seeds indicates improved germination performance with long-term effects. J. Appl. Phys. **129**, 044904 (2021)
39. S. Bußler, J. Ehlbeck, O. Schlüter, Pre-drying treatment of plant related tissues using plasma processed air: Impact on enzyme activity and quality attributes of cut apple and potato. Innov. Food Sci. Emerg. Technol. **40**, 78–86 (2017)
40. S. Bußler, W.B. Herppich, S. Neugart, M. Schreiner, J. Ehlbeck, S. Rohn, O. Schlüter, Impact of cold atmospheric pressure plasma on physiology and flavonol glycoside profile of peas (*Pisum sativum* 'Salamanca'). Food Res. Int. **76**, 132–141 (2015)

41. S. Bußler, B.A. Rumpold, A. Fröhling, E. Jander, H.M. Rawel, O.K. Schlüter, Cold atmospheric pressure plasma processing of insect flour from *Tenebrio molitor*: Impact on microbial load and quality attributes in comparison to dry heat treatment. Innov. Food Sci. Emerg. Technol. **36**, 277–286 (2016)
42. D. Butscher, D. Zimmermann, M. Schuppler, P.R. von Rohr, Plasma inactivation of bacterial endospores on wheat grains and polymeric model substrates in a dielectric barrier discharge. Food Control **60**, 636–645 (2016)
43. S. Chaple, C. Sarangapani, J. Jones, E. Carey, L. Causeret, A. Genson, B. Duffy, P. Bourke, Effect of atmospheric cold plasma on the functional properties of whole wheat (*Triticum aestivum* L.) grain and wheat flour. Innov. Food Sci. Emerg. Technol. **66**, 10 (2020)
44. J. Chavan, S. Kadam, D. Salunkhe, Changes in tannin, free amino acids, reducing sugars, and starch during seed germination of low and high tannin cultivars of sorghum. J. Food Sci. **46**, 638–639 (1981)
45. H.H. Chen, H.C. Chang, Y.K. Chen, C.L. Hung, S.Y. Lin, Y.S. Chen, An improved process for high nutrition of germinated brown rice production: Low-pressure plasma. Food Chem. **191**, 120–127 (2016)
46. E.J. Choi, H.W. Park, S.B. Kim, S. Ryu, J. Lim, E.J. Hong, Y.S. Byeon, H.H. Chun, Sequential application of plasma-activated water and mild heating improves microbiological quality of ready-to-use shredded salted kimchi cabbage (*Brassica pekinensis* L.). Food Control **98**, 501–509 (2019)
47. M.S. Choi, E.B. Jeon, J.Y. Kim, E.H. Choi, J.S. Lim, J. Choi, K.S. Ha, J.Y. Kwon, S.H. Jeong, S.Y. Park, Virucidal effects of dielectric barrier discharge plasma on human Norovirus infectivity in fresh oysters (*Crassostrea gigas*). Foods **9**, 13 (2020)
48. S. Choi, P. Puligundla, C. Mok, Corona discharge plasma jet for inactivation of *Escherichia coli* O157:H7 and *Listeria monocytogenes* on inoculated pork and its impact on meat quality attributes. Ann. Microbiol. **66**, 685–694 (2016)
49. Y.-J. Chou, K.-C. Cheng, F.-C. Hsu, J.S.-B. Wu, Y. Ting, Producing high quality mung bean sprout using atmospheric cold plasma treatment: better physical appearance and higher γ-aminobutyric acid (GABA) content. J. Sci. Food Agric. **101**, 6463–6471 (2021)
50. S. Colejo, A. Alvarez-Ordóñez, M. Prieto, M. González-Raurich, M. López, Evaluation of ultraviolet light (UV), non-thermal atmospheric plasma (NTAP) and their combination for the control of foodborne pathogens in smoked salmon and their effect on quality attributes. Innov. Food Sci. Emerg. Technol. **50**, 84–93 (2018)
51. R.R. Colwell, Viable but not cultivable bacteria, in *Uncultivated Microorganisms*, ed. by S. Epstein (Springer, Berlin, Heidelberg, 2009), pp. 121–129
52. E. Commission, Commission regulation (EU 2018/1500) Concerning the non-renewal of approval of the active substance thiram, and prohibiting the use and sale of seeds treated with plant protection products containing thiram, in accordance with Regulation (EC) No 1107/2009 of the European Parliament and of the Council concerning the placing of plant protection products on the market, and amending Commission Implementing Regulation (EU) No 540/2011. E Union, editor Official J. Euro. Union Brussels **3** (2018)
53. M. Cooper, G. Fridman, D. Staack, A.F. Gutsol, V.N. Vasilets, S. Anandan, Y.I. Cho, A. Fridman, A. Tsapin, Decontamination of surfaces from extremophile organisms using nonthermal atmospheric-pressure plasmas. IEEE Trans. Plasma Sci. **37**, 866–871 (2009)
54. D.J. Cui, Y. Yin, J.Q. Wang, Z.W. Wang, H.B. Ding, R.N. Ma, Z. Jiao, Research on the physio-biochemical mechanism of non-thermal plasma-regulated seed germination and early seedling development in *Arabidopsis*. Front. Plant Sci. **10**, 12 (2019)
55. J. Cui, T. Zhao, L. Zou, X. Wang, Y. Zhang, Molecular dynamics simulation of *S. cerevisiae* glucan destruction by plasma ROS based on ReaxFF. J. Phys. D: Appl. Phys. **51**, 355401 (2018)
56. B.G. Dasan, I.H. Boyaci, M. Mutlu, Inactivation of aflatoxigenic fungi (*Aspergillus* spp.) on granular food model, maize, in an atmospheric pressure fluidized bed plasma system. Food Control **70**, 1–8 (2016a)

57. B.G. Dasan, M. Mutlu, I.H. Boyaci, Decontamination of *Aspergillus flavus* and *Aspergillus parasiticus* spores on hazelnuts via atmospheric pressure fluidized bed plasma reactor. Int. J. Food Microbiol. **216**, 50–59 (2016b)
58. B.G. Dasan, I.H. Boyaci, M. Mutlu, Nonthermal plasma treatment of *Aspergillus* spp. spores on hazelnuts in an atmospheric pressure fluidized bed plasma system: impact of process parameters and surveillance of the residual viability of spores. J. Food Eng. **196**, 139–149 (2017)
59. A.R.M. da Silva, M.L. Farias, D.L.S. da Silva, J.O. Vitoriano, R.C. de Sousa, C. Alves-Junior, Using atmospheric plasma to increase wettability, imbibition and germination of physically dormant seeds of *Mimosa caesalpiniafolia*. Colloids Surf. B **157**, 280–285 (2017)
60. D.L.S. da Silva, M.D. Farias, J.D. Vitoriano, C. Alves, S.B. Torres, Use of atmospheric plasma in germination of *Hybanthus calceolaria* (L.) Schulze-Menz seeds. Rev. Caatinga Mossoró **31**, 632–639 (2018)
61. R. Dauwe, R. Roulard, M. Ramos, B. Thiombiano, F. Mesnard, E. Gontier, A. Jamali, Etching of the seed cuticle by cold plasma shortens imbibitional leakage in *Linum usitatissimum* L. Ind. Crops Prod. **167**, 113536 (2021)
62. D. Dave, A.E. Ghaly, Meat spoilage mechanisms and preservation techniques: a critical review. Am. J. Agric. Biol. Sci. **6**, 486–510 (2011)
63. M.J. Davies, Protein oxidation and peroxidation. Biochem. J. **473**, 805–825 (2016)
64. N. Dawood, Effect of RF plasma on Moringa seeds germination and growth. J. Taibah Univ. Sci. **14**, 279–284 (2020)
65. I. Debeaujon, K.M. Léon-Kloosterziel, M. Koornneef, Influence of the testa on seed dormancy, germination, and longevity in *Arabidopsis*. Plant Physiol. **122**, 403–414 (2000)
66. G. de Groot, A. Hundt, A.B. Murphy, M.P. Bange, A. Mai-Prochnow, Cold plasma treatment for cotton seed germination improvement. Sci. Rep. **8**, 10 (2018)
67. L. Degutytė-Fomins, G. Paužaitė, R. Žūkienė, V. Mildažienė, K. Koga, M. Shiratani, Relationship between cold plasma treatment-induced changes in radish seed germination and phytohormone balance. Jpn. J. Appl. Phys. **59**, SH1001 (2020)
68. Y. Devi, R. Thirumdas, C. Sarangapani, R.R. Deshmukh, U.S. Annapure, Influence of cold plasma on fungal growth and aflatoxins production on groundnuts. Food Control **77**, 187–191 (2017)
69. M. Dhayal, S.-Y. Lee, S.-U. Park, Using low-pressure plasma for *Carthamus tinctorium* L. seed surface modification. Vacuum **80**, 499–506 (2006)
70. B.P. Dirks, D. Dobrynin, G. Fridman, Y. Mukhin, A. Fridman, J.J. Quinlan, Treatment of raw poultry with nonthermal dielectric barrier discharge plasma to reduce *Campylobacter jejuni* and *Salmonella enterica*. J. Food Prot. **75**, 22–28 (2012)
71. D. Dobrin, M. Magureanu, N.B. Mandache, M.D. Ionita, The effect of non-thermal plasma treatment on wheat germination and early growth. Innov. Food Sci. Emerg. Technol. **29**, 255–260 (2015)
72. E. Dolezalova, P. Lukes, Membrane damage and active but nonculturable state in liquid cultures of *Escherichia coli* treated with an atmospheric pressure plasma jet. Bioelectrochemistry **103**, 7–14 (2015)
73. T. Dufour, Q. Gutierrez, Cold plasma treatment of seeds: deciphering the role of contact surfaces through multiple exposures, randomizing and stirring. J. Phys. D Appl. Phys. **54**, 505202 (2021)
74. T. Dufour, Q. Gutierrez, C. Bailly, Sustainable improvement of seeds vigor using dry atmospheric plasma priming: evidence through coating wettability, water uptake, and plasma reactive chemistry. J. Appl. Phys. **129**, 084902 (2021)
75. M.-T. Ebadi, S. Abbasi, A. Harouni, F. Sefidkon, Effect of cold plasma on essential oil content and composition of lemon verbena. Food Sci. Nutr. **7**, 1166–1171 (2019)
76. Y. Elad, I. Pertot, Climate change impact on plant pathogens and plant diseases. J. Crop Improv. **28**, 99–139 (2014)
77. S.A. Fadhlalmawla, A.-A.H. Mohamed, J.Q.M. Almarashi, T. Boutraa, The impact of cold atmospheric pressure plasma jet on seed germination and seedlings growth of fenugreek (*Trigonella foenum-graecum*). Plasma Sci. Technol. **21**, 105503 (2019)

78. M. Farooq, M. Usman, F. Nadeem, H. Rehman, A. Wahid, S.M.A. Basra, K.H.M. Siddique, Seed priming in field crops: potential benefits, adoption and challenges. Crop Pasture Sci. **70**, 731–771 (2019)
79. F. Feng, H. Mei, P. Fan, Y. Li, X. Xu, H. Wei, M. Yan, L. Luo, Dynamic transcriptome and phytohormone profiling along the time of light exposure in the mesocotyl of rice seedling. Sci. Rep. **7**, 11961 (2017)
80. J. Feng, D. Wang, C. Shao, L. Zhang, X. Tang, Effects of cold plasma treatment on alfalfa seed growth under simulated drought stress. Plasma Sci. Technol. **20**, 035505 (2018)
81. I. Filatova, V. Azharonok, M. Kadyrov, V. Beljavsky, A. Gvozdov, A. Shik, A. Antonuk, The effect of plasma treatment of seeds of some grain and legumes on their sowing quality and productivity. Rom. J. Phys. **56**, 139–143 (2011)
82. I. Filatova, V. Azharonok, A. Shik, A. Antoniuk, N. Terletskaya, Fungicidal effects of plasma and radio-wave pre-treatments on seeds of grain crops and legumes, in *Plasma for Bio-Decontamination, Medicine and Food Security. NATO Science for Peace and Security Series A: Chemistry and Biology*, eds. by Z. Machala et al., (Springer Science + Business Media, Berlin, Heidelberg, 2012), pp. 469–479
83. I.I. Filatova, V.V. Azharonok, S.V. Goncharik, V.A. Lushkevich, A.G. Zhukovsky, G.I. Gadzhieva, Effect of RF plasma treatment on the germination and phytosanitary state of seeds. J. Appl. Spectrosc. **81**, 250–256 (2014)
84. I. Filatova, V. Lyushkevich, S. Goncharik, A. Zhukovsky, N. Krupenko, J. Kalatskaja, The effect of low-pressure plasma treatment of seeds on the plant resistance to pathogens and crop yields. J. Phys. D Appl. Phys. **53**, 244001 (2020)
85. E.G.A. Filho, P.J. Cullen, J.M. Frias, P. Bourke, B.K. Tiwari, E.S. Brito, S. Rodrigues, F.A.N. Fernandes, Evaluation of plasma, high-pressure and ultrasound processing on the stability of fructooligosaccharides. Int. J. Food Sci. Technol. **51**, 2034–2040 (2016)
86. A. Filipić, G. Primc, R. Zaplotnik, N. Mehle, I. Gutierrez-Aguirre, M. Ravnikar, M. Mozetič, J. Žel, D. Dobnik, Cold atmospheric plasma as a novel method for inactivation of potato virus Y in water samples. Food Environ. Virol. **11**, 220–228 (2019)
87. R. Förch, H. Schöherr, A. Jenkins, Appendix C: contact angle goniometry, in *Surface Design: applications in Bioscience and Nanotechnology*, eds. by R. Fçrch, H. Schçnherr, A. Tobias, A. Jenkins. (Wiley-VCH, Weinheim, 2009), pp. 471–473.
88. A. Frohling, J. Durek, U. Schnabel, J. Ehlbeck, J. Bolling, O. Schluter, Indirect plasma treatment of fresh pork: decontamination efficiency and effects on quality attributes. Innov. Food Sci. Emerg. Technol. **16**, 381–390 (2012)
89. X.T. Gao, A. Zhang, P. Heroux, W. Sand, Z.Y. Sun, J.X. Zhan, C.H. Wang, S.Y. Hao, Z.Y. Li, Z.Y. Li, Y. Guo, Y.N. Liu, Effect of dielectric barrier discharge cold plasma on pea seed growth. J. Agric. Food Chem. **67**, 10813–10822 (2019)
90. G.G. Gebremical, S.A. Emire, T. Berhanu, Effects of multihollow surface dielectric barrier discharge plasma on chemical and antioxidant properties of peanut. J. Food Qual. **2019**, 3702649 (2019)
91. M. Ghaemi, A. Majd, A. Iranbakhsh, Transcriptional responses following seed priming with cold plasma and electromagnetic field in *Salvia nemorosa* L. J. Theor. Appl. Phys. **14**, 323–328 (2020)
92. A. Gholami, N.N. Safa, M. Khoram, J. Hadian, H. Ghomi, Effect of low-pressure radio frequency plasma on ajwain seed germination. Plasma Meed. **6**, 389–396 (2016)
93. M. Giannoglou, P. Stergiou, P. Dimitrakellis, E. Gogolides, N.G. Stoforos, G. Katsaros, Effect of cold atmospheric plasma processing on quality and shelf-life of ready-to-eat rocket leafy salad. Innov. Food Sci. Emerg. Technol. **66**, 6 (2020)
94. S.-M. Go, M.-R. Park, H.-S. Kim, W.S. Choi, R.-D. Jeong, Antifungal effect of non-thermal atmospheric plasma and its application for control of postharvest *Fusarium oxysporum* decay of paprika. Food Control **98**, 245–252 (2019)
95. V. Gök, S. Aktop, M. Özkan, O. Tomar, The effects of atmospheric cold plasma on inactivation of *Listeria monocytogenes* and *Staphylococcus aureus* and some quality characteristics of pastırma—A dry-cured beef product. Innov. Food Sci. Emerg. Technol. **56**, 102188 (2019)

96. A. Gómez-Ramírez, C. López-Santos, M. Cantos, J.L. García, R. Molina, J. Cotrino, J.P. Espinós, A.R. González-Elipe, Surface chemistry and germination improvement of Quinoa seeds subjected to plasma activation. Sci. Rep. **7**, 5924 (2017)
97. M.N. Groot, T. Abee, H. van Bokhorst-van de Veen, Inactivation of conidia from three *Penicillium* spp. isolated from fruit juices by conventional and alternative mild preservation technologies and disinfection treatments. Food Microbiol. **81**, 108–114 (2019)
98. F. Grzegorzewski, J. Ehlbeck, O. Schluter, L.W. Kroh, S. Rohn, Treating lamb's lettuce with a cold plasma—Influence of atmospheric pressure Ar plasma immanent species on the phenolic profile of *Valerianella locusta*. LWT-Food Sci. Technol. **44**, 2285–2289 (2011)
99. Q. Guo, Y. Wang, H. Zhang, G. Qu, T. Wang, Q. Sun, D. Liang, Alleviation of adverse effects of drought stress on wheat seed germination using atmospheric dielectric barrier discharge plasma treatment. Sci. Rep. **7**, 16680 (2017)
100. L. Guo, R. Xu, L. Gou, Z. Liu, Y. Zhao, D. Liu, L. Zhang, H. Chen, M.G. Kong, Mechanism of virus inactivation by cold atmospheric-pressure plasma and plasma-activated water. Appl. Environ. Microbiol. **84**, e00726-e818 (2018a)
101. Q. Guo, Y.R. Meng, G.Z. Qu, T.C. Wang, F.N. Yang, D.L. Liang, S.B. Hu, Improvement of wheat seed vitality by dielectric barrier discharge plasma treatment. Bioelectromagnetics **39**, 120–131 (2018b)
102. C. Hallmann, M. Sorg, E. Jongejans, H. Siepel, N. Hofland, H. Schwan, W. Stenmans, A. Müller, H. Sumser, T. Hörren, More than 75% decline over 27 years in total flying insect biomass in protected areas. PLoS ONE **12**(10), e0185809 (2017)
103. L. Han, D. Boehm, E. Amias, V. Milosavljevic, P.J. Cullen, P. Bourke, Atmospheric cold plasma interactions with modified atmosphere packaging inducer gases for safe food preservation. Innov. Food Sci. Emerg. Technol. **38**, 384–392 (2016)
104. S.E. Hanbal, K. Takashima, S. Miyashita, S. Ando, K. Ito, M.M. Elsharkawy, T. Kaneko, H.J. Takahashi, Atmospheric-pressure plasma irradiation can disrupt tobacco mosaic virus particles and RNAs to inactivate their infectivity. Adv. Virol. **163**, 2835–2840 (2018)
105. Y. Hara, Calculation of population parameters using richards function and application of indices of growth and seed vigor to rice plants. Plant Prod. Sci. **2**, 129–135 (1999)
106. S. Hati, M. Patel, D. Yadav, Food bioprocessing by non-thermal plasma technology. Curr. Opin. Food Sci. **19**, 85–91 (2018)
107. N. Hayashi, Y. Yagyu, A. Yonesu, M. Shiratani, Sterilization characteristics of the surfaces of agricultural products using active oxygen species generated by atmospheric plasma and UV light. Jpn. J. Appl. Phys.**53**, 05FR03 (2014)
108. N. Hayashi, R. Ono, M. Shiratani, A. Yonesu, Antioxidative activity and growth regulation of *Brassicaceae* induced by oxygen radical irradiation. Jpn. J. Appl. Phys. **54**, 06GD01 (2015a)
109. N. Hayashi, R. Ono, S. Uchida, Growth enhancement of plant by plasma and UV light irradiation to seeds. J. Photopolym. Sci. Technol. **28**, 445–448 (2015b)
110. M. Henselová, Ľ Slováková, M. Martinka, A. Zahoranová, Growth, anatomy and enzyme activity changes in maize roots induced by treatment of seeds with low-temperature plasma. Biologia **67**, 490–497 (2012)
111. C. Hertwig, A. Leslie, N. Meneses, K. Reineke, C. Rauh, O. Schluter, Inactivation of *Salmonella enteritidis* PT30 on the surface of unpeeled almonds by cold plasma. Innov. Food Sci. Emerg. Technol. **44**, 242–248 (2017)
112. C. Hertwig, K. Reineke, J. Ehlbeck, D. Knorr, O. Schluter, Decontamination of whole black pepper using different cold atmospheric pressure plasma applications. Food Control **55**, 221–229 (2015)
113. N. Hojnik, M. Modic, G. Tavcar-Kalcher, J. Babic, J.L. Walsh, U. Cvelbar, Mycotoxin decontamination efficacy of atmospheric pressure air plasma. Toxins **11**, 219 (2019)
114. N. Hojnik, M. Modic, D. Zigon, J. Kovac, A. Jurov, A. Dickenson, J.L. Walsh, U. Cvelbar, Cold atmospheric pressure plasma-assisted removal of aflatoxin B-1 from contaminated corn kernels. Plasma Process. Polym. **18**, 12 (2021)
115. Ľ Holubová, R. Švubová, Ľ Slováková, B. Bokor, V. Chobotová Kročková, J. Renčko, F. Uhrin, V. Medvecká, A. Zahoranová, E. Gálová, Cold atmospheric pressure plasma treatment

of maize grains—Induction of growth, enzyme activities and heat shock proteins. Int. J. Mol. Sci. **22**, 8509 (2021)
116. T. Homola, V. Prukner, A. Artemenko, J. Hanuš, O. Kylián, M. Šimek, Direct treatment of pepper (*Capsicum annuum* L.) and melon (*Cucumis melo*) seeds by amplitude-modulated dielectric barrier discharge in air. J. Appl. Phys. **129**, 193303 (2021)
117. L. Hoppanová, V. Medvecká, J. Dylíková, D. Hudecová, B. Kaliňáková, S. Kryštofová, A. Zahoranová, Low-temperature plasma applications in chemical fungicide treatment reduction. Acta Chim. Slovaca **13**, 26–33 (2020)
118. S.I. Hosseini, N. Farrokhi, K. Shokri, M.R. Khani, B. Shokri, Cold low pressure O(2) plasma treatment of Crocus sativus: an efficient way to eliminate toxicogenic fungi with minor effect on molecular and cellular properties of saffron. Food Chem. **257**, 310–315 (2018a)
119. S.I. Hosseini, S. Mohsenimehr, J. Hadian, M. Ghorbanpour, B. Shokri, Physico-chemical induced modification of seed germination and early development in artichoke (*Cynara scolymus* L.) using low energy plasma technology. Phys. Plasmas **25**, 9 (2018b)
120. C.-C. Huang, J.S.-B. Wu, J.-S. Wu, Y. Ting, Effect of novel atmospheric-pressure jet pretreatment on the drying kinetics and quality of white grapes. J. Sci. Food Agric. **99**, 5102–5111 (2019a)
121. H. Huang, F. Ullah, D.-X. Zhou, M. Yi, Y. Zhao, Mechanisms of ROS regulation of plant development and stress responses. Front. Plant Sci. **10**, 800 (2019b)
122. Y.M. Huang, W.C. Chang, C.A.L. Hsu, Inactivation of norovirus by atmospheric pressure plasma jet on salmon sashimi. Food Res. Int. **141**, 7 (2021)
123. Y. Hui, D. Wan, Y. You, C. Shao, C. Zhong, H. Wang, Effect of low temperature plasma treatment on biological characteristics and yield components of wheat seeds (*Triticum aestivum* L.). Plasma Chem. Plasma Process. **40**, 1555–1570 (2020)
124. A. Iranbakhsh, M. Ghoranneviss, Z. Oraghi Ardebili, N. Oraghi Ardebili, S. Hesami Tackallou, H. Nikmaram, Non-thermal plasma modified growth and physiology in *Triticum aestivum* via generated signaling molecules and UV radiation. Biol. Plant. **61**, 702–708 (2017)
125. A. Iranbakhsh, Z.O. Ardebili, N.O. Ardebili, M. Ghoranneviss, N. Safari, Cold plasma relieved toxicity signs of nano zinc oxide in *Capsicum annuum* cayenne via modifying growth, differentiation, and physiology. Acta Physiol. Plant. **40**, 11 (2018)
126. A. Iranbakhsh, Z. Oraghi Ardebili, H. Molaei, N. Oraghi Ardebili, M. Amini, Cold Plasma Up-Regulated Expressions of WRKY1 Transcription Factor and Genes Involved in Biosynthesis of Cannabinoids in Hemp (*Cannabis sativa* L.). Plasma Chem. Plasma Process. **40**, 527–537 (2020)
127. S. Iseki, H. Hashizume, F. Jia, K. Takeda, K. Ishikawa, T. Ohta, M. Ito, M. Hori, Inactivation of Penicillium digitatum spores by a high-density ground-state atomic oxygen-radical source employing an atmospheric-pressure plasma. Appl. Phys. Express **4**, 116201 (2011)
128. N. Ishida, M. Koizumi, H. Kano, The NMR microscope: a unique and promising tool for plant science. Ann. Bot. **86**, 259–278 (2000)
129. S. Islam, F.B. Omar, S.A. Sajib, N.C. Roy, A. Reza, M. Hasan, M.R. Talukder, A.H. Kabir, Effects of LPDBD plasma and plasma activated water on germination and growth in rapeseed (*Brassica napus*). Gesunde Pflanzen **71**, 175–185 (2019)
130. A. Ivankov, Z. Naučienė, L. Degutytė-Fomins, R. Žūkienė, I. Januškaitienė, A. Malakauskienė, V. Jakštas, L. Ivanauskas, D. Romanovskaja, A. Šlepetienė, I. Filatova, V. Lyushkevich, V. Mildažienė, Changes in agricultural performance of common buckwheat induced by seed treatment with cold plasma and electromagnetic field. Appl. Sci. **11**, 4391 (2021a)
131. A. Ivankov, Z. Nauciene, R. Zukiene, L. Degutyte-Fomins, A. Malakauskiene, P. Kraujalis, P.R. Venskutonis, I. Filatova, V. Lyushkevich, V. Mildaziene, Changes in growth and production of non-psychotropic cannabinoids induced by pre-sowing treatment of hemp seeds with cold plasma vacuum and electromagnetic field. Appl. Sci. **10**, 8519 (2020)
132. A. Ivankov, R. Zukiene, Z. Nauciene, L. Degutyte-Fomins, I. Filatova, V. Lyushkevich, V. Mildaziene, The effects of red clover seed treatment with cold plasma and electromagnetic field on germination and seedling growth are dependent on seed color. Appl. Sci. **11**, 4676 (2021b)

133. P.E. Jameson, P. Dhandapani, O. Novak, J. Song, Cytokinins and expression of SWEET, SUT, CWINV and AAP genes increase as pea seeds germinate. Int. J. Mol. Sci. **17**, 2013 (2016)
134. D.D. Jayasena, H.J. Kim, H.I. Yong, S. Park, K. Kim, W. Choe, C. Jo, Flexible thin-layer dielectric barrier discharge plasma treatment of pork butt and beef loin: effects on pathogen inactivation and meat-quality attributes. Food Microbiol. **46**, 51–57 (2015)
135. S.H. Ji, T. Kim, K. Panngom, Y. Hong, A. Pengkit, D.H. Park, M.H. Kang, S.H. Lee, J.S. Im, J.S. Kim, H.S. Uhm, E.H. Choi, G. Park, Assessment of the effects of nitrogen plasma and plasma-generated nitric oxide on early development of *Coriandum sativum*. Plasma Process. Polym. **12**, 1164–1173 (2015)
136. S.-H. Ji, K.-H. Choi, A. Pengkit, J.S. Im, J.S. Kim, Y.H. Kim, Y. Park, E.J. Hong, S.K. Jung, E.-H. Choi, G. Park, Effects of high voltage nanosecond pulsed plasma and micro DBD plasma on seed germination, growth development and physiological activities in spinach. Arch. Biochem. Biophys. **605**, 117–128 (2016)
137. S.H. Ji, S.H. Ki, J.H. Ahn, J.H. Shin, E.J. Hong, Y.J. Kim, E.H. Choi, Inactivation of escherichia coli and staphylococcus aureus on contaminated perilla leaves by dielectric barrier discharge (DBD) plasma treatment. Arch. Biochem. Biophys. **643**, 32–41 (2018a)
138. S.H. Ji, S.H. Ki, M.H. Kang, J.S. Choi, Y. Park, J. Oh, S.B. Kim, S.J. Yoo, E.H. Choi, G. Park, Characterization of physical and biochemical changes in plasma treated spinach seed during germination. J. Phys. D Appl. Phys. **51**, 145205 (2018b)
139. J. Jiang, X. He, L. Li, J. Li, H. Shao, Q. Xu, R. Ye, Y. Dong, Effect of cold plasma treatment on seed germination and growth of wheat. Plasma Sci. Technol **16**, 54–58 (2014a)
140. J. Jiang, Y. Lu, J. Li, L. Li, X. He, H. Shao, Y. Dong, Effect of seed treatment by cold plasma on the resistance of tomato to *Ralstonia solanacearum* (bacterial wilt). PLoS ONE **9**, e97753 (2014b)
141. Y. Jiang, K. Sokorai, G. Pyrgiotakis, P. Demokritou, X. Li, S. Mukhopadhyay, T. Jin, X. Fan, Cold plasma-activated hydrogen peroxide aerosol inactivates *Escherichia coli* O157:H7, *Salmonella* Typhimurium, and *Listeria innocua* and maintains quality of grape tomato, spinach and cantaloupe. Int. J. Food Microbiol. **249**, 53–60 (2017)
142. J. Jiang, J. Li, Y. Dong, Effect of cold plasma treatment on seedling growth and nutrient absorption of tomato. Plasma Sci. Technol **20**, 044007 (2018)
143. Y.-K. Jo, J. Cho, T.-C. Tsai, D. Staack, M.-H. Kang, J.-H. Roh, D.-B. Shin, W. Cromwell, D. Gross, A non-thermal plasma seed treatment method for management of a seedborne fungal pathogen on rice seed. Crop Sci. **54**, 796–803 (2014)
144. S. Jung, J. Lee, Y. Lim, W. Choe, H.I. Yong, C. Jo, Direct infusion of nitrite into meat batter by atmospheric pressure plasma treatment. Innov. Food Sci. Emerg. Technol. **39**, 113–118 (2017)
145. J.-P. Kamseu-Mogo, G. Kamgang-Youbi, S.A. Djepang, B.S. Tamo, S. Laminsi, Treatment of maize seeds (*Zea mays*) by nonthermal plasma generated by gliding electric discharge for application in agriculture. IEEE Trans. Plasma Sci. **49**, 2318–2328 (2021)
146. M.H. Kang, A. Pengkit, K. Choi, S.S. Jeon, H.W. Choi, D.B. Shin, E.H. Choi, H.S. Uhm, G. Park, Differential inactivation of fungal spores in water and on seeds by ozone and arc discharge plasma. PLoS ONE **10**, e0139263 (2015)
147. K. Kartaschew, S. Baldus, M. Mischo, E. Bründermann, P. Awakowicz, M. Havenith, Cold atmospheric-pressure plasma and bacteria: understanding the mode of action using vibrational microspectroscopy. J. Phys. D Appl. Phys. **49**, 374003 (2016)
148. S.J. Kays, Preharvest factors affecting appearance. Postharvest Biol. Technol. **15**, 233–247 (1999)
149. F. Khalili, B. Shokri, M.-R. Khani, M. Hasani, F. Zandi, A. Aliahmadi, A study of the effect of gliding arc non-thermal plasma on almonds decontamination. AIP Adv. **8**, 105024 (2018)
150. N. Khamsen, D. Onwimol, N. Teerakawanich, S. Dechanupaprittha, W. Kanokbannakorn, K. Hongesombut, S. Srisonphan, Rice (*Oryza sativa* L.) seed sterilization and germination enhancement via atmospheric hybrid nonthermal discharge plasma. ACS Appl. Mater. Interfaces **8**, 19268–19275 (2016)

151. S. Khatami, A. Ahmadinia, Increased germination and growth rates of pea and Zucchini seed by FSG plasma. J. Theor. Appl. Phys. **12**, 33–38 (2018)
152. A. Kilonzo-Nthenge, S. Liu, S. Yannam, A. Patras, Atmospheric cold plasma inactivation of *Salmonella* and *Escherichia coli* on the surface of golden delicious apples. Front. Nutr. **5**, 120–120 (2018)
153. B. Kim, H. Yun, S. Jung, Y. Jung, H. Jung, W. Choe, C. Jo, Effect of atmospheric pressure plasma on inactivation of pathogens inoculated onto bacon using two different gas compositions. Food Microbiol. **28**, 9–13 (2011)
154. H.-J. Kim, H. Yong, S. Park, W. Choe, C. Jo, Effects of dielectric barrier discharge plasma on pathogen inactivation and the physicochemical and sensory characteristics of pork loin. Curr. Appl. Phys. **13**, 1420–1425 (2013)
155. J.E. Kim, D.U. Lee, S.C. Min, Microbial decontamination of red pepper powder by cold plasma. Food Microbiol. **38**, 128–136 (2014)
156. J.E. Kim, Y.J. Oh, M.Y. Won, K.S. Lee, S.C. Min, Microbial decontamination of onion powder using microwave-powered cold plasma treatments. Food Microbiol. **62**, 112–123 (2017a)
157. J.W. Kimm, P. Puligundla, C. Mok, Effect of corona discharge plasma jet on surface-borne microorganisms and sprouting of broccoli seeds. J. Sci. Food Agric. **97**, 128–134 (2017b)
158. S. Kitazaki, K. Koga, M. Shiratani, N. Hayashi, Growth enhancement of radish sprouts induced by low pressure O_2 radio frequency discharge plasma irradiation. Jpn. J. Appl. Phys. **51**, 01AE01 (2012)
159. S. Kitazaki, T. Sarinont, K. Koga, N. Hayashi, M. Shiratani, Plasma induced long-term growth enhancement of *Raphanus sativus* L. using combinatorial atmospheric air dielectric barrier discharge plasmas. Curr. Appl. Phys. **14**, S149–S153 (2014)
160. K. Koga, S. Thapanut, T. Amano, H. Seo, N. Itagaki, N. Hayashi, M. Shiratani, Simple method of improving harvest by nonthermal air plasma irradiation of seeds of *Arabidopsis thaliana* (L.). Appl. Phys. Express **9**, 016201 (2016)
161. L. Kordas, W. Pusz, T. Czapka, R. Kacprzyk, The effect of low-temperature plasma on fungus colonization of winter wheat grain and seed quality. Pol. J. Environ. Stud. **24**, 433–438 (2015)
162. T.G. Kronn, K.C. Lawrence, H. Zhuang, K.L. Hiett, M.J. Rothrock, Y.W. Huang, K.M. Keener, Z. Abdo, Nonthermal plasma system for extending shelf life of raw broiler breast fillets. Trans. ASABE **58**, 493–500 (2015)
163. B. Kucera, M.A. Cohn, G. Leubner-Metzger, Plant hormone interactions during seed dormancy release and germination. Seed Sci. Res. **15**, 281–307 (2005)
164. A. Lacombe, B. Niemira, J. Gurtler, X. Fan, J. Sites, G. Boyd, H. Chen, Atmospheric cold plasma inactivation of aerobic microorganisms on blueberries and effects on quality attributes. Food Microbiol. **46**, 479–484 (2015)
165. M. Laroussi, Nonthermal decontamination of biological media by atmospheric-pressure plasmas: review, analysis, and prospects. IEEE Trans. Plasma Sci. **30**, 1409–1415 (2002)
166. M. Laroussi, F. Leipold, Evaluation of the roles of reactive species, heat, and UV radiation in the inactivation of bacterial cells by air plasmas at atmospheric pressure. Int. J. Mass Spectrom. **233**, 81–86 (2004)
167. M. Laroussi, D. Mendis, M. Rosenberg, Plasma interaction with microbes. New J. Phys. **5**, 41 (2003)
168. K.-Y. Law, Definitions for hydrophilicity, hydrophobicity, and superhydrophobicity: getting the basics right. J. Phys. Chem. Lett. **5**, 686–688 (2014)
169. E.S. Lee, C.-I. Cheigh, J.H. Kang, S.Y. Lee, S.C. Min, Evaluation of in-package atmospheric dielectric barrier discharge cold plasma treatment as an intervention technology for decontaminating bulk ready-to-eat chicken breast cubes in plastic containers. Appl. Sci. **10**, 6301 (2020)
170. H. Lee, J.E. Kim, M.-S. Chung, S.C. Min, Cold plasma treatment for the microbiological safety of cabbage, lettuce, and dried figs. Food Microbiol. **51**, 74–80 (2015)
171. H. Lee, H.I. Yong, H.-J. Kim, W. Choe, S.J. Yoo, E.J. Jang, C. Jo, Evaluation of the microbiological safety, quality changes, and genotoxicity of chicken breast treated with flexible thin-layer dielectric barrier discharge plasma. Food Sci. Biotechnol. **25**, 1189–1195 (2016)

172. H.J. Lee, H. Jung, W. Choe, J.S. Ham, J.H. Lee, C. Jo, Inactivation of *Listeria monocytogenes* on agar and processed meat surfaces by atmospheric pressure plasma jets. Food Microbiol. **28**, 1468–1471 (2011)
173. S.Y. Lee, J. In, M.S. Chung, S.C. Min, Microbial decontamination of particulate food using a pilot-scale atmospheric plasma jet treatment system. J. Food Eng. **294**, 8 (2021a)
174. Y. Lee, Y.Y. Lee, Y.S. Kim, K. Balaraju, Y.S. Mok, S.J. Yoo, Y. Jeon, Enhancement of seed germination and microbial disinfection on ginseng by cold plasma treatment. J. Ginseng Res. **45**, 519–526 (2021b)
175. L. Li, J. Jiang, J. Li, M. Shen, X. He, H. Shao, Y. Dong, Effects of cold plasma treatment on seed germination and seedling growth of soybean. Sci. Rep. **4**, 5859 (2014)
176. L. Li, J. Li, H. Shao, Y. Dong, Effects of low-vacuum helium cold plasma treatment on seed germination, plant growth and yield of oilseed rape. Plasma Sci. Technol **20**, 095502 (2018)
177. L. Li, J. Li, M. Shen, J. Hou, H. Shao, Y. Dong, J. Jiang, Improving seed germination and peanut yields by cold plasma treatment. Plasma Sci. Technol. **18**, 1027 (2016)
178. L. Li, J. Li, M. Shen, C. Zhang, Y. Dong, Cold plasma treatment enhances oilseed rape seed germination under drought stress. Sci. Rep. **5**, 13033 (2015)
179. Y.J. Li, T.C. Wang, Y.R. Meng, G.Z. Qu, Q.H. Sun, D.L. Liang, S.B. Hu, Air atmospheric dielectric barrier discharge plasma induced germination and growth enhancement of wheat seed. Plasma Chem. Plasma Process. **37**, 1621–1634 (2017)
180. X. Liao, D. Liu, Q. Xiang, J. Ahn, S. Chen, X. Ye, T. Ding, Inactivation mechanisms of non-thermal plasma on microbes: a review. Food Control **75**, 83–91 (2017)
181. X.Y. Liao, J. Li, A.I. Muhammad, Y.J. Suo, S.G. Chen, X.Q. Ye, D.H. Liu, T. Ding, Application of a dielectric barrier discharge atmospheric cold plasma (Dbd-Acp) for eshcerichia coli inactivation in apple juice. J. Food Sci. **83**, 401–408 (2018)
182. K.A. Lis, A. Boulaaba, S. Binder, Y. Li, C. Kehrenberg, J.L. Zimmermann, G. Klein, B. Ahlfeld, Inactivation of *Salmonella* Typhimurium and *Listeria monocytogenes* on ham with nonthermal atmospheric pressure plasma. PLoS ONE **13**, e0197773 (2018)
183. C.-M. Liu, Y. Nishida, K. Iwasaki, K. Ting, Prolonged preservation and sterilization of fresh plants in controlled environments using high-field plasma. IEEE Trans. Plasma Sci. **39**, 717–724 (2011)
184. Z.R. Liu, W.Q. Zhao, Q.A. Zhang, G.T. Gao, Y.H. Meng, Effect of cold plasma treatment on sterilizing rate and quality of kiwi turbid juice. J. Food Process Eng. **44**, 7 (2021)
185. R. Lokeswari, P.S. Sharanyakanth, S. Jaspin, R. Mahendran, Cold plasma effects on changes in physical, nutritional, hydration, and pasting properties of pearl millet (*Pennisetum glaucum*). IEEE Trans. Plasma Sci. **49**, 1745–1751 (2021)
186. L. Lopez Del Egido, D. Navarro-Miró, V. Martinez-Heredia, P.E. Toorop, P.P.M. Iannetta, A spectrophotometric assay for robust viability testing of seed batches using 2,3,5-Triphenyl Tetrazolium chloride: using *Hordeum vulgare* L. as a model. Front. Plant Sci. **8**, 747 (2017)
187. M. López, T. Calvo, M. Prieto, R. Múgica-Vidal, I. Muro-Fraguas, F. Alba-Elías, A. Alvarez-Ordóñez, A review on non-thermal atmospheric plasma for food preservation: mode of action, determinants of effectiveness, and applications. Front. Microbiol. **10**, 622 (2019)
188. C. Lo Porto, L. Sergio, F. Boari, A.F. Logrieco, V. Cantore, Cold plasma pretreatment improves the germination of wild asparagus (*Asparagus acutifolius* L.) seeds. Sci. Hortic. **256**, 108554 (2019)
189. A. Los, D. Ziuzina, S. Akkermans, D. Boehm, P.J. Cullen, J. Van Impe, P. Bourke, Improving microbiological safety and quality characteristics of wheat and barley by high voltage atmospheric cold plasma closed processing. Food Res. Int. **106**, 509–521 (2018)
190. A. Los, D. Ziuzina, D. Boehm, P.J. Cullen, P. Bourke, Investigation of mechanisms involved in germination enhancement of wheat (*Triticum aestivum*) by cold plasma: effects on seed surface chemistry and characteristics. Plasma Process. Polym. **16**, 1800148 (2019)
191. K. Lotfy, Effects of cold atmospheric plasma jet treatment on the seed germination and enhancement growth of watermelon. Open J. Appl. Sci. **07**, 705–719 (2017)
192. K. Lotfy, N.A. Al-Harbi, H. Abd El-Raheem, Cold atmospheric pressure nitrogen plasma jet for enhancement germination of wheat seeds. Plasma Chem. Plasma Process. **39**, 897–912 (2019)

193. Q. Lu, D. Liu, Y. Song, R. Zhou, J. Niu, Inactivation of the tomato pathogen Cladosporium fulvum by an atmospheric-pressure cold plasma jet. Plasma Process. Polym. **11**, 1028–1036 (2014)
194. F. Ma, E. Cholewa, T. Mohamed, C.A. Peterson, M. Gijzen, Cracks in the palisade cuticle of soybean seed coats correlate with their permeability to water. Ann. Bot. **94**, 213–228 (2004)
195. N.K. Mahnot, C.L. Mahanta, K.M. Keener, N.N. Misra, Strategy to achieve a 5-log Salmonella inactivation in tender coconut water using high voltage atmospheric cold plasma (HVACP). Food Chem. **284**, 303–311 (2019)
196. M. Măgureanu, R. Sirbu, D. Dobrin, M. Gidea, Stimulation of the germination and early growth of tomato seeds by non-thermal plasma. Plasma Chem. Plasma Process. **38**, 989–1001 (2018)
197. A. Mai-Prochnow, M. Bradbury, K. Ostrikov, A.B. Murphy, *Pseudomonas aeruginosa* biofilm response and resistance to cold atmospheric pressure plasma is linked to the redox-active molecule phenazine. PLoS ONE **10**, e0130373 (2015)
198. A. Mai-Prochnow, M. Clauson, J. Hong, A.B. Murphy, Gram positive and Gram negative bacteria differ in their sensitivity to cold plasma. Sci. Rep. **6**, 1–11 (2016)
199. M. Makari, M. Hojjati, S. Shahbazi, H. Askari, Elimination of *Aspergillus flavus* from pistachio nuts with dielectric barrier discharge (DBD) cold plasma and its impacts on biochemical indices. J. Food Qual. **2021**, 12 (2021)
200. D. Manoharan, J. Stephen, M. Radhakrishnan, Study on low-pressure plasma system for continuous decontamination of milk and its quality evaluation. J. Food Process. Preserv. **45**, 9 (2021)
201. S. Marin, A.J. Ramos, G. Cano-Sancho, V. Sanchis, Mycotoxins: occurrence, toxicology, and exposure assessment. Food Chem. Toxicol. **60**, 218–237 (2013)
202. K. Matra, Non-thermal plasma for germination enhancement of radish seeds. Procedia Comput. Sci. **86**, 132–135 (2016)
203. K. Matra, Atmospheric non-thermal argon–oxygen plasma for sunflower seedling growth improvement. Jpn. J. Appl. Phys.**57**, 01AG03 (2018)
204. J.D. McClurkin-Moore, K.E. Ileleji, K.M. Keener, The effect of high-voltage atmospheric cold plasma treatment on the shelf-life of distillers wet grains. Food Bioprocess Technol. **10**, 1431–1440 (2017)
205. N. Mehle, M. Ravnikar, Plant viruses in aqueous environment-survival, water mediated transmission and detection. Water Res. **46**, 4902–4917 (2012)
206. D. Mehta, N. Sharma, V. Bansal, R.S. Sangwan, S.K. Yadav, Impact of ultrasonication, ultraviolet and atmospheric cold plasma processing on quality parameters of tomato-based beverage in comparison with thermal processing. Innov. Food Sci. Emerg. Technol. **52**, 343–349 (2019)
207. Y.R. Meng, G.Z. Qu, T.C. Wang, Q.H. Sun, D.L. Liang, S.B. Hu, Enhancement of germination and seedling growth of wheat seed using dielectric barrier discharge plasma with various gas sources. Plasma Chem. Plasma Process. **37**, 1105–1119 (2017)
208. A.L. Mihai, D. Dobrin, M. Magureanu, M.E. Popa, Positive effect of non-thermal plasma treatment on radish seeds. Roman. Rep. Phys. **66**, 1110–1117 (2014)
209. V. Mildažienė, V. Aleknavičiūtė, R. Žūkienė, G. Paužaitė, Z. Naučienė, I. Filatova, V. Lyushkevich, P. Haimi, I. Tamošiūnė, D. Baniulis, Treatment of common sunflower (*Helianthus annus* L.) seeds with radio-frequency electromagnetic field and cold plasma induces changes in seed phytohormone balance, seedling development and leaf protein expression. Sci. Rep. **9**, 12 (2019)
210. V. Mildažienė, G. Pauzaite, Z. Nauciene, A. Malakauskiene, R. Zukiene, I. Januskaitiene, V. Jakstas, L. Ivanauskas, I. Filatova, V. Lyushkevich, Pre-sowing seed treatment with cold plasma and electromagnetic field increases secondary metabolite content in purple coneflower (*Echinacea purpurea*) leaves. Plasma Process. Polym. **15**, 11 (2018)
211. V. Mildažienė, G. Paužaitė, Z. Naučienė, R. Žūkienė, A. Malakauskienė, E. Norkevičienė, A. Šlepetienė, V. Stukonis, V. Olšauskaitė, A. Padarauskas, I. Filatova, V. Lyuskevich, Effect of seed treatment with cold plasma and electromagnetic field on red clover germination, growth and content of major isoflavones. J. Phys. D Appl. Phys. **53**, 264001 (2020)

212. S. Milusheva, L. Nacheva, E. Benova, P. Marinova, N. Dimitrova, A. Georgieva-Hristeva, Experiments on Plum pox virus inactivation from micropropagated plum plants through non-thermal plasma treatment. Bitki Koruma Bülteni **60**, 83–90 (2020)
213. S.C. Min, S.H. Roh, B.A. Niemira, J.E. Sites, G. Boyd, A. Lacombe, Dielectric barrier discharge atmospheric cold plasma inhibits *Escherichia coli* O157:H7, *Salmonella, Listeria monocytogenes*, and Tulane virus in Romaine lettuce. Int. J. Food Microbiol. **237**, 114–120 (2016)
214. N.N. Misra, S. Patil, T. Moiseev, P. Bourke, J.P. Mosnier, K.M. Keener, P.J. Cullen, In-package atmospheric pressure cold plasma treatment of strawberries. J. Food Eng. **125**, 131–138 (2014)
215. N.N. Misra, X. Yepez, L. Xu, K. Keener, In-package cold plasma technologies. J. Food Eng. **244**, 21–31 (2019)
216. A. Mitra, Y.-F. Li, T.G. Klämpfl, T. Shimizu, J. Jeon, G.E. Morfill, J.L. Zimmermann, Inactivation of surface-borne microorganisms and increased germination of seed specimen by cold atmospheric plasma. Food Bioprocess Technol. **7**, 645–653 (2014)
217. M. Moisan, J. Barbeau, M.-C. Crevier, J. Pelletier, N. Philip, B. Saoudi, Plasma sterilization. methods and mechanisms. Pure Appl. Chem. **74**, 349–358 (2002)
218. M. Moisan, J. Barbeau, S. Moreau, J. Pelletier, M. Tabrizian, L.H. Yahia, Low-temperature sterilization using gas plasmas: a review of the experiments and an analysis of the inactivation mechanisms. Int. J. Pharm. **226**, 1–21 (2001)
219. R. Molina, C. López-Santos, A. Gómez-Ramírez, A. Vílchez, J.P. Espinós, A.R. González-Elipe, Influence of irrigation conditions in the germination of plasma treated *Nasturtium* seeds. Sci. Rep. **8**, 16442 (2018)
220. R. Molina, A. Lalueza, C. López-Santos, R. Ghobeira, P. Cools, R. Morent, N. de Geyter, A.R. González-Elipe, Physicochemical surface analysis and germination at different irrigation conditions of DBD plasma-treated wheat seeds. Plasma Process. Polym. **18**, 2000086 (2021)
221. A.Y. Moon, S. Noh, S.Y. Moon, S. You, Feasibility study of atmospheric-pressure plasma treated air gas package for grape's shelf-life improvement. Curr. Appl. Phys. **16**, 440–445 (2016)
222. F.J. Morales, History and current distribution of begomoviruses in Latin America. Adv. Virus Res. **67**, 127–162 (2006)
223. M. Moreau, M. Feuilloley, W. Veron, T. Meylheuc, S. Chevalier, J.-L. Brisset, N. Orange, Gliding arc discharge in the potato pathogen *Erwinia carotovora* subsp. *atroseptica*: mechanism of lethal action and effect on membrane-associated molecules. Appl. Environ. Microbiol. **73**, 5904–5910 (2007)
224. M. Moreau, M.G. Feuilloley, N. Orange, J.L. Brisset, Lethal effect of the gliding arc discharges on *Erwinia* spp. J. Appl. Microbiol. **98**, 1039–1046 (2005)
225. M. Moritz, C. Wiacek, M. Koethe, P.G. Braun, Atmospheric pressure plasma jet treatment of *Salmonella* Enteritidis inoculated eggshells. Int. J. Food Microbiol. **245**, 22–28 (2017)
226. S. Mosovska, V. Medvecka, N. Halaszova, P. Durina, L. Valik, A. Mikulajova, A. Zahoranova, Cold atmospheric pressure ambient air plasma inhibition of pathogenic bacteria on the surface of black pepper. Food Res. Int. **106**, 862–869 (2018)
227. A. Motyka, A. Dzimitrowicz, P. Jamroz, E. Lojkowska, W. Sledz, P. Pohl, Rapid eradication of bacterial phytopathogens by atmospheric pressure glow discharge generated in contact with a flowing liquid cathode. Biotechnol. Bioeng. **115**, 1581–1593 (2018)
228. J. Mravlje, M. Regvar, P. Starič, M. Mozetič, K. Vogel-Mikuš, Cold plasma affects germination and fungal community structure of buckwheat seeds. Plants **10**, 851 (2021a)
229. J. Mravlje, M. Regvar, K. Vogel-Mikuš, Development of cold plasma technologies for surface decontamination of seed fungal pathogens: present status and perspectives. J. Fungi **7**, 650 (2021b)
230. I. Mraz, P. Beran, B. Šerá, B. Gavril, E. Hnatiuc, Effect of low-temperature plasma treatment on the growth and reproduction rate of some plant pathogenic bacteria. J. Plant Pathol. **96**, 63–67 (2014)
231. A. Muhammad, L. Yang, X. Liao, D. Liu, X. Ye, S. Chen, Y. Hu, J. Wang, T. Ding, Effect of dielectric barrier discharge plasma on background microflora and physicochemical properties of tiger nut milk. Food Control **96**, 119–127 (2019)

232. Y.H. Na, G. Park, E.H. Choi, H.S. Uhm, Effects of the physical parameters of a microwave plasma jet on the inactivation of fungal spores. Thin Soild Films **547**, 125–131 (2013)
233. R. Nakano, K. Tashiro, R. Aijima, N. Hayashi, Effect of oxygen plasma irradiation on gene expression in plant seeds induced by active oxygen species. Plasma Med. **6**, 303–313 (2016)
234. C. Nalwa, A.K. Thakur, A. Vikram, R. Rane, A. Vaid, Studies on plasma treatment and priming of seeds of bell pepper (*Capsicum annuum* L.). J. Appl. Nat. Sci. **9**, 1505–1509 (2017)
235. P. Narayanasamy, *Soilborne Microbial Plant Pathogens and Disease Management, Volume Two: management of Crop Diseases*. (CRC Press, Boca Raton , 2019)
236. R. Nastase, J.-M. Tatibouët, E. Fourré, Depolymerization of inulin in the highly reactive gas phase of a non thermal plasma at atmospheric pressure. Plasma Process. Polym. **15**, 1800067 (2018)
237. C.Z. Nie, X.P. Qin, Z.Q. Duan, S.S. Huang, X. Yu, Q.C. Deng, Q.S. Xiang, F. Geng, Comparative structural and techno-functional elucidation of full-fat and defatted flaxseed extracts: implication of atmospheric pressure plasma jet. J. Sci. Food Agric. **102**, 823–835 (2022)
238. I. Niedzwiedz, A. Wasko, J. Pawlat, M. Polak-Berecka, The state of research on antimicrobial activity of cold plasma. Pol. J. Microbiol. **68**, 153–164 (2019)
239. B.A. Niemira, Cold plasma reduction of *Salmonella* and *Escherichia coli* O157:H7 on Almonds using ambient pressure gases. J. Food Sci. **77**, M171–M175 (2012)
240. T.M.C. Nishime, N. Wannicke, S. Horn, K.D. Weltmann, H. Brust, A coaxial dielectric barrier discharge reactor for treatment of winter wheat seeds. Appl. Sci. **10**, 19 (2020)
241. T. Nishioka, Y. Takai, M. Kawaradani, K. Okada, H. Tanimoto, T. Misawa, S. Kusakari, Seed disinfection effect of atmospheric pressure plasma and low pressure plasma on *Rhizoctonia solani*. Biocontrol Sci. **19**, 99–102 (2014)
242. T. Nishioka, Y. Takai, T. Mishima, M. Kawaradani, H. Tanimoto, K. Okada, T. Misawa, S. Kusakari, Low-pressure plasma application for the inactivation of the seed-borne pathogen *Xanthomonas campestris*. Biocontrol Sci. **21**, 37–43 (2016)
243. H. Nojima, R.-E. Park, J.-H. Kwon, I. Suh, J. Jeon, E. Ha, H.-K. On, H.-R. Kim, K. Choi, K.-H. Lee, Novel atmospheric pressure plasma device releasing atomic hydrogen: reduction of microbial-contaminants and OH radicals in the air. J. Phys. D Appl. Phys. **40**, 501 (2007)
244. H. Nonogaki, G.W. Bassel, J.D. Bewley, Germination—Still a mystery. Plant Sci. **179**, 574–581 (2010)
245. E. Noriega, G. Shama, A. Laca, M. Diaz, M.G. Kong, Cold atmospheric gas plasma disinfection of chicken meat and chicken skin contaminated with *Listeria innocua*. Food Microbiol. **28**, 1293–1300 (2011)
246. A. Ochi, H. Konishi, S. Ando, K. Sato, K. Yokoyama, S. Tsushima, S. Yoshida, T. Morikawa, T. Kaneko, H. Takahashi, Management of bakanae and bacterial seedling blight diseases in nurseries by irradiating rice seeds with atmospheric plasma. Plant. Pathol. **66**, 67–76 (2017)
247. T. Okumura, Y. Saito, K. Takano, K. Takahashi, K. Takaki, N. Satta, T. Fujio, Inactivation of bacteria using discharge plasma under liquid fertilizer in a hydroponic culture system. Plasma Med. **6**, 247–254 (2016)
248. S.A. Ouf, A.H. Basher, A.-A.H. Mohamed, Inhibitory effect of double atmospheric pressure argon cold plasma on spores and mycotoxin production of Aspergillus niger contaminating date palm fruits. J. Sci. Food Agric. **95**, 3204–3210 (2015)
249. S.K. Pankaj, Z.F. Wan, K.M. Keener, Effects of cold plasma on food quality: a review. Foods **7**, 21 (2018)
250. K. Panngom, S.H. Lee, D.H. Park, G.B. Sim, Y.H. Kim, H.S. Uhm, G. Park, E.H. Choi, Non-thermal plasma treatment diminishes fungal viability and up-regulates resistance genes in a plant host. PLoS ONE **9**, e99300 (2014)
251. S. Paparella, S.S. Araújo, G. Rossi, M. Wijayasinghe, D. Carbonera, A. Balestrazzi, Seed priming: state of the art and new perspectives. Plant Cell Rep. **34**, 1281–1293 (2015)
252. B.J. Park, D. Lee, J.-C. Park, I.-S. Lee, K.-Y. Lee, S. Hyun, M.-S. Chun, K.-H. Chung, Sterilization using a microwave-induced argon plasma system at atmospheric pressure. Phys. Plasmas **10**, 4539–4544 (2003)

253. B.J. Park, K. Takatori, Y. Sugita-Konishi, I.H. Kim, M.H. Lee, D.W. Han, K.H. Chung, S.O. Hyun, J.C. Park, Degradation of mycotoxins using microwave-induced argon plasma at atmospheric pressure. Surf. Coat. Technol. **201**, 5733–5737 (2007)
254. J.-C. Park, B.-J. Park, D.-W. Han, D.-H. Lee, I.-S. Lee, S.-O. Hyun, M.-S. Chun, K.-H. Chung, A. Maki, T. Kosuke, Fungal sterilization using microwave-induced argon plasma at atmospheric pressure. J. Microbiol. Biotechnol. **14**, 188–192 (2004)
255. S.Y. Park, S.-D. Ha, Application of cold oxygen plasma for the reduction of *Cladosporium cladosporioides* and *Penicillium citrinum* on the surface of dried filefish (*Stephanolepis cirrhifer*) fillets. Int. J. Food Sci. Technol. **50**, 966–973 (2015)
256. Y. Park, K.S. Oh, J. Oh, D.C. Seok, S.B. Kim, S.J. Yoo, M.-J. Lee, The biological effects of surface dielectric barrier discharge on seed germination and plant growth with barley. Plasma Process. Polym. **15**, 1600056 (2018)
257. F. Pasquali, A.C. Stratakos, A. Koidis, A. Berardinelli, C. Cevoli, L. Ragni, R. Mancusi, G. Manfreda, M. Trevisani, Atmospheric cold plasma process for vegetable leaf decontamination: a feasibility study on radicchio (red chicory, *Cichorium intybus* L.). Food Control **60**, 552–559 (2016)
258. G. Pauzaite, A. Malakauskiene, Z. Nauciene, R. Zukiene, I. Filatova, V. Lyushkevich, I. Azarko, V. Mildaziene, Changes in Norway spruce germination and growth induced by pre-sowing seed treatment with cold plasma and electromagnetic field: short-term versus long-term effects. Plasma Process. Polym. **15**, 11 (2018)
259. J. Pawłat, A. Starek, A. Sujak, M. Kwiatkowski, P. Terebun, M. Budzeń, Effects of atmospheric pressure plasma generated in GlidArc reactor on *Lavatera thuringiaca* L. seeds' germination. Plasma Process. Polym. **15**, 1700064 (2018a)
260. J. Pawłat, A. Starek, A. Sujak, P. Terebun, M. Kwiatkowski, M. Budzeń, D. Andrejko, Effects of atmospheric pressure plasma jet operating with DBD on *Lavatera thuringiaca* L. seeds' germination. PLoS One **13**, e0194349 (2018b)
261. C. Pelosi, C. Bertrand, G. Daniele, M. Coeurdassier, P. Benoit, S. Nélieu, F. Lafay, V. Bretagnolle, S. Gaba, E. Vulliet, C. Fritsch, Residues of currently used pesticides in soils and earthworms: a silent threat? *Agriculture*. Ecosyst. Environ. **305**, 107167 (2021)
262. K.N.M. Penado, C.L.S. Mahinay, I.B. Culaba, Effect of atmospheric plasma treatment on seed germination of rice (*Oryza sativa* L.). Jpn. J. Appl. Phys. **57**, 01AG08 (2017)
263. S. Penfield, Revealing the water uptake pathways of seeds with high resolution magnetic resonance imaging. New Phytol. **216**, 965–966 (2017)
264. M.C. Pérez Pizá, L. Prevosto, C. Zilli, E. Cejas, H. Kelly, K. Balestrasse, Effects of non–thermal plasmas on seed-borne *Diaporthe/Phomopsis* complex and germination parameters of soybean seeds. Innov. Food Sci. Emerg. Technol. **49**, 82–91 (2018)
265. M.C. Pérez-Pizá, E. Cejas, C. Zilli, L. Prevosto, B. Mancinelli, D. Santa-Cruz, G. Yannarelli, K. Balestrasse, Enhancement of soybean nodulation by seed treatment with non-thermal plasmas. Sci. Rep. **10**, 4917 (2020)
266. M.C. Pérez-Pizá, V.N. Ibañez, A. Varela, E. Cejas, M. Ferreyra, J.C. Chamorro-Garcés, C. Zilli, P. Vallecorsa, B. Fina, L. Prevosto, C.F. Marfil, K.B. Balestrasse, Non-thermal plasmas affect plant growth and DNA methylation patterns in *Glycine max*. J. Plant Growth Regul. (2021a). https://doi.org/10.1007/s00344-021-10470-8
267. M. Pětková, R. Švubová, S. Kyzek, V. Medvecká, Ľ Slováková, A. Ševčovičová, E. Gálová, The effects of cold atmospheric pressure plasma on germination parameters, enzyme activities and induction of DNA damage in barley. Int. J. Mol. Sci. **22**, 2833 (2021b)
268. M.C. Pina-Perez, D. Rodrigo, C. Ellert, M. Beyrer, Surface micro discharge-cold atmospheric pressure plasma processing of common house cricket *Acheta domesticus* powder: antimicrobial potential and lipid-quality preservation. Front. Bioeng. Biotechnol. **9**, 7 (2021)
269. N. Puač, Z.L. Petrović, S. Živković, Z. Giba, D. Grubišić, A.R. Đorđević, Low-temperature plasma treatment of dry empress-tree seeds, in *Plasma Processes and Polymers*, eds. by R. d'Agostino, P. Favia, C. Oehr, M.R. Wertheimer. (Wiley, Hoboken, NJ, USA, 2005), pp. 193–203

270. P. Puligundla, J.-W. Kim, C. Mok, Effect of corona discharge plasma jet treatment on decontamination and sprouting of rapeseed (*Brassica napus* L.) seeds. Food Control **71**, 376–382 (2017a)
271. P. Puligundla, J.-W. Kim, C. Mok, Effects of nonthermal plasma treatment on decontamination and sprouting of radish (*Raphanus sativus* L.) seeds. Food Bioprocess Technol. **10**, 1093–1102 (2017b)
272. P. Puligundla, T. Lee, C. Mok, Effect of intermittent corona discharge plasma treatment for improving microbial quality and shelf life of kumquat (*Citrus japonica*) fruits. LWT **91**, 8–13 (2018)
273. P. Puligundla, T. Lee, C. Mok, Effect of corona discharge plasma jet treatment on the degradation of aflatoxin B1 on glass slides and in spiked food commodities. LWT **124**, 108333 (2020)
274. P. Putnik, Ž Kresoja, T. Bosiljkov, A. Režek Jambrak, F.J. Barba, J.M. Lorenzo, S. Roohinejad, D. Granato, I. Žuntar, D. Bursać Kovačević, Comparing the effects of thermal and non-thermal technologies on pomegranate juice quality: a review. Food Chem. **279**, 150–161 (2019)
275. J. Qian, C. Wang, H. Zhuang, M.M. Nasiru, J.H. Zhang, W.J. Yan, Evaluation of meat-quality and myofibrillar protein of chicken drumsticks treated with plasma-activated lactic acid as a novel sanitizer. LWT-Food Sci. Technol. **138**, 9 (2021)
276. M.M. Rahman, S.A. Sajib, M.S. Rahi, S. Tahura, N.C. Roy, S. Parvez, M.A. Reza, M.R. Talukder, A.H. Kabir, Mechanisms and signaling associated with LPDBD plasma mediated growth improvement in wheat. Sci. Rep. **8**, 10498 (2018)
277. T. Ramamurthy, A. Ghosh, G.P. Pazhani, S. Shinoda, Current perspectives on viable but non-culturable (VBNC) pathogenic bacteria. Front. Public Health **2**, 103 (2014)
278. P. Ranieri, N. Sponsel, J. Kizer, M. Rojas-Pierce, R. Hernández, L. Gatiboni, A. Grunden, K. Stapelmann, Plasma agriculture: review from the perspective of the plant and its ecosystem. Plasma Process. Polym. **18**, 2000162 (2021)
279. F. Rashid, Y.W. Bao, Z. Ahmed, J.Y. Huang, Effect of high voltage atmospheric cold plasma on extraction of fenugreek galactomannan and its physicochemical properties. Food Res. Int. **138**, 10 (2020)
280. Z. Rasooli, G. Barzin, T.D. Mahabadi, M. Entezari, Stimulating effects of cold plasma seed priming on germination and seedling growth of cumin plant. S. Afr. J. Bot. **142**, 106–113 (2021)
281. N. Recek, M. Holc, A. Vesel, R. Zaplotnik, P. Gselman, M. Mozetič, G. Primc, Germination of *Phaseolus vulgaris* L. Seeds after a short treatment with a powerful RF plasma. Int. J. Mol. Sci. **22**, 6672 (2021)
282. C.R. Ren, G.Q. Huang, S.Q. Wang, J.X. Xiao, X.B. Xiong, H.F. Wang, Y.L. Liu, Influence of atmospheric pressure argon plasma treatment on the quality of peanut oil. Eur. J. Lipid Sci. Technol. **119**, 7 (2017)
283. F.J. Richards, A flexible growth function for empirical use. J. Exp. Bot. **10**, 290–301 (1959)
284. E. Robilotti, S. Deresinski, B.A. Pinsky, Norovirus. Clin. Microbiol. Rev. **28**, 134–164 (2015)
285. N.C. Roy, M.M. Hasan, M.R. Talukder, M.D. Hossain, A.N. Chowdhury, Prospective applications of low frequency glow discharge plasmas on enhanced germination, growth and yield of wheat. Plasma Chem. Plasma Process. **38**, 13–28 (2018)
286. C.L. Rüntzel, J.R. da Silva, B.A. da Silva, E.S. Moecke, V.M. Scussel, Effect of cold plasma on black beans (*Phaseolus vulgaris* L.), fungi inactivation and micro-structures stability. Emir. J. Food Agric. **31**, 864–873 (2019)
287. S. Sadhu, R. Thirumdas, R.R. Deshmukh, U.S. Annapure, Influence of cold plasma on the enzymatic activity in germinating mung beans (*Vigna radiate*). LWT-Food Sci. Technol. **78**, 97–104 (2017)
288. N. Safari, A. Iranbakhsh, Z.O. Ardebili, Non-thermal plasma modified growth and differentiation process of *Capsicum annuum* PP805 Godiva in *in vitro* conditions. Plasma Sci. Technol. **19**, 055501 (2017)
289. A. Sakudo, Y. Yagyu, Application of a roller conveyor type plasma disinfection device with fungus-contaminated citrus fruits. AMB Express **11**, 16 (2021)

290. A. Sakudo, Y. Yagyu, T. Onodera, Disinfection and sterilization using plasma technology: fundamentals and future perspectives for biological applications. Int. J. Mol. Sci. **20**, 5216 (2019)
291. T. Sarinont, T. Amano, S. Kitazaki, K. Koga, G. Uchida, M. Shiratani, N. Hayashi, Growth enhancement effects of radish sprouts: atmospheric pressure plasma irradiation versus heat shock. J. Phys.: Conf. Ser.**518**, 012017 (2014)
292. T. Sarinont, T. Amano, K. Koga, M. Shiratani, N. Hayashi, Multigeneration effects of plasma irradiation to seeds of *Arabidopsis thaliana* and *Zinnia* on their growth. MRS Proc. **1723**
293. T. Sarinont, T. Amano, P. Attri, K. Koga, N. Hayashi, M. Shiratani, Effects of plasma irradiation using various feeding gases on growth of *Raphanus sativus* L. Arch. Biochem. Biophys. **605**, 129–140 (2016)
294. K.S. Sastry, *Seed-Borne Plant Virus Diseases* (Springer, Berlin, Heidelberg, 2013)
295. U. Schnabel, O. Handorf, H. Winter, T. Weihe, C. Weit, J. Schafer, J. Stachowiak, D. Boehm, H. Below, P. Bourke, J. Ehlbeck, The effect of plasma treated water unit processes on the food quality characteristics of fresh-cut endive. Front. Nutr. **7**, 14 (2021)
296. U. Schnabel, C. Schmidt, J. Stachowiak, A. Bosel, M. Andrasch, J. Ehlbeck, Plasma processed air for biological decontamination of PET and fresh plant tissue. Plasma Process. Polym. **15**, 9 (2018)
297. U. Schnabel, R. Niquet, U. Krohmann, J. Winter, O. Schlüter, K.-D. Weltmann, J. Ehlbeck, Decontamination of microbiologically contaminated specimen by direct and indirect plasma treatment. Plasma Process. Polym. **9**, 569–575 (2012)
298. F. Schottroff, A. Fröhling, M. Zunabovic-Pichler, A. Krottenthaler, O. Schlüter, H. Jäger, Sublethal injury and viable but non-culturable (VBNC) state in microorganisms during preservation of food and biological materials by non-thermal processes. Front. Microbiol. **9**, 2773 (2018)
299. F.S. Seddighinia, A. Iranbakhsh, Z.O. Ardebili, T.N. Satari, S. Soleimanpour, Seed priming with cold plasma and multi-walled carbon nanotubes modified growth, tissue differentiation, anatomy, and yield in bitter melon (*Momordica charantia*). J. Plant Growth Regul. **39**, 87–98 (2020)
300. A. Segat, N.N. Misra, P.J. Cullen, N. Innocente, Effect of atmospheric pressure cold plasma (ACP) on activity and structure of alkaline phosphatase. Food Bioprod. Process. **98**, 181–188 (2016)
301. M. Selcuk, L. Oksuz, P. Basaran, Decontamination of grains and legumes infected with *Aspergillus* spp. and *Penicillum* spp. by cold plasma treatment. Biores. Technol. **99**, 5104–5109 (2008)
302. Y. Sen, B. Onal-Ulusoy, M. Mutlu, *Aspergillus* decontamination in hazelnuts: Evaluation of atmospheric and low-pressure plasma technology. Innov. Food Sci. Emerg. Technol. **54**, 235–242 (2019a)
303. Y. Sen, B. Onal-Ulusoy, M. Mutlu, Detoxification of hazelnuts by different cold plasmas and gamma irradiation treatments. Innov. Food Sci. Emerg. Technol. **54**, 252–259 (2019b)
304. B. Šerá, M. Šerý, V. Štraňák, P. Špatenka, M. Tichý, Does cold plasma affect breaking dormancy and seed germination? A study on seeds of lamb's quarters (*Chenopodium album* agg.). Plasma Sci. Technol. **11**, 750–754 (2009)
305. B. Šerá, P. Špatenka, M. Šerý, N. Vrchotová, I. Hrušková, Influence of plasma treatment on wheat and oat germination and early growth. IEEE Trans. Plasma Sci. **38**, 2963–2968 (2010)
306. B. Šerá, M. Šerý, B. Gavril, I. Gajdova, Seed germination and early growth responses to seed pre-treatment by non-thermal plasma in hemp cultivars (*Cannabis sativa* L.). Plasma Chem. Plasma Process. **37**, 207–221 (2017)
307. B. Šerá, A. Zahoranová, H. Bujdáková, M. Šerý, Disinfection from pine seeds contaminated with *Fusarium circinatum* Nirenberg & O'Donnell using non-thermal plasma treatment. Roman. Rep. Phys. **71**, 701 (2019)
308. B. Šerá, M. Šerý, A. Zahoranová, J. Tomeková, Germination improvement of three pine species (Pinus) after diffuse coplanar surface barrier discharge plasma treatment. Plasma Chem. Plasma Process. **41**, 211–226 (2021)

309. M. Šerý, A. Zahoranová, A. Kerdík, B. Šerá, Seed germination of black pine (*Pinus nigra* Arnold) after diffuse coplanar surface barrier discharge plasma treatment. IEEE Trans. Plasma Sci. **48**, 939–945 (2020)
310. C. Shao, D. Wang, X. Tang, L. Zhao, Y. Li, Stimulating effects of magnetized arc plasma of different intensities on the germination of old spinach seeds. Math. Comput. Model. **58**, 814–818 (2013)
311. M.S. Sheteiwy, J. An, M. Yin, X. Jia, Y. Guan, F. He, J. Hu, Cold plasma treatment and exogenous salicylic acid priming enhances salinity tolerance of *Oryza sativa* seedlings. Protoplasma **256**, 79–99 (2019)
312. D.D. Shukla, C.W. Ward, A.A. Brunt, *The Potyviridae*. (CAB International, Wallingford, UK Wallingford, UK, 1994)
313. I. Siciliano, D. Spadaro, A. Prelle, D. Vallauri, M.C. Cavallero, A. Garibaldi, M.L. Gullino, Use of cold atmospheric plasma to detoxify hazelnuts from aflatoxins. Toxins **8**, 10 (2016)
314. I. Sifuentes-Nieves, G. Mendez-Montealvo, P.C. Flores-Silva, M. Nieto-Perez, G. Neira-Velazquez, O. Rodriguez-Fernandez, E. Hernandez-Hernandez, G. Velazquez, Dielectric barrier discharge and radio-frequency plasma effect on structural properties of starches with different amylose content. Innov. Food Sci. Emerg. Technol. **68**, 9 (2021)
315. D. Sikorska, E. Papierowska, J. Szatyłowicz, P. Sikorski, K. Suprun, R.J. Hopkins, Variation in leaf surface hydrophobicity of wetland plants: the role of plant traits in water retention. Wetlands **37**, 997–1002 (2017)
316. J. Šimončicová, B. Kaliňáková, D. Kováčik, V. Medvecká, B. Lakatoš, S. Kryštofová, L. Hoppanová, V. Palušková, D. Hudecová, P. Ďurina, Cold plasma treatment triggers antioxidative defense system and induces changes in hyphal surface and subcellular structures of *Aspergillus flavus*. Appl. Microbiol. Biotechnol. **102**, 6647–6658 (2018)
317. R. Singh, P. Prasad, R. Mohan, M.K. Verma, B. Kumar, Radiofrequency cold plasma treatment enhances seed germination and seedling growth in variety CIM-Saumya of sweet basil (*Ocimum basilicum* L.). J. Appl. Res. Med. Arom. Plants **12**, 78–81 (2019)
318. F. Sohbatzadeh, S. Mirzanejhad, H. Shokri, M. Nikpour, Inactivation of *Aspergillus flavus* spores in a sealed package by cold plasma streamers. J. Theor. Appl. Phys. **10**, 99–106 (2016)
319. J.-S. Song, M.J. Lee, J.E. Ra, K.S. Lee, S. Eom, H.M. Ham, H.Y. Kim, S.B. Kim, J. Lim, Growth and bioactive phytochemicals in barley (*Hordeum vulgare* L.) sprouts affected by atmospheric pressure plasma during seed germination. J. Phys. D: Appl. Phys. **53**, 314002 (2020)
320. H. Soušková, V. Scholtz, J. Julák, D. Savická, The fungal spores survival under the low-temperature plasma, in *Plasma for Bio-Decontamination, Medicine and Food Security, NATO Science for Peace and Security Series A: chemistry and Biology*. (Springer, Berlin, Heidelberg, 2012), pp. 57–66
321. K. Srakaew, A. Chingsungnoen, W. Sutthisa, A. Lakhonchai, P. Poolcharuansin, P. Chunpeng, C. Rojviriya, K. Thumanu, S. Tunmee, Development of a multihole atmospheric plasma jet for growth rate enhancement of broccoli seeds. Processes **9**, 1134 (2021)
322. S. Srisonphan, Tuning surface wettability through hot carrier initiated impact ionization in cold plasma. ACS Appl. Mater. Interfaces. **10**, 11297–11304 (2018)
323. A. Starek, J. Pawlat, B. Chudzik, M. Kwiatkowski, P. Terebun, A. Sagan, D. Andrejko, Evaluation of selected microbial and physicochemical parameters of fresh tomato juice after cold atmospheric pressure plasma treatment during refrigerated storage. Sci. Rep. **9**, 8407 (2019)
324. A. Starek, A. Sagan, D. Andrejko, B. Chudzik, Z. Kobus, M. Kwiatkowski, P. Terebun, J. Pawlat, Possibility to extend the shelf life of NFC tomato juice using cold atmospheric pressure plasma. Sci. Rep. **10**, 13 (2020)
325. V. Štěpánová, P. Slavíček, J. Kelar, J. Prášil, M. Smékal, M. Stupavská, J. Jurmanová, M. Černák, Atmospheric pressure plasma treatment of agricultural seeds of cucumber (*Cucumis sativus* L.) and pepper (*Capsicum annuum* L.) with effect on reduction of diseases and germination improvement. Plasma Process. Polym. **15**, e1700076 (2018)
326. W.A. Stirk, J.D. Gold, O. Novák, M. Strnad, J.V. Staden, Changes in endogenous cytokinins during germination and seedling establishment of *Tagetes minuta* L. Plant Growth Regul. **47**, 1–7 (2005)

327. W.A. Stirk, O. Novák, E. Žižková, V. Motyka, M. Strnad, J. van Staden, Comparison of endogenous cytokinins and cytokinin oxidase/dehydrogenase activity in germinating and thermoinhibited *Tagetes minuta* achenes. J. Plant Physiol. **169**, 696–703 (2012a)
328. W.A. Stirk, K. Václavíková, O. Novák, S. Gajdošová, O. Kotland, V. Motyka, M. Strnad, J. van Staden, Involvement of cis-zeatin, dihydrozeatin, and aromatic cytokinins in germination and seedling establishment of maize, oats, and lucerne. J. Plant Growth Regul. **31**, 392–405 (2012b)
329. S.D. Stoev, Food safety and increasing hazard of mycotoxin occurrence in foods and feeds. Crit. Rev. Food Sci. Nutr. **53**, 887–901 (2013)
330. T. Stolárik, M. Henselová, M. Martinka, O. Novák, A. Zahoranová, M. Černák, Effect of Low-Temperature Plasma on the Structure of Seeds, Growth and Metabolism of Endogenous Phytohormones in Pea (*Pisum sativum* L.). Plasma Chem. Plasma Process. **35**, 659–676 (2015)
331. E. Stoffels, Y. Sakiyama, D.B. Graves, Cold atmospheric plasma: charged species and their interactions with cells and tissues. IEEE Trans. Plasma Sci. **36**, 1441–1457 (2008)
332. K. Suhem, N. Matan, M. Nisoa, N. Matan, Inhibition of *Aspergillus flavus* on agar media and brown rice cereal bars using cold atmospheric plasma treatment. Int. J. Food Microbiol. **161**, 107–111 (2013)
333. F.S. Sun, X.X. Xie, Y.F. Zhang, M.Y. Ma, Y.Q. Wang, J.W. Duan, X.P. Lu, G.X. Yang, G.Y. He, Wheat gliadin in ethanol solutions treated using cold air plasma at atmospheric pressure. Food Biosci. **39**, 11 (2021)
334. R. Švubová, S. Kyzek, V. Medvecká, Ľ Slováková, E. Gálová, A. Zahoranová, Novel insight at the effect of cold atmospheric pressure plasma on the activity of enzymes essential for the germination of pea (*Pisum sativum* L. cv. Prophet) seeds. Plasma Chem. Plasma Process. **40**, 1221–1240 (2020)
335. R. Švubová, Ľ Slováková, Ľ Holubová, D. Rovňanová, E. Gálová, J. Tomeková, Evaluation of the impact of cold atmospheric pressure plasma on soybean seed germination. Plants **10**, 177 (2021)
336. I. Tamošiūnė, D. Gelvonauskienė, P. Haimi, V. Mildažienė, K. Koga, M. Shiratani, D. Baniulis, Cold plasma treatment of sunflower seeds modulates plant-associated microbiome and stimulates root and lateral organ growth. Front. Plant Sci. **11**, 568924 (2020)
337. Y. Tanakaran, V. Luang-In, K. Matra, Effect of atmospheric pressure multicorona air plasma and plasma-activated water on germination and growth of rat-tailed radish seeds. IEEE Trans. Plasma Sci. **49**, 563–572 (2021)
338. S. Tappi, G. Gozzi, L. Vannini, A. Berardinelli, S. Romani, L. Ragni, P. Rocculi, Cold plasma treatment for fresh-cut melon stabilization. Innov. Food Sci. Emerg. Technol. **33**, 225–233 (2016)
339. L. Ten Bosch, K. Pfohl, G. Avramidis, S. Wieneke, W. Viöl, P. Karlovsky, Plasma-based degradation of mycotoxins produced by *Fusarium, Aspergillus* and *Alternaria* species. Toxins **9**, 97 (2017)
340. P. Terebun, M. Kwiatkowski, A. Starek, S. Reuter, Y.S. Mok, J. Pawłat, Impact of short time atmospheric plasma treatment on onion seeds. Plasma Chem. Plasma Process. **41**, 559–571 (2021)
341. M. Thisawech, O. Saritnum, W. Phakham, K. Prakrajang, S. Sarapirom, Effects of plasma technique and gamma irradiation on seed germination and seedling growth of chili pepper. Chiang Mai J. Sci. **47**, 73–82 (2020)
342. J. Tomeková, S. Kyzek, V. Medvecká, E. Gálová, A. Zahoranová, Influence of cold atmospheric pressure plasma on pea seeds: DNA damage of seedlings and optical diagnostics of plasma. Plasma Chem. Plasma Process. **40**, 1571–1584 (2020)
343. J. Tong, R. He, X. Zhang, R. Zhan, W. Chen, S. Yang, Effects of atmospheric pressure air plasma pretreatment on the seed germination and early growth of *Andrographis paniculata*. Plasma Sci. Technol. **16**, 260–266 (2014)
344. Y. Toyokawa, Y. Yagyu, T. Misawa, A. Sakudo, A new roller conveyer system of non-thermal gas plasma as a potential control measure of plant pathogenic bacteria in primary food production. Food Control **72**, 62–72 (2017)

345. N. Ulbin-Figlewicz, E. Brychcy, A. Jarmoluk, Effect of low-pressure cold plasma on surface microflora of meat and quality attributes. J. Food Sci. Technol. **52**, 1228–1232 (2015)
346. C. Varilla, M. Marcone, G.A. Annor, Potential of cold plasma technology in ensuring the safety of foods and agricultural produce: a review. Foods **9**, 1435 (2020)
347. I. Velichko, I. Gordeev, A. Shelemin, D. Nikitin, J. Brinar, P. Pleskunov, A. Choukourov, K. Pazderů, J. Pulkrábek, Plasma jet and dielectric barrier discharge treatment of wheat seeds. Plasma Chem. Plasma Process. **39**, 913–928 (2019)
348. A.G. Volkov, J.S. Hairston, J. Marshall, A. Bookal, A. Dholichand, D. Patel, Plasma seeds: cold plasma accelerates *Phaseolus vulgaris* seed imbibition, germination, and speed of seedling growth. Plasma Med. **10**, 139–158 (2020)
349. A.G. Volkov, J.S. Hairston, D. Patel, R.P. Gott, K.G. Xu, Cold plasma poration and corrugation of pumpkin seed coats. Bioelectrochemistry **128**, 175–185 (2019)
350. Z. Wan, S.K. Pankaj, C. Mosher, K.M. Keener, Effect of high voltage atmospheric cold plasma on inactivation of *Listeria innocua* on Queso Fresco cheese, cheese model and tryptic soy agar. LWT **102**, 268–275 (2019)
351. Z.F. Wan, N.N. Misra, G. Li, K.M. Keener, High voltage atmospheric cold plasma treatment of *Listeria innocua* and *Escherichia coli* K-12 on Queso Fresco (fresh cheese). LWT-Food Sci. Technol. **146**, 10 (2021)
352. J. Wang, D. Cui, L. Wang, M. Du, Y. Yin, R. Ma, H. Sun, Z. Jiao, Atmospheric pressure plasma treatment induces abscisic acid production, reduces stomatal aperture and improves seedling growth in *Arabidopsis thaliana*. Plant Biol. **23**, 564–573 (2021)
353. J.M. Wang, H. Zhuang, A. Hinton, J.H. Zhang, Influence of in-package cold plasma treatment on microbiological shelf life and appearance of fresh chicken breast fillets. Food Microbiol. **60**, 142–146 (2016)
354. Y. Wang, T. Wang, Y. Yuan, Y. Fan, K. Guo, T. Yue, Inactivation of yeast in apple juice using gas-phase surface discharge plasma treatment with a spray reactor. LWT **97**, 530–536 (2018)
355. N. Wannicke, R. Wagner, J. Stachowiak, T.M.C. Nishime, J. Ehlbeck, K.-D. Weltmann, H. Brust, Efficiency of plasma-processed air for biological decontamination of crop seeds on the premise of unimpaired seed germination. Plasma Process. Polym. **18**, 2000207 (2021)
356. A. Waskow, J. Betschart, D. Butscher, G. Oberbossel, D. Klöti, A. Büttner-Mainik, J. Adamcik, P.R. von Rohr, M. Schuppler, Characterization of efficiency and mechanisms of cold atmospheric pressure plasma decontamination of seeds for sprout production. Front. Microbiol. **9**, 3164 (2018)
357. A. Waskow, D. Butscher, G. Oberbossel, D. Klöti, P.R. von Rohr, A. Büttner-Mainik, D. Drissner, M. Schuppler, Low-energy electron beam has severe impact on seedling development compared to cold atmospheric pressure plasma. Sci. Rep. **11**, 16373 (2021)
358. A. Wiktor, B. Hrycak, M. Jasinski, K. Rybak, M. Kieliszek, K. Krasniewska, D. Witrowa-Rajchert, Impact of atmospheric pressure microwave plasma treatment on quality of selected spices. Appl. Sci. **10**, 6815 (2020)
359. M.Y. Won, S.J. Lee, S.C. Min, Mandarin preservation by microwave-powered cold plasma treatment. Innov. Food Sci. Emerg. Technol. **39**, 25–32 (2017)
360. M. Wu, C. Liu, C. Chiang, Y. Lin, Y. Lin, Y. Chang, J. Wu, Inactivation effect of *Colletotrichum gloeosporioides* by long-lived chemical species using atmospheric-pressure corona plasma-activated water. IEEE Trans. Plasma Sci. **47**, 1100–1104 (2019a)
361. T.-Y. Wu, C.-R. Chang, T.-J. Chang, Y.-J. Chang, Y. Liew, C.-F. Chau, Changes in physicochemical properties of corn starch upon modifications by atmospheric pressure plasma jet. Food Chem. **283**, 46–51 (2019b)
362. Q. Xia, B.D. Green, Z. Zhu, Y. Li, S.M.T. Gharibzahedi, S. Roohinejad, F.J. Barba, Innovative processing techniques for altering the physicochemical properties of wholegrain brown rice (*Oryza sativa* L.)—Opportunities for enhancing food quality and health attributes. Crit. Rev. Food Sci. Nutr. **59**, 3349–3370 (2019)
363. Q. Xiang, X. Liu, S. Liu, Y. Ma, C. Xu, Y. Bai, Effect of plasma-activated water on microbial quality and physicochemical characteristics of mung bean sprouts. Innov. Food Sci. Emerg. Technol. **52**, 49–56 (2019)

364. H. Xu, Y. Zhu, M. Du, Y. Wang, S. Ju, R. Ma, Z. Jiao, Subcellular mechanism of microbial inactivation during water disinfection by cold atmospheric-pressure plasma. Water Res. **188**, 116513 (2021)
365. L. Xu, A.L. Garner, B. Tao, K.M. Keener, Microbial inactivation and quality changes in orange juice treated by high voltage atmospheric cold plasma. Food Bioprocess Technol. **10**, 1778–1791 (2017)
366. Y.Y. Xu, Y. Tian, R.N. Ma, Q.H. Liu, J. Zhang, Effect of plasma activated water on the postharvest quality of button mushrooms, *Agaricus bisporus*. Food Chem. **197**, 436–444 (2016)
367. Z. Xu, C. Cheng, J. Shen, Y. Lan, S. Hu, W. Han, P.K. Chu, In vitro antimicrobial effects and mechanisms of direct current air-liquid discharge plasma on planktonic *Staphylococcus aureus* and *Escherichia coli* in liquids. Bioelectrochemistry **121**, 125–134 (2018)
368. B. Yadav, A.C. Spinelli, B.N. Govindan, Y.Y. Tsui, L.M. McMullen, M.S. Roopesh, Cold plasma treatment of ready-to-eat ham: Influence of process conditions and storage on inactivation of *Listeria innocua*. Food Res. Int. **123**, 276–285 (2019)
369. B. Yadav, A.C. Spinelli, N.N. Misra, Y.Y. Tsui, L.M. McMullen, M.S. Roopesh, Effect of in-package atmospheric cold plasma discharge on microbial safety and quality of ready-to-eat ham in modified atmospheric packaging during storage. J. Food Sci. **85**, 1203–1212 (2020)
370. Y. Yagyu, Y. Hatayama, N. Hayashi, T. Mishima, T. Nishioka, A. Sakudo, T. Ihara, T. Ohshima, H. Kawasaki, Y. Suda, Direct plasma disinfection of green mold spore on citrus by atmospheric pressure dielectric barrier discharge for agricultural applications. Trans. Mater. Res. Soc. Jpn **41**, 127–130 (2016)
371. S. Yawirach, S. Sarapirom, K. Janpong, The effects of dielectric barrier discharge atmospheric air plasma treatment to germination and enhancement growth of sunflower seeds. J. Phys: Conf. Ser. **1380**, 012148 (2019)
372. X.V. Yepez, K.M. Keener, High-voltage atmospheric atmospheric cold plasma (HVACP) hydrogenation of soybean oil without trans-fatty acids. Innov. Food Sci. Emerg. Technol. **38**, 169–174 (2016)
373. S. Yodpitak, S. Mahatheeranont, D. Boonyawan, P. Sookwong, S. Roytrakul, O. Norkaew, Cold plasma treatment to improve germination and enhance the bioactive phytochemical content of germinated brown rice. Food Chem. **289**, 328–339 (2019)
374. H.I. Yong, H.-J. Kim, S. Park, A.U. Alahakoon, K. Kim, W. Choe, C. Jo, Evaluation of pathogen inactivation on sliced cheese induced by encapsulated atmospheric pressure dielectric barrier discharge plasma. Food Microbiol. **46**, 46–50 (2015a)
375. H.I. Yong, H.J. Kim, S. Park, W. Choe, M.W. Oh, C. Jo, Evaluation of the treatment of both sides of raw chicken breasts with an atmospheric pressure plasma jet for the inactivation of *Escherichia coli*. Foodborne Pathog. Dis. **11**, 652–657 (2014)
376. H.I. Yong, H.J. Kim, S. Park, K. Kim, W. Choe, S.J. Yoo, C. Jo, Pathogen inactivation and quality changes in sliced cheddar cheese treated using flexible thin-layer dielectric barrier discharge plasma. Food Res. Int. **69**, 57–63 (2015a)
377. H.I. Yong, H. Lee, S. Park, J. Park, W. Choe, S. Jung, C. Jo, Flexible thin-layer plasma inactivation of bacteria and mold survival in beef jerky packaging and its effects on the meat's physicochemical properties. Meat Sci. **123**, 151–156 (2017)
378. H.I. Yong, S.H. Lee, S.Y. Kim, S. Park, J. Park, W. Choe, C. Jo, Color development, physiochemical properties, and microbiological safety of pork jerky processed with atmospheric pressure plasma. Innov. Food Sci. Emerg. Technol. **53**, 78–84 (2019)
379. A. Zahoranová, M. Henselová, D. Hudecová, B. Kaliňáková, D. Kováčik, V. Medvecká, M. Černák, Effect of cold atmospheric pressure plasma on the wheat seedlings vigor and on the inactivation of microorganisms on the seeds surface. Plasma Chem. Plasma Process. **36**, 397–414 (2016)
380. A. Zahoranová, L. Hoppanová, J. Šimončicová, Z. Tučeková, V. Medvecká, D. Hudecová, B. Kaliňáková, D. Kováčik, M. Černák, Effect of cold atmospheric pressure plasma on maize seeds: enhancement of seedlings growth and surface microorganisms inactivation. Plasma Chem. Plasma Process. **38**, 969–988 (2018)

381. B. Zhang, R. Li, J. Yan, Study on activation and improvement of crop seeds by the application of plasma treating seeds equipment. Arch. Biochem. Biophys. **655**, 37–42 (2018)
382. J.J. Zhang, J.O. Jo, D.L. Huynh, R.K. Mongre, M. Ghosh, A.K. Singh, S.B. Lee, Y.S. Mok, P. Hyuk, D.K. Jeong, Growth-inducing effects of argon plasma on soybean sprouts via the regulation of demethylation levels of energy metabolism-related genes. Sci. Rep. **7**, 41917 (2017)
383. Q.Z. Zhang, Z.Z. Cheng, J.H. Zhang, M.M. Nasiru, Y.B. Wang, L.L. Fu, Atmospheric cold plasma treatment of soybean protein isolate: insights into the structural, physicochemical, and allergenic characteristics. J. Food Sci. **86**, 68–77 (2021)
384. Y. Zhou, Y. Yan, M. Shi, Y. Liu, Effect of an atmospheric pressure plasma jet on the structure and physicochemical properties of waxy and normal maize starch. Polymers **11**, 8 (2019)
385. M. Zhu, S. Dai, Q. Ma, S. Li, Identification of the initial water-site and movement in *Gleditsia sinensis* seeds and its relation to seed coat structure. Plant Methods **17**, 55 (2021)
386. H. Zhuang, M.J. Rothrock Jr., K.L. Hiett, K.C. Lawrence, G.R. Gamble, B.C. Bowker, K.M. Keener, In-package air cold plasma treatment of chicken breast meat: treatment time effect. J. Food Qual. **2019**, 1837351 (2019)
387. D. Ziuzina, N.N. Misra, P.J. Cullen, K.M. Keener, J.P. Mosnier, I. Vilaró, E. Gaston, P. Bourke, Demonstrating the potential of industrial scale in-package atmospheric cold plasma for decontamination of cherry tomatoes. Plasma Med. **6**, 397–412 (2016)
388. D. Ziuzina, S. Patil, P.J. Cullen, K.M. Keener, P. Bourke, Atmospheric cold plasma inactivation of *Escherichia coli*, *Salmonella enterica* serovar Typhimurium and *Listeria monocytogenes* inoculated on fresh produce. Food Microbiol. **42**, 109–116 (2014)
389. S. Živković, N. Puač, Z. Giba, D. Grubišić, Z.L. Petrović, The stimulatory effect of non-equilibrium (low temperature) air plasma pretreatment on light-induced germination of *Paulownia tomentosa* seeds. Seed Sci. Technol. **32**, 693–701 (2004)
390. R. Zukiene, Z. Nauciene, I. Januskaitiene, G. Pauzaite, V. Mildaziene, K. Koga, M. Shiratani, Dielectric barrier discharge plasma treatment-induced changes in sunflower seed germination, phytohormone balance, and seedling growth. Appl. Phys. Express **12**, 126003 (2019)

Chapter 7
Plasma Devices for Cosmetic and Aesthetic Treatment

Ihn Han

7.1 Plasma Devices

Plasma medicine is a relatively new scientific discipline that employs nonthermal atmospheric plasmas for a variety of applications including sterilization, dental care, cosmetics and skin disorders, wound therapy, blood coagulation, and, more recently, cancer treatment. Plasma sources for plasma medicine are commonly operated in air, argon, or helium with little air contaminants. As a result, they produce reactive oxygen and nitrogen species (RONS), such as OH, O_2•, O, NO, H_2O_2, NO_2, NO_3, ONOO, NO_2, ONOO, which are thought to be responsible for plasma's biological effects. Although CAP appears to be a potential anti-cancer drug, direct irradiation of cancer cells or tissue by CAP has significant limitations, including the necessity for a consistent plasma supply and the possibility of a rapid temperature rise during therapy. Furthermore, it is inconvenient for many cancer kinds throughout the body.

There are four types of plasma discharge: floating electrode technique with ground electrode, non-floating electrode method without ground electrode, jet type, and DBD type. A plasma jet is a basic plasma source that uses a hollow cathode construction. The driving frequency is roughly 10 kHz, while the high frequency ranges from 10 to 100 MHz, and microwave plasma jet stations are categorized based on frequency. The DBD technique involves covering both electrodes with a dielectric substance and creating a discharge in the gap between the two electrodes. The discharge at the surface DBD is constructed using two electrodes on a substrate and a high current delivered to the electrodes to create plasma discharge. The floating electrode technique is a discharge method that uses skin as a ground electrode instead of the ground electrode mentioned above.

I. Han (✉)
Plasma Bioscience Research Center, Department of Electrical and Biological Physics, Kwangwoon University, Seoul, Korea
e-mail: hanihn@kw.ac.kr

E. H. Choi (ed.), *Plasma Biosciences and Medicine*, Topics in Applied Physics 148,
https://doi.org/10.1007/978-981-19-7935-4_7

7.2 Opportunities for Plasma Devices in Cosmetics/Aesthetics Applications

Nonthermal atmospheric pressure plasma is an innovative technique that has opened up new research avenues in cancer therapy and other medical sectors such as dental brightening, wound treatments, and skin care. The plasma device produces plasma by combining various gases or by employing a high voltage and current. Ionized gas molecules, electrons, excited atoms, ultraviolet (UV) radiation, electromagnetic fields, and other particles are among the plasma particles released. Plasma devices caused reactive atoms to connect with one another, as well as reactive oxygen and nitrogen species (RONS). Plasma-generated reactive species have physiological features that may encourage therapeutic uses for skin therapy.

Since the 1990s, plasma has been employed as one of the sterilizing procedures for medical devices. Plasma sterilization has the benefit of providing safe and non-toxic dry pasteurization. Following sterilization, the charged particles produce active oxygen via ion reactions while producing water and oxygen. When free oxygen is provided to microorganisms, bacteria, viruses, and fungus can perish effectively. Plasma has recently been shown to have anticancer properties in several tumors such as the brain, breast, prostate, ovaries, and lungs. Several medical gadgets that utilize plasma needles have also been reported.

A technique of processing plasma particles under three-dimensional settings may be constructed with appropriate physical components and mechanical cavity elements in the case of cosmetics and aesthetics. Skin treatment method development is also critical for effective beauty and medicinal applications. It was recently suggested that plasma might be to blame. Plasma-induced RONS production causes skin cell differentiation and proliferation. There is, however, no evidence to support plasma therapy for skin cell differentiation.

The skin's barrier system not only protects against antigens and hazardous chemicals, but it also prevents medications and cosmetics from penetrating the dermis. Several technologies, including the use of chemicals and skin ablation devices, have been developed to improve the absorption capacity of skin for cosmetics. The cost and inconvenience of these measures, on the other hand, underline the need for an unique and safe means of enhancing skin absorption.

The influence of low temperature atmospheric pressure plasma (LTAPP) on drug penetration through the skin and its mode of action [1]. HaCaT human keratinocytes and hairless mice were treated with LTAPP, and cellular and tissue gene expression, as well as morphological alterations, were studied [1]. They discovered that LTAPP exposure decreased E-cadherin expression in skin cells, resulting in the loss of cell-cell connections. LTAPP also inhibited E-cadherin expression and impeded intercellular connection formation inside the tissue, resulting in increased absorption of hydrophilic substances, eosin, and epidermal growth factor. Within 3 h of LTAPP exposure, the drop in E-cadherin expression and epidermal barrier function had entirely restored. These findings suggest that LTAPP can cause a transient reduction in skin barrier function by altering E-cadherin-mediated intercellular contacts, resulting in improved transdermal medication and cosmetic delivery.

The combination of nonthermal atmospheric pressure plasma with 15% CP is more effective for teeth bleaching than traditional light sources. The temperature of the tooth surface was kept about 37 °C, showing that the plasma did not cause any thermal damage to the tooth. The use of plasma had no structural effects on the bleached surface. Nonthermal atmospheric pressure plasma has been shown to be harmless to bleached enamel. As a result, plasma tooth bleaching treatments constitute a cosmetic dentistry technology with several potential practical uses [2].

Foster presented the first commercial plasma system, the Portrait_ PSR, and reviewed early in vivo therapy outcomes. The Portrait_ PSR is essentially a nitrogen-generated radio frequency plasma jet. A high energy therapy (3–4 J) resulted in regulated skin damage. After 10 days, the epidermis had entirely recovered. They also demonstrated continued collagen formation, elastosis decrease, and gradual skin rejuvenation one year following the therapy. Patients experienced a 60% improvement in skin texture, including wrinkle reduction and skin tone enhancement [3].

The application of plasma to the nail surface via a specially constructed prototype enables the longer-lasting nail polish touted by many traditional nail varnish makers in commercials. This innovative, working prototype is simple to use and low-cost; it is also portable due to its battery power. The goal of this paper is to investigate the changes in fingernails caused by plasma utilizing surface analysis methodologies. This article also looks into the benefits of plasma therapy for nail varnish adherence and drying times [4].

Through plasma pulses generated by radiofrequency energy being applied to nitrogen gas, plasma skin regeneration provides energy to the skin. High-energy, single-pass treatments have been shown to produce positive outcomes with a great safety record. A total of three full-face treatments with energy settings ranging from 1.2 to 1.8 J were given to eight individuals every three weeks. The quality of the regenerated epidermis, the length of downtime, and erythema were noted prior to each successive treatment. Six individuals had full-thickness skin biopsy samples taken both before and 90 days after their final treatment. Four days after each treatment, 30 days after the second, and 90 days after the third, patients were seen for follow-up visits.

Researchers discovered a 37% decrease in face rhytides three months following treatment, and trial participants reported a 68% improvement in overall facial attractiveness. In 4 days, re-epithelialization was finished. Patients reported that the erythema persisted for an average of 6 days following therapy. The first treatment's duration of epidermal regeneration was greater than that of the subsequent treatments (9 vs 4 and 5 days, respectively). Following the initial treatment, one patient experienced localized hyperpigmentation, which disappeared by the follow-up appointment on day 30. There was no hypopigmentation or scarring. A histopathological assessment A ring of new collagen at the dermo epidermal junction and less thick elastin in the upper dermis were visible 3 months after therapy. The new collagen was 72.3 m deep on average. With little recovery time, photodamaged facial skin can be successfully treated using plasma skin regeneration and the multiple low-energy treatment technique. Results are similar to one high-energy treatment, but recovery takes less time [5].

Plasma is designed to sculpt the pigments and tones of the skin. This non-invasive form provides high utilization for clinical applications, the behavior of plasma is multi-directional stimulating, uses low levels of energy to directly irritate the skin, while using custom options to explain in a distinct way. This can be compared to existing technologies such as laser irradiation or LED irradiation, or the main differentiation between emerging technologies is the size of the treatment area.

In addition to the use of cosmetics, new physical technologies have recently emerged in the field of skin treatment, and various devices have been proposed for skin care, anti-aging and rejuvenation applications (mechanical cleansing devices, massage devices, ultrasonic waves, light, radio frequency, and cooling). Among these innovations, plasma devices were developed mainly for medical use [6] to treat skin, but recently for cosmetic use. In addition, some small plasma devices can be found mainly in the cosmetics/esthetic market for skin rejuvenation, but so far most of them lack a scientific foundation. As a result, problems arise not only in the user but also in the effectiveness and safety of these devices when managing the skin after repeated procedures.

It is required to specify the region enclosing the cosmetics in order to deal with the possibilities of using cold plasma in cosmetics. It is challenging to precisely draw the line between aesthetics and medical (cortical) (including reconstruction molding and non-surgical procedures). Typically, aesthetics and medicine refer to modifications to the body or its function. Because they simply need to take care of their look without altering their body or function, cosmetics appear to be different from medicine or aesthetics.

7.3 Trends of Plasma Technology in Cosmetics and Marketing

Plasma is designed to sculpt the pigments and tones of the skin. This non-invasive form provides high utilization for clinical applications, the behavior of plasma is multi-directional stimulating, uses low levels of energy to directly irritate the skin, while using custom options to explain in a distinct way. This can be compared to existing technologies such as laser irradiation or LED irradiation, or the main differentiation between emerging technologies is the size of the treatment area.

Besides the use of cosmetic products, new physical technologies have recently emerged in the field of skin treatment and various devices are proposed for skin care, anti-aging and rejuvenation applications (mechanical cleansing devices, massager instruments, ultrasounds, light, radiofrequencies, cool sculpting,). Among these innovations, plasma devices have been developed to treat skin, mainly for medical use [8] but more recently for cosmetics applications.

Moreover, some small plasma devices can now be found on the market for cosmetics/aesthetics applications, mainly for skin rejuvenation, but, up to now, mostly with a lack of scientific fundament. This raises the problem of the effectiveness of such devices in the care of the skin and their safety of use, both for the user but also for the skin after repeated treatments.

To address the possibilities of using cold plasma in cosmetics, it is necessary to define which area covers the cosmetics. The boundaries between medicine (dermatology), aesthetic medicine (including reconstructive and plastic surgery and non-surgical procedures) and cosmetics are difficult to define precisely. Usually, medicine and aesthetic medicine imply a modification of the body or of its functions. Cosmetics seems to be different from medicine and aesthetic medicine as it should not modify the body or its functions but only improve its external aspect.

According to the European Union Cosmetics Directive, a cosmetic "any substance or preparation designed to be applied to the skin, hair, nails, lips, external genital organs, teeth, and mucous membranes of the oral cavity with the sole or primary goals of cleaning, perfuming, altering appearance, reducing body odor, protecting, or maintaining the health of the various external parts of the human body (epidermis, hair system, nails, lips, and external genital organs). When used under typical or fairly anticipated situations, they must not harm human health." (Regulation of the European Community, CE No. 1223/2009; available at: http://eur-lex.europa.eu/legal-content/EN/ALL/?uri=CELEX%3A32009R1223) Cosmetics are defined as "materials intended to be rubbed, poured, sprinkled, or sprayed on, introduced into, or otherwise applied to the human body...for washing, beautifying, enhancing attractiveness, or altering the look" under the US Federal Food, Drug & Cosmetic Act (FD&C Act).

The definition is less clear when it comes to cosmetic gadgets. According to the FDA (https://www.fda.gov/medical-devices/products-and-medical-procedures/cosmetic-devices), cosmetic devices are used in the US "to improve appearance and do not impart any health benefits." The situation is changing in Europe, where the majority of medical and cosmetic/aesthetic devices will be regulated more strictly under one group (EU Regulation 2017/745; available at https://eur-lex.europa.eu/legal-content/EN/TXT/?uri=CELEX%3A32017R0745).

So, it remains to debate whether plasma devices can be used for skin care and whether they can be considered as cosmetic or medical devices according to their effects on skin and the respective regulations of various countries.

In 2017, the world's cosmetics exports scale was about 94.2 billion and especially in France cosmetics exports scale was amounted to 15.2 billion dollar which accounting for 16% of the world's cosmetic export market and taking first place. In 2017, France's Trade Specification Index (TSI) appears at 0.6 and followed by Korea (0.5) and Italy (0.4).

In Korea, skin care equipment is classified managed as beauty device and industrial products. As for medical device items, low-frequency stimulation, ultraviolet irradiator, infrared irradiator, light irradiator, ultra-high frequency stimulation, high-frequency stimulation, ultrasonic stimulation, paraffin bath, treatment steam machine, treatment heater medical treatment, limb circulation, treatment laser irradiator, medicine inhaler, and vibrators for medical treatment etc.

According to the medical equipment, the skin care equipment also needs approval from the Korea Food and Drug Administration (KFDA). Skin care and aesthetic equipment are categorized under one of four medical device classification systems used in Korea, which is most similar to the EU system. Despite efforts to create a market for the sale of plasma cosmetics/aesthetics devices, there is currently no item category or management regulation for atmospheric plasma devices. Cosmetic devices that use plasma technology must be divided into groups based on whether they use high-pressure, atmospheric-pressure, or low-pressure plasma with or without a thermal effect.

It is recommended that the skin care devices in Korea be presented in the "Guideline for Evaluation Equipment for Clinical Treatment for Wound healing Using Plasma" published by KFED. The guidelines include measurements of plasma density (electron density), ozone, nitrogen monoxide, nitrogen dioxide production, OH radical optical emission production, ultraviolet radiation production, photo-biological safety, plasma degree, etc. of plasma equipment.

Floating electrode plasma sources are currently used by the majority of plasma beauty devices in Korea. Human skin serves as the ground electrode in a floating-electrode plasma discharge, and discharge takes place between the electrode of the device and the skin. Based on the non-floating electrode jet plasma whitening machine, this device was created.

Recently it has been reported that plasma devices have functions mainly for wrinkles, acne, wound healing, wound removal, and skin regeneration. This product can helps rejuvenate the skin to promotes collagen formation and can cure the acne and atopy to kills microbes on skin.

Target	Plasma source	Species	Feed gas	References
Skin decontamination	Jet and dielectric barrier discharge (DBD)	Human	Argon	Daeschlein et al. [96]
Wound healing	Jet	Rat	Helium	Nastuta et al. [125]
Wound healing	Jet	Mouse	Argon	Xu et al. [126]
Root canal of tooth	Jet	Human	Argon	Bussiahn et al. [127]
Wound healing	Jet	Human	Argon	Heinlin et al. [128]
Chronic wound	Torch	Human	Argon	Isbary et al. [8]
Wound healing	Jet	Rat	Argon	Salehi et al. [129]

(continued)

(continued)

Target	Plasma source	Species	Feed gas	References
Skin decontamination	Jet	Human	Argon	Daeschlein et al. [98]
Wound healing	Jet	Human	Argon	Isbary et al. [8]
Wound healing	Jet	Mouse	Argon, Helium	García-Alcantara et al. [130]
Sterilization of surface wound	Floating Electrode (FE)-DBD	Mouse Human	Room air	Fridman et al. [131]
Tooth whitening	Micro-jet	Human	Air	Pan et al. [132]
Wound healing	Jet	Mouse	Argon	Schmidt et al. [133]
Bactericidal effect in biofilm of infected wounds	Torch	Rat	Argon	Ermolaeva et al. [134]
Wound healing in diabetes	Jet	Rat	Helium	Fathollah et al. [135]
Wound healing	DBD	Rat	Argon, oxygen mixture	Hung et al. [60]
Acne	DBD	Human	Room air	Chutsirimongkol et al. [136]
Rejuvenation	Jet	Human	Helium	Heinlin et al. [13]
Transdermal drug delivery	DBD	Human	Argon	Shimizu et al. (2015)
Actinic keratoses	Jet	Human	Argon	Wirtz et al. [137]
Transdermal delivery	DBD	Porcine	Room air	Kalghatgi et al. [138]
Skin penetration	Jet	Mouse	Air	Liu et al. [139]

The plasma discharge can be distinguished into a floating electrode method with a ground electrode, a non-floating electrode method without a ground electrode, a jet and a DBD according to the form of the electrode. The plasma jet is a hollow cathode electrode structure. A driving frequency of 10 kHz, a high frequency of 10–100 MHz, and microwave plasma jet stations of GHz are classified by frequency. The DBD method involves discharging in the area between two electrodes that are separated by a dielectric layer. The surface eruption Two electrodes are created on a substrate for the DBD, and after a high electric current is delivered to the electrodes to discharge plasma, a dielectric is created on the surface of the dielectric. The floating electrode method is a discharge technique in which the ground electrode in the non-floating electrode approach previously mentioned is replaced by the skin or a similar material [7].

7.4 Optimization of Plasma Dose for Wound Healing and Cancer Treatment

Several studies investigated the effectiveness of non-thermal bio-compatible plasma (NBP) to heal acute and chronic wounds. However, Initial studies addressed the safety concerns and bacterial load decrease in protracted wounds [8–11]. How can a single application express two opposite characters? "Sola dosis facit venenum" according to Paracelsus in 1538, "means just the dose makes the poison. The case is the same for plasma, the dose determines its outcomes which means a low dose can be beneficial, however, a high dose may cause destruction. Thus, for wound healing low dose is recommended however, a low dose may not have required killing impact while treating cancer. Moreover, the term dose is not uniform or well defined because there are numerous plasma devices with various operating characteristics. Therefore, one of the key issues requiring future clinical studies to pay attention is the optimization of doses in relation to an application or clinical trials, it will help to understand the mechanism of NBP at the molecular level. Although research on NBP mechanisms in cancer treatment and wound healing has progressed well, however, there is still, a lot that needs to be explored. Thus, the molecular mechanism of NBP has been under investigation to find the optimum procedure for clinical practices. Although, the main mechanism of NBP in wound and cancer treatment is explained briefly (Fig. 7.1).

Additionally, reactive oxygen and nitrogen species (RONS), electromagnetic fields, and ultraviolet radiations are all linked to the biological effectiveness of NBP. Chemical reactive oxygen and nitrogen species play a part in enhancing biological applications and are directly related to the ability of wound healing and tumor inhibition. The impact of UV light during NBP administration is most likely minimal. Compared to its natural sun exposure, the UV intensity during NBP treatment for wound healing is significantly reduced [12–15].

7.5 NBP Anti-Cancer Effect During in Vivo and in Vitro Application

NBP's anti-cancer mechanism in vivo treatment is debatable. Phosphate-buffered saline (PBS) after plasma exposure is associated with the production of H_2O_2, NO_2^-, and OH, which are covered by a 1 mm gelatin film [16]. This indicates that ROS diffusion across the skin is possible. Another possibility is to control tumors by NBP through immune response activation [17–19]. The uniform nanosecond pulsed DBD (nspDBD) has been shown to activate macrophages and improve healing in artificial wounds [19]. Thus, nspDBD is shown to augment the anti-tumor efficacy through both, tumor cell death and activation of macrophage function [18].

In the NBP treatment, a layer of cell culture media was used in melanoma cells in in vitro studies [20–22]. The natural phenomena of metabolism engage in the

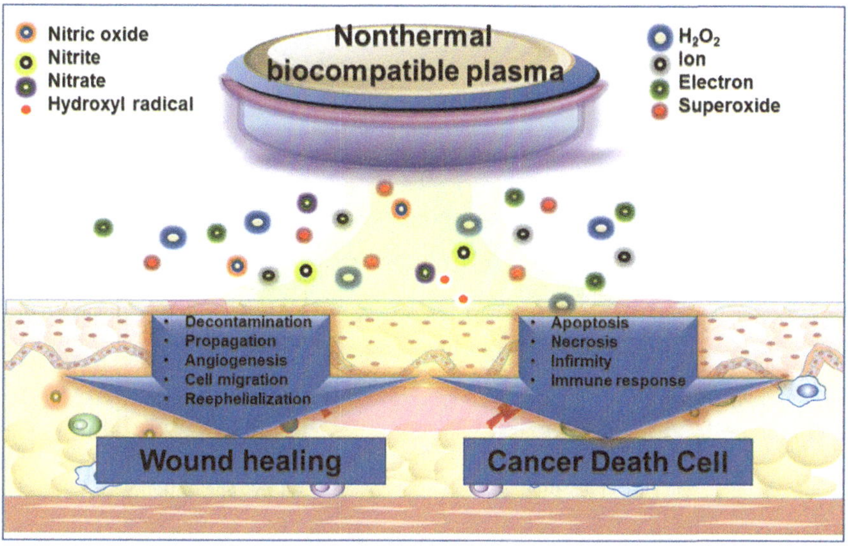

Fig. 7.1 Non-thermal bio-compatible plasma (NBP) for the treatment of cancer and wounds treatment. Low or optimal NBP therapy promotes tissue growth, wound healing, and purification. Reactive oxygen and nitrogen species acting separately or combined are thought to have an impact on the reported results. Long-term cancer therapy increases oxidative stress, which causes cancer cells to die as a result

generation of long-lived and short-lived reactive species which is congruent to the production of RONS during NBP. The role of RONS has already been reported in the propagation of physiological and pathological mechanisms for example wound regeneration, cell propagation, apoptosis, and immunogenic responsive element [23]. A schematic example of the interaction of NBP with cells in vitro and in vivo is provided illustration is shown in (Fig. 7.2).

7.6 Human Skin Anatomy

The human body consists of an external layer, skin that collects approximately, 1/3 of circulating blood. On the basis of its morphological and physiological features categorized for instance protective, homeostatic, or thermo and osmoregulation, etc. The epidermis, dermis, and hypodermis are the three primary integuments that makeup skin as shown in Fig. 7.3 [24, 25].

Epidermis

Depending on the number of layers and cell size, the thickness of the epidermis varies from one person to the next, as well as from one region to the next. The epidermis is

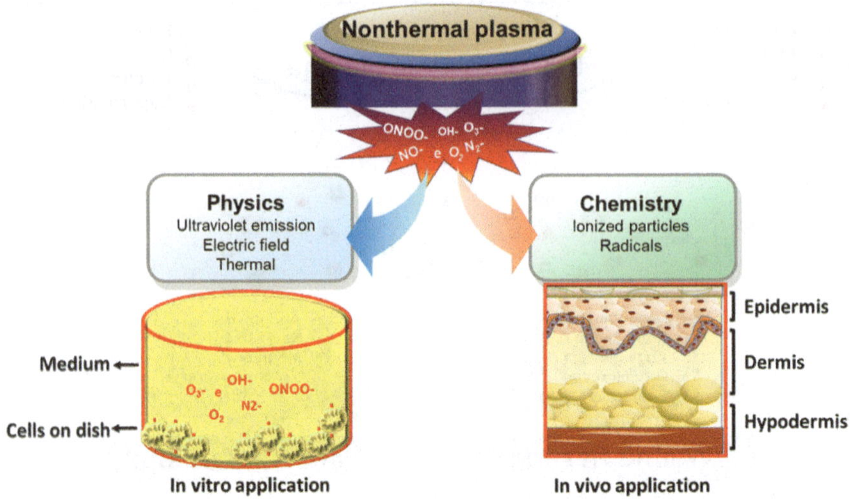

Fig. 7.2 NBP and cell interaction in vitro and in vivo are schematically illustrated. The primary causes of cancer cells dying in vitro have been identified as reactive oxygen and nitrogen species in the culture media. Furthermore, it remains a perplexing mystery in plasma medicine how tumor tissues internal growth might be prevented by the administration of NBP above the skin

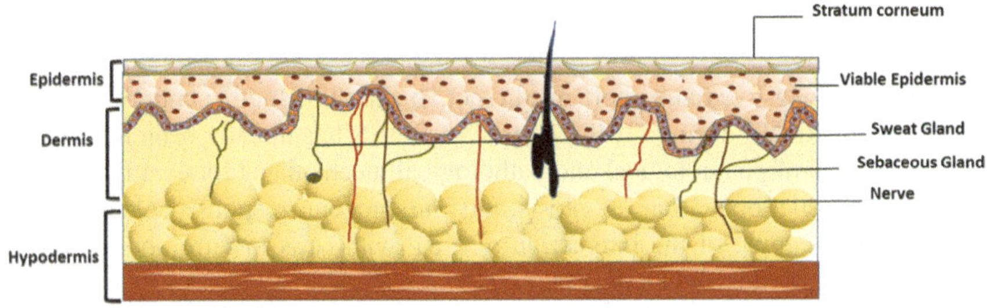

Fig. 7.3 Human skin anatomy. Structure of skin layer which has three layers, outermost is epidermis, middle layer dermis and inner layer hypodermis also called subcutaneous layer

comprised of numerous layered and distinguished into sections: (i) non-viability of the epidermis and (ii) viability of the epidermis.

Non-viability of epidermis

Non-viability of the epidermis of human skin also known as the horny layer and increases many times in thickness when hydrated which is roughly 10 mm thick in dry conditions [26, 27]. The non–viable epidermis consists of Corneocytes, keratinized cells organized in numerous lipid bilayer structures. In a structure where corneocytes are represented as bricks and bilipid layers are thought to be mortar used for the perception of drugs work as an amalgamation hurdle with its specific ‘brick and mortar, just as it is like a wall in between [28, 29].

Viability of epidermis

Underneath the SC, the viability of the epidermis is divided into four layers, viz., stratum lucidum, stratum granulosum, stratum spinosum, and stratum germinativum [26]. Keratinocytes that are flat, clear, and dead make up stratum lucidum. Two to four keratinocyte cell layers with a 3 m thickness make up the stratum granulosum. The desmosomes group the stratum spinosum also known as the prickle cell layer, which has a thickness of 50–150 m and is composed of 8–10 layers of polygonal keratinocytes. It is located right above the basal cell layer. The innermost layer of the epidermis, known as the stratum basale is made up entirely of basal cells. It aids in hydration retention, but as it ages, it loses this function. [30].

Dermis

The 3–4 mm thick dermis is a matrix of connective tissues, blood vessels, sweat glands, hair follicles, lymph vessels, and nerves in a matrix, its thickness of 3–4 mm. Collagen and elastic fibers are connective tissues that provide skin strength and flexibility, respectively [30].

Hypodermis/subcutaneous layer

The dermis and epidermis are supported by the larger blood and lymph vessels in the hypodermis, which also function as a zone for storing fat. It contributes to mechanical strength and controls body temperature. Additionally, this layer is made up of blood vessels, skin nerves, and pressure- sensitive organs [28].

7.7 Methods of Enhancing Skin Permeability

Various methods are used to control the SCs barrier-like nature. There are divided into passive/chemical or active/physical approaches [31, 32]. The following is a brief demonstration of the many methodologies and ways based on their principles and mechanisms of action.

Passive/chemical strategies

Penetration enhancers, supersaturated systems, prodrugs, liposomes, and other vesicles are all products of passive methods. Chemical penetration enhancers could serve through one or greater of the subsequent mechanisms [33]: (i) drug operation across the intercellular pathway is increased by interfering with the SCs highly ordered bilipid structure (e.g. terpenes and azones), (ii) Through an intracellular pathway, interacting with protein shape of corneocytes (e.g. pyrrolidones, dimethylformamide, dimethyl sulphoxide) (iii) Improving the distribution of the solute throughout the SC (e.g. propylene glycol, polyethylene glycol) [30].

Active/physical techniques

The active physical techniques mainly describe the thermal ablation technique. Large MW (>500 Da) hydrophilic molecules, such as proteins and peptides have been attained via active methods. However, Active/assisted strategies are under progress for the active transport of larger biomolecules. Many physical and active methods have been demonstrated across the formation of biomolecules.

Thermal ablation

The thermal ablation procedure, also known as thermophoresis increases the drug perception across the skin and involves the depletion or elimination of the SC [34, 35]. Additionally, the thermal exposure rate should be shorter to sustain a significant skin surface temperature compared to the underneath viable epidermis [36]. It would be attained by two methods: (i) For an extended period of time at a moderate temperature (100 °C), and (ii) during a shorter time at a relatively high temperature (100 °C). The following procedures can be used to achieve thermal ablation: (i) chemical heating based on thermal ablation (ii) Miroporation based on thermal ablation

Chemical heating based on thermal ablation

Chemical methods have been employed to promote medication penetration through the skin. The local body temperature rose as a result of chemical substances. They determined the intensity of heat production. The majority of commercial transdermal patches use two initiators, such as oxygen and water, are used [37, 38]. The patch comprises an iron powder based on heat generating chemical together with 70 mg of lidocaine and tetracaine. An alternative method known as Eutectic Mixture Local Anaesthetics (EMLA®)–based cream, which contained 2.5% of lidocaine and prilocaine. In a double-blind, randomized study with 82 adult human subjects, the EMLA® cream was applied to one antecubital surface prior to a vascular access procedure and the Synera® patch was applied to the other [39].

Miroporation based on thermal ablation

Thermoporation, commonly referred to as microporation, is a method for forming aqueous channels across the SC to improve the permeability of active substances passing through the skin into the systemic circulation. In this method, a variety of metallic filaments are briefly kept in contact with the skin surface. The passage of electric current along these filaments causes them to heat up, resulting in localized disintegration and vaporization of the SC. Resultantly, microchannels formed on the skin's surface. Afterward, the use of transdermal formulations such as gels, creams, patches, or vaccinations will increase the permeability of the medications that have been included [40, 41]. Due to the use of sterile and disposable metal filaments, the microporation device offers the advantage of reducing the danger of the transfer of blood-borne pathogens [42].

Thermoporation is a technique that uses controlled thermal radiation to increase drug absorption through the skin. passport and Tixel, two FDA-approved devices, can create a corporation. With their patented patch method, known as PassPort™,

Altea Therapeutics Corporation has made an significant advancement in the delivery of drugs and vaccine via the skin (Altea Therapeutics Corp., Atlanta, GA). Aqueous micropores are used for the ablation of the SC by heat. The micropore/microchannels are reported to have a width of 50–200 μm and a depth of 30–50 μm. This technology permits the non-invasive, economical, and regulated delivery of drugs of numerous therapeutic types. This method has the advantage of avoiding the usage of needles, pumps, and costly devices which are employed in other techniques [36, 43]. Furthermore, the patient's application of the patch can be recorded by this device along with the date and time [44].

7.8 Skin and Its Microenvironment

Skin is made up of numerous layers of defense. It has a built-in defense system. Hair and hair follicles, as well as a sebaceous gland and a sweat gland, are distinctive skin appendages. It is possible to experience external stimuli due to some sensory organs, such as the Pacini corpuscle. The layers of epidermis are made up of keratinocyte cells. The palms of the hands and the soles of the feet have specific types of skin layers called stratum corneum and stratum lucidum. The lamina basalis separates the epidermis from the dermis. This epidermis also contains a variety of other cell types, including Merkel, Langerhans, and melanocytes. The extracellular matrix fibers and a small number of cell types, including fibroblasts and mast cells, make up the vascularized dermis layer that regulates skin moisture levels. The skin's stratum corneum, which has an acidic surface layer, and the accompanying bacteria that live there and in hair follicles. For the transportation of oxygenation by oxygen coming on one side from the atmosphere and on the other side from dermis blood vessels, the skin layer maintains an oxygen gradient [45] (Fig. 7.4).

7.9 Chitosan Biocompatible Material as Skin Rejuvenation

Chitin is converted into chitosan, a natural, new polyheterosaccharide copolymer, using an alkaline deacetylation process. It has been demonstrated that chitosan and its derivatives are efficient sources for boosting musical and transmucosal deliveries [46]. Chitosan's biological characteristics, such as biocompatibility, non-toxicity, and biodegradability, have opened up new possibilities for the treatment of skin conditions and bone regeneration. The improved biomaterial properties of chitosan, such as mucosal adherence and absorption, are mostly due to its surface chemistry. Chitosan's positive charge and ability to adhere to various epithelial surfaces open up new possibilities for medication interactions with mucus layer. Future skin treatments might derive from these features. Due to its stimulation of osteoblast cell proliferation and attachment as well as the production of mineralized bone matrix, chitosan is used as a bone scaffold material and may open up new therapy options for skin in

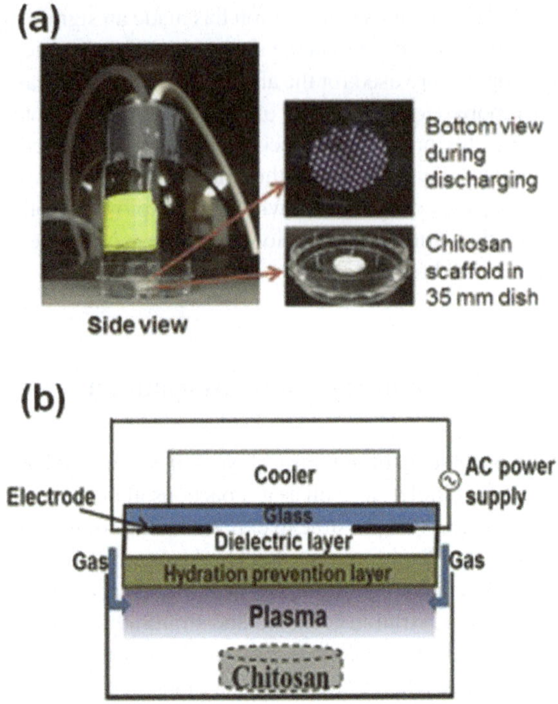

Fig. 7.4 **a** Experimental setup of NBP device for chitosan treatment. **b** Diagrammatic representation of a dielectric barrier discharge (DBD) plasma generator

the future [47, 48]. A promising method to improve affectivity toward cells, NBP treatment of 3D chitosan scaffolds have the potential to augment biological effects. The 3D chitosan scaffolds treated with NBP, however, primarily have a sporadic effect [49–52] (Fig. 7.5).

7.10 Skin Treatment by Using Nonthermal Plasma

Reactive species production plays major role inside body by metabolic activity. For maintained of metabolic activity inside cells it is necessary to absorb reactive species after its generation because it causes many side effects while it interrupts in other normal cells function. So, in our body there are some antioxidant enzymes are present to absorb these free radicals like catalase. Therefore, stem cell self-renewal and differentiation are greatly influenced by the balance of intracellular reduction-oxidation (redox) homeostasis [54–56]. The most effective method for cell rejuvenation is the introduction of non-thermal biocompatible plasma. Plasma contains reactive species that are crucial for activating certain pathways that are highly beneficial for cells [57, 58]. Plasma produces reactive oxygen and reactive nitrogen species, which could indicate a crucial role for the second a participant in the cell's active antioxidant system and signaling network. These reactive species are important for both treating

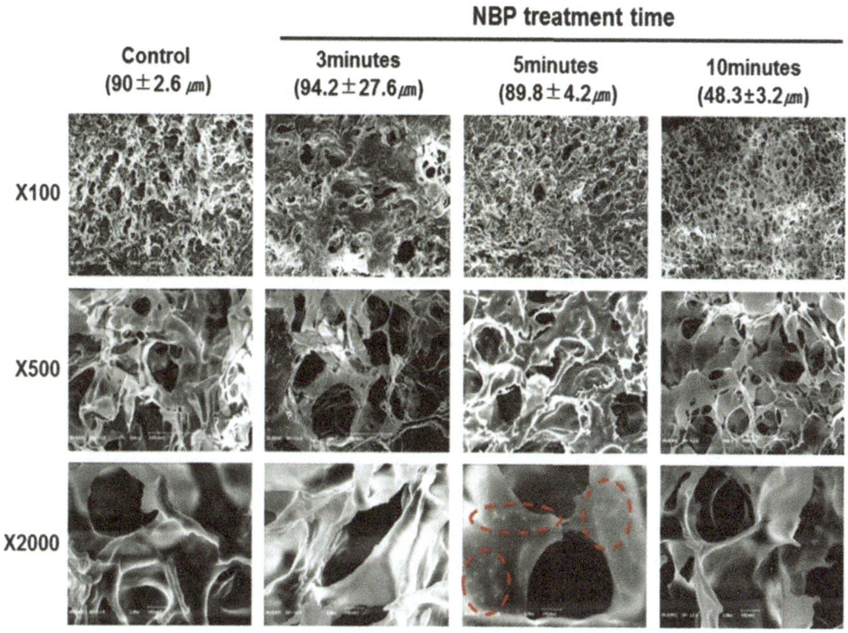

Fig. 7.5 Chitosan morphology was examined using scanning electron microscopy (SEM) following plasma treatment [53]. Chitosan pore size demonstrates and gives significance enhancement of the cell's attachment on scaffold

and rejuvenating the skin. The current understanding of the processes by which NBP regulates cell proliferation and differentiation through redox change is provided in this review [58]. Understanding the role of redox homeostasis in stem cell differentiation control and carefully elaborating the underlying molecular mechanisms would offer important new approaches that NBP stimulated stem cell differentiation for skin treatment [59, 60]. Non-thermal plasma also has certain antibacterial properties that can be used as a therapeutic technique for the management of skin diseases and chronic wounds. This study [61, 62] evaluated the effects of plasma on the healing of a rat model of a full-thickness acute skin wound. Skin wounds from a study were analyzed by histological and gene expression analyses three, seven, and fourteen days after the wounding. The wounds were exposed to three daily plasma treatments for one or two minutes [63]. When compared to untreated wounds, plasma therapy effectively increased epithelization and wound contraction on day 7. In conclusion, plasma treatment improved acute skin wound healing effectiveness while minimizing side effects. Since it is the largest organ in the human body and occupies a unique position, the skin plays a significant role in many different ways. It acts as a bridge between the interior of the organism and the external world [64, 65]. The entire body is protected from external aggressions and significant water loss by its keratinized tegument [66–68]. The skin serves as a significant physical barrier against pathogens and the environment, but it also produces vitamin D, regulates temperature, senses humidity and mechanically, absorbs, excretes, and secretes molecules, among other things [69, 70]. Maintaining its integrity is crucial to stop the loss of function [71–73]. Ancient civilizations used a variety of techniques to conduct cosmetic skin care. With an increase in life expectancy in the 21st century, people are putting more emphasis on skin care to seem younger [74, 75]. Because of this, there is an increasing need for skin care products on a global scale to meet consumer demand. The global market for cosmetics was worth 508 billion dollars in 2018. By 2025, the market is anticipated to be worth roughly 758 billion US dollars. The variety of modern skin care options includes both physical and chemical treatments. While skin peeling treatments are frequently prescribed by licensed beauticians, creams, serum, and oils are frequently utilized as DIY home remedies for skin maintenance [76, 77]. Some of the equipment is also available and used commercially for skin care. One of them that is frequently used for rejuvenation is lasers and LED lights. By physically removing the outer layers of the skin and triggering skin cell metabolism for a more thorough skin rejuvenation, these light sources interact with skin cells to enhance their ability for renewal. For skin treatment, more intrusive and pricey methods like cosmetic surgery may occasionally be needed. Ionized gases are currently used in revolutionary technology-based physicochemical methods for non-surgical skin treatments [78, 79]. Dermatology has long employed cold atmospheric pressure plasma-mediated skin treatments to promote wound healing. It is demonstrating quicker healing abilities and increased cell renewal capabilities. Today, plasma offers a fresh approach to the beauty industry. Several studies are currently being conducted to apply plasma with various materials, such as microneedles. One method for allowing plasma to easily penetrate the skin is the use of microneedles. As plasma has a variety of reactive properties, these properties play a significant role when skin cells are penetrated [80–82]. However, research is ongoing to identify the precise mechanisms of action of cold plasma effects on skin as well as their scientific underpinnings at the cellular and molecular levels. Plasma offers new directions in the field of study and is a well-known skin-treatment technology [83–85].

7.11 Plasma Activated Water Play Important Role in Skin Rejuvenation

Non-small cell lung cancer (NSCLC) is the most prevalent form of lung cancer, accounting for 85% of all cases. In addition to being extremely difficult to treat, there is currently no advanced approach for treating lung cancer. Nonthermal plasma opens up new therapeutic possibilities in medicine. This study's results have shown that oral administration of plasma-treated water (PTW) is effective in treating NSCLC [86, 87]. Oral administration of this mixture to mice demonstrated no toxicities even at the highest dose of PTW, after a single dose and repeated doses for 28 d in mice [88–90]. Cold plasma in water produces a variety of reactive species. PTW shown promising anticancer effects on chemo-resistant lung cancer cells, according to in vivo investigations. There are several plasma treated liquids which are also useful for medication purposes. Plasma activated water one of the promising tools for skin treatment purpose. As it contains high number of reactive species and it is easy to store water for longer time [91–93]. The PTW anticancer strategy appears to be complex in preventing angiogenesis and proliferation of cancer cells while promoting apoptosis. Oral administration of plasma-activated water may prove to be a promising kind of therapy for skin conditions. As a result, plasma activated liquids may present a unique approach to treating skin conditions and promoting skin cell renewal in the future [94, 95].

7.12 Plasma Skin Regeneration Treatment in the Dermo-Cosmetic Application

There are various skin diseases and the severity of skin diseases can vary from benign, over-disturbing (i.e. minor eczema or ichthyosis), and painful (i.e. infected chronic wounds) to lethal (i.e. malignant melanoma). Mild inflammatory skin diseases are usually treated with topical cremes made up of disinfected substances and steroids. In addition, severe forms of antibiotic treatment are inevitable, whether it is topical application as a component of cream or consistently. In particular, in the case of chronic skin diseases superinfected patients suffer from painful treatments and side effects of medications.

7.12.1 Epithelialized Skin Diseases that Are Highly Contaminated with Germs

7.12.1.1 Atopic Eczema Treatment

Atopic eczema is a highly prevalent form of eczema (3–5% of the population is affected). Rash, inflammation, and dry and itchy skin are some of the signs and symptoms. Patients are normally treated with a moisturizer followed by anti-inflammatory and topical antimicrobial treatments. An anti-inflammatory and topical microbial medication is typically given to patients after a moisturizer. Mertens and his colleagues reported a case study in 2009 about atopic eczema in a patient. In this study, the patient's left arm was exposed to plasma, while the right arm was treated with hydrating cream. They used DBD device with an output of 0.2 W. The plasma treatment time was 1 min per day for 30 days. The energy density per day can be calculated at around 1 J/cm^2.

After 30 days, a decrease in some symptoms such as swelling and redness on the upper arm, which was uncovered by the plasma, can be observed. In addition, the patient reported a reduction in itching from 8 to 3 points. The point scale ranged from

0 to 10, with 10 being the most severe itch. Eczema improved by two points overall (scale of −5 to 0–5 points, indicating complete cure and severe worsening of eczema). No side effects were observed during and after the study [96]. Moreover, agar plates were pressed onto the skin of the patients and the results showed a reduction of 1 log level in the bacterial load of the plasma-treated skin after two days (Staphylococcus aureus).

Daeschlein and others also reported high inactivation rates for plasma treatment using a plasma jet in another study. The plasma jet was made of argon gas and operated at a frequency of 1.5 MHz and voltage supply of 1–5 kV [97]. A decline in Staphylococcus aureus colonies grown on agar media was measured by 2.7 log levels after plasma treatment for 2 min (RF 2.7). The efficiency of plasma inactivation varied among the five different species tested in this study, indicating that plasma doses are appropriate for the efficient eradication of wound harmful microbes. While the lowest reduction factor, 1.9 log steps, was found for Enterococcus faecium, this still represents a pronounced inactivation of bacteria, which seems to be suitable for inactivating nearly all types of realistic microbial contamination, colonization, and also infection.in vivo. Pseudomonas aeruginosa, extended-spectrum beta-lactamases (ESBLs) multidrug resistance pathogens, E.coli, Staphylococcus aureus, Staphylococcus epidermidis, hygienic disinfection, preoperative skin antisepsis, and highly gentamicin-resistant enterococci (HLGR), which offer new perspectives on skin disinfection and wound decontamination, were also shown to be significantly reduced by plasma in vitro [98]. In addition, plasma therapy may be effective in curing dermatological conditions involving parasites, such as dermatitis, as a strong killing effect against Demodex folliculorum has been observed [99].

7.12.2 Wounded Epidermis and Germ-Contaminated Skin Diseases Treatment

7.12.2.1 Chronic Wounds with Plasma Treatment

Many people have wound infections on their feet and legs, which are caused by aortic illnesses (15%), hyperglycemia (5%), and other arteriolar disorders. Venous ulcers affect the elderly in particular, and their treatment consumes a significant percentage of the healthcare budget necessitating the development of cost-effective alternative ulcer remedies [100, 101]. Plasma has the potential to be such an alternative, not only because it can be produced at a low cost, but also because it is easy to handle. In addition, plasma treatment combines several mechanisms of action that are beneficial to the wound healing process including (i) the potent antibacterial impact might reduce the number of bacteria present in the injured areas, preventing the recovery process from being slowed down by invasive colonization [102]; (ii) plasma is known to stimulate the proliferation of endothelial cells [103]; and (iii) plasma treatment leads to a decrease in pH, which would also support the healing process since the hyperacidity of wounds is also a natural response of the body [104].

In vivo tests, Isbary and colleagues investigated the effects of argon plasma on infected wounds [105]. In 2010, they published the results of the first clinical trial using the indirect plasma technology MicroPlaSter. Additionally, for routine wound care, 36 patients with 38 ulcers (mainly venous, traumatic, arterial, and diabetic causes) were treated with plasma for 5 min daily.

Patients acted as their controls in the control areas, which were merely treated with normal wound care and were around 3 cm2 in diameter. There were 291 treatments carried out, with the findings revealing a 34% reduction in bacterial load in the wounds ($p < 10–6$) [106]. A year later, another study by Isbary and co-workers found that plasma applications could help people with the genetic disorder Hailey-Hailey disease [107]. Severe outbreaks of this disease often cause blisters and rashes that often lead to chronic, infected sores when they burst. One patient with Hailey-Hailey was treated using MicroPlaSter, a newer version of the MicroPlaSter device that is more convenient to use due to its smaller size and flexible treatment arm with four joints. The torch was held at a distance of 2 cm from the target region for 5 min. As a result, the plasma treatments significantly improved the healing process in both the patient's right axilla and groin. After plasma treatment a sustained positive effect was also observed; the patient remained symptom-free for several months. In the recent clinical trial on chronic wounds conducted by the same research group, two MicroPlaster devices were evaluated using two minutes of treatment time with the same experimental conditions (except for the period of the therapy session) [108].

A considerable decrease in microbial contamination was observed in the patients who received plasma treatments for the injuries. A large decrease in microbial infections of 40% ($p < 0.016$) was noted in the wounds and a decrease of 23.5% ($p < 0.008$) with the use of MicroPlaSter β. Overall, no side effects occurred in these studies by Isbary, and the applications were also well tolerated by patients. MicroPlaSter α, a major reduction in bacterial load of 40% ($p < 0.016$) and a reduction of 23.5% was observed in wounds. It is safe to suppose that in all of this research on chronic wounds, the plasma and tissues of deeper skin layers, such as keratinocytes in the proliferative basal layer or fibroblasts in the dermis, were given access to wound pathogens. Although not entirely appreciated, the direct microbiological actions by the plasma over eukaryotic cells are of tremendous interest. To utterly make sure that no long-term damage or side effects will occur, possible genotoxic effects of plasma treatment must be investigated genetically.

To this end, a risk analysis is going to be performed for each plasma device (kinpen MED and our DBD device), including the investigation of cell damage at the DNA and cell membrane levels. For this purpose, common genotoxicity and cytotoxicity tests such as the Ames test and various host cell reactivation tests are performed. The Ames test takes a look at may be used to decide the mutagenic capacity of chemical substances (in our case this will be plasma). Different strains of Salmonella typhimurium with a mutation in histidine biosynthesis are used for the test so that the auxotrophic mutants require the addition of histidine for their growth.

Treatment with the mutagenic substance can generate revertants, which can grow on a histidine-free medium.

This method has been successfully used to detect the mutagenicity of various substances metabolized by the cytochrome P450 enzyme system [109]. To replace animal testing two plasmid DNA vector assay systems are planned as methods and the infected tissues renewal tests use gene sequences to quantitatively analyze the recovery of DNA rate of cells [107, 110]. Plasma is used to treat a non-replicating reporter gene plasmid that codes for an enzyme before it is transfected into host cells. The expression level of the reporter gene would be reflected in the plasma treatment causing DNA lesions that are repaired by the host cells.

Consequently, the enzymatic expression would be an indirect indicator of plasma's mutagenic potential. The Plasmid-Shuttle-Vector-Mutagenesis-Assay is another test system that has been successfully used to detect age-related DNA repair capacity in various cells. [111]. This assay is based on a plasmid that has E. coli microbial repressor tRNA genes (supF genes) which serve as a mutation indicator [112]. Plasma will be used to treat the plasmid DNA (pSP189), which will then be transfected into host cells, isolated after a few days, and transformed into bacteria. Light blue or white colonies indicate a mutation in the supF gene, and the number of these colonies indicates the mutation frequency.

The mutant plasmids may also be used for mutation spectra and sequence analysis. Finally, these assays are useful for the characterization of the two different plasma sources as well as the standardization of experimental parameters and criteria for medical applications.

7.12.3 The Effect of Plasma on the Skin Surface

7.12.3.1 CAP Treatment Restores the Physiological pH Barrier

RONS generated by CAP also induced acidification in the target as well as oxidizing and stimulating effects. It is common to observe decreasing the initial pH in moist three-dimensional structures and semi or poorly buffered fluids [113]. It might be explained by the existence of acidic substances in liquids that develop from the progenitor NO•, which results in the production of nitrous (HNO_2) and nitric (HNO_3) acids [114]. The plasma exposure time is proportional to its acidification. The rapid pH decline and seems stable the pH values around 3.5 and 2.5 as a result of the temporary development of the HONO/ONO buffer and the production of nitrous acid [113–115]. Human triglycerides and porcine epidermal sebaceous may both become much more acidic, after exposure to CAP.

The medical experiments using healthy human epidermis confirmed CAP-induced acidification. [116, 117]. Due to its acidifying characteristics, CAP therapy may help in protecting healthy skin. Cold plasma may enhance and improve skin rejuvenation by reducing the pH. Physiological acidification has been demonstrated to boost potency as well as enhance fibroblast growth in chronic wound infections [118].

Physiological values can cause pathologies when skin pH is greater, and strong acidic pH might damage the outer tissues. Skin contact with CAP must be strictly regulated to prevent chemical burns [117]. Chemical peels, also known as chemexfoliation, are used in cosmetics to gently exfoliate the epidermis's outer layer and force skin renewal. To decrease pH and remove the outer layers of the epidermis, organic acids are frequently utilized in this technique. The successful plasma application might provide a comparable non-invasive peeling effect. Furthermore, CAP therapies may also be able to restore the physiological pH barrier and promote the aged skin because the rise of alkaline pH with age weakens the threshold [119].

7.12.3.2 The Plasma Effect Improves Skin Hydration and Acidification

The proper amount of water is required for healthy and functional skin. Strong water-absorbing GAGs like hyaluronic acid keep the dermis hydrated. The epidermal has a relative humidity that ranges from 15 to 30% in the outer skin to 70% in the vital portion. The inner layer can detect humidity in the environment and adjust the metabolic processes [120]. The highly porous chemicals that make up Organic Hydration components and keratinocytes, the dying cells that make up skin barriers called corneocytes retain moisture [121, 122]. Strong adhesion between corneocytes prevents significant moisture loss. In addition, ceramides and other intercellular lipids further provide hydrophilic barriers that prevent dehydration [123]. Skin moisture may be influenced in two different ways by cold plasma therapies. In the beginning, CAP might weaken the layer of the skin and stop flowing the epidermal outer layer. A brief, temporary moisture depletion was noticed inside the human skin surface following plasma treatment. [124]. The goal of plasma skin rejuvenation is to preserve the thermal wounded tissues during the healing process with the non-ablated, dry skin [124]. The skin may attract more water molecules following plasma therapy because CAP can release ions on the surface layer. Within the first few seconds of plasma therapy, the human epithelial tissue becomes much more wettable [124]. The plasma therapy enhances the adherence of nail polish for aesthetic purposes, and fingernail hydrophilicity has also been found to improve [124].

Acknowledgements This research was supported by Basic Science Research Program through the National Research Foundation of Korea (NRF) funded by the Ministry of Education (2021R1A6A1A03038785 and 2020R1I1A1A01073071).

References

1. J.H. Choi et al., Treatment with low-temperature atmospheric pressure plasma enhances cutaneous delivery of epidermal growth factor by regulating E-cadherin-mediated cell junctions. Arch. Dermatol. Res. **306**(7), 635–643 (2014)
2. S.H. Nam et al., Efficacy of nonthermal atmospheric pressure plasma for tooth bleaching. Sci. World J. **2015**, 581731 (2015)

3. R. Tiede et al., Plasma applications: a dermatological view. Contrib. Plasma Phys. **54**(2), 118–130 (2014)
4. C. Kaemling et al., Plasma treatment on finger nails prior to coating with a varnish. Surf. Coat. Technol. **200**(1–4), 668–671 (2005)
5. M.A. Bogle, K.A. Arndt, J.S. Dover, Evaluation of plasma skin regeneration technology in low-energy full-facial rejuvenation. Arch. Dermatol. **143**(2), 168–174 (2007)
6. G. Isbary et al., Cold atmospheric plasma for local infection control and subsequent pain reduction in a patient with chronic post-operative ear infection. New Microb. New Infect **1**(3), 41–43 (2013a)
7. G. Isbary, T. Shimizu, Y.F. Li, W. Stolz, H.M. Thomas, G.E. Morfill, J.L. Zimmermann, Cold atmospheric plasma devices for medical issues. Expert Rev. Med. Devices **10**(3), 367–377 (2013b). https://doi.org/10.1586/erd.13.4
8. G. Isbary, G. Morfill, H. Schmidt, M. Georgi, K. Ramrath, J. Heinlin, S. Karrer, M. Landthaler, T. Shimizu, B. Steffes et al., A first prospective randomized controlled trial to decrease bacterial load using cold atmospheric argon plasma on chronic wounds in patients. Br. J. Dermatol. **163**, 78–82 (2010)
9. G. Isbary, J. Heinlin, T. Shimizu, J. Zimmermann, G. Morfill, H.-U. Schmidt, R. Monetti, B. Steffes, W. Bunk, Y. Li et al., Successful and safe use of 2 min cold atmospheric argon plasma in chronic wounds: results of a randomized controlled trial. Br. J. Dermatol. **167**, 404–410 (2012)
10. M. Klebes, C. Ulrich, F. Kluschke, A. Patzelt, S. Vandersee, H. Richter, A. Bob, J. Von Hutten, J.T. Krediet, A. Kramer et al., Combined antibacterial effects of tissue-tolerable plasma and a modern conventional liquid antiseptic on chronic wound treatment. J. Biophotonics **8**, 382–391 (2015)
11. C. Ulrich, F. Kluschke, A. Patzelt, S. Vandersee, M.V.A. Czaika, H. Richter, A. Bob, J. Von Hutten, C. Painsi, R. Huge et al., Clinical use of cold atmospheric pressure argon plasma in chronic leg ulcers: a pilot study. J. Wound Care **24**, 196–203 (2015)
12. J. Lademann, H. Richter, A. Alborova, D. Humme, A. Patzelt, A. Kramer, K.-D. Weltmann, B. Hartmann, C. Ottomann, J.W. Fluhr et al., Risk assessment of the application of a plasma jet in dermatology. J. Biomed. Opt. **14**, 054025 (2009). [CrossRef] [PubMed]
13. J. Heinlin, G. Isbary, W. Stolz, G. Morfill, M. Landthaler, T. Shimizu, B. Steffes, T. Nosenko, J. Zimmermann, S. Karrer, Plasma applications in medicine with a special focus on dermatology. J. Eur. Acad. Dermatol. Venereol. **25**, 1–11 (2010). [CrossRef] [PubMed]
14. S. Bekeschus, A. Schmidt, K.-D. Weltmann, T. Von Woedtke, The plasma jet kINPen—A powerful tool for wound healing. Clin. Plasma Med. **4**, 19–28 (2016). [CrossRef]
15. N. Gaur, E.J. Szili, J.-S. Oh, S.-H. Hong, A. Michelmore, D.B. Graves, A. Hatta, R.D. Short, Combined effect of protein andoxygen on reactive oxygen and nitrogen species in the plasma treatment of tissue. Appl. Phys. Lett. **107**(10), 103703 (2015)
16. V. Miller, A. Lin, A. Fridman, Why target immune cells for plasma treatment of cancer. Plasma Chem. Plasma Process. **36**(1), 259–268 (2016)
17. A. Lin, B. Truong, A. Pappas, L. Kirifides, A. Oubarri, S. Chen, S. Lin, D. Dobrynin, G. Fridman, A. Fridman, Uniform nanosecond pulsed dielectric barrier discharge plasma enhances anti-tumor effects by induction of immunogenic cell death in tumors and stimulation of macrophages. Plasma Process. Polym. **12**(12), 1392–1399 (2015)
18. V. Miller, A. Lin, G. Fridman, D. Dobrynin, A. Fridman, Plasma stimulation of migration of macrophages. Plasma Process. Polym. **11**(12), 1193–1197 (2014)
19. S.U. Kang, J.H. Cho, J.W. Chang, Y.S. Shin, K.I. Kim, J.K. Park, S.S. Yang, J.S. Lee, E. Moon, K. Lee, C.H. Kim, Nonthermal plasma induces head and neck cancer cell death: the potential involvement of mitogen-activated protein kinasedependent mitochondrial reactive oxygen species. Cell Death Dis. **5**, e1056 (2014)
20. N.K. Kaushik, N. Kaushik, D. Park, E.H. Choi, Altered antioxidant system stimulates dielectric barrier discharge plasma-induced cell death for solid tumor cell treatment. PLoS ONE **9**(7), e103349 (2014)

21. A.R. Gibson, H.O. McCarthy, A.A. Ali, D. O'Connell, W.G. Graham, Interactions of a non-thermal atmospheric pressure plasma effluent with PC-3 prostate cancer cells. Plasma Process. Polym. **11**(12), 1142–1149 (2014)
22. J. Roy, J.-M. Galano, T. Durand, J.-Y. Le Guennec, J.C.-Y. Lee, Physiological role of reactive oxygen species as promoters of natural defenses. FASEB J. **31**, 3729–3745 (2017)
23. T. Senyigit, O. Ozer, Corticosteroids for skin delivery: challenges and new formulation opportunities: IntechOpen (2012). https://doi.org/10.5772/53909
24. L. Latheeshjlal, P. Phanitejaswini, Y. Soujanya, U. Swapna, V. Sarika, G. Moulika, Transdermal drug delivery systems: an overview. Int. J. Pharm. Technol. Res. **3**, 2140–2148 (2011)
25. N.R. Jawale, C.D. Bhangale, M.A. Chaudhari, T.A. Deshmukh, Physical approach to transdermal drug delivery: a review. J. Drug Deliv. Ther. **7**, 28–35 (2017)
26. P. Bala, S. Jathar, S. Kale, K. Pal, Transdermal drug delivery system (TDDS)-a multifaceted approach for drug delivery. J. Pharm. Res. **8**, 1805–1835 (2014)
27. V.B. Kumbhar, P.S. Malpure, Y.M. More, A Review on transdermal drug delivery system. **7**, 1258–1269 (2018)
28. I.A. Aljuffali, C.-F. Lin, J.-Y. Fang, Skin ablation by physical techniques for enhancing dermal/transdermal drug delivery. J. Drug Deliv. Sci. Technol. **24**, 277–287 (2014)
29. R. Parhi, P. Suresh, S. Mondal, P.M. Kumar, Novel penetration enhancers for skin applications: a review. Curr. Drug Deliv. **9**, 219–230 (2012)
30. W.I. Choi, J.H. Lee, J.-Y. Kim, J.-C. Kim, Y.H. Kim, G. Tae, Efficient skin permeation of soluble proteins via flexible and functional nano-carrier. J. Control. Release **157**, 272–278 (2012)
31. M.B. Brown, G.P. Martin, S.A. Jones, F.K. Akomeah, Dermal and transdermal drug delivery systems: current and future prospects. Drug Deliv. **13**, 175–187 (2006)
32. A.C. Williams, B.W. Barry, Terpenes and the lipid-proteinpartitioning theory of skin penetration enhancement. Pharm. Res. **8**, 17–24 (1991)
33. A.Z. Alkilani, M.T.C. McCrudden, R.F. Donnelly, Transdermal drug delivery: innovative pharmaceutical developments based on disruption of the barrier properties of the stratum corneum. Pharm. J. **7**, 438–470 (2015)
34. J.W. Lee, P. Gadiraju, J. Park, M.G. Allen, M.R. Prausnitz, Microsecond thermal ablation of skin for transdermal drug delivery. J Control. Release. **154**, 58–66 (2011)
35. A. Hussain, G.M.K.A. Wahab, M.A.S. ur Rahman, H. Altaf, N. Akhtar, M.I. Qayyum, Potential enhancers for transdermal drug delivery: a review. Int. J. Basic Med. Sci. Pharm. **4**, 19–22 (2014)
36. Y. Shahzad, R. Louw, M. Gerber, J. du Plessis, Breaching the skin barrier through temperature modulations. J. Control. Release **202**, 1–13 (2015)
37. D.G. Wood, M.B. Brown, S.A. Jones, Controlling barrier penetration via exothermic iron oxidation. Int. J. Pharm. **404**, 42–48 (2011)
38. J. Sawyer, S. Febbraro, S. Masud, M.A. Ashburn, J.C. Campbell, Heated lidocaine/tetracaine patch (Synera™, Rapydan™) compared with lidocaine/prilocaine cream (EMLA®) for topical anaesthesia before vascular access. Br. J. Anaesth. **102** (2009)
39. A. Herwadkar, A.K. Banga, Peptide and protein transdermal drug delivery. Drug Discov. Today Technol. **9** (2012)
40. S. Lakshmanan, G.K. Gupta, P. Avci, R. Chandran, M. Sadasivam, A.E.S. Jorge et al., Physical energy for drug delivery; poration, concentration and activation. Adv. Drug Deliv. Rev. **71**, 98–114 (2014)
41. S. Garg, M. Hoelscher, J.A. Belser, C. Wang, L. Jayashankar, Z. Guo et al., Needle-free skin patch delivery of a vaccine for a potentially pandemic influenza virus provides protection against lethal challenge in mice. Clin. Vaccine Immunol. **14**, 92 (2007)
42. A. Ahad, A.A. Al-Saleh, N. Akhtar, A.M. Al-Mohizea, F.I. Al-Jenoobi, Transdermal delivery of antidiabetic drugs: formulation and delivery strategies. Drug Discov. Today **20**, 1217–1227 (2015)

43. Y.R. Patel, S. Damon, PassPort™ apomorphine HCL patch: meeting unmet needs in management of Parkinson's disease Altea therapeutics. https://pdfs.semanticscholar.org/3acc/0f6ae72ced8672215635659035e60b835350.pdf
44. G. Busco, E. Robert, N. Chettouh-Hammas, J.-M. Pouvesle, C. Grillon, The emerging potential of cold atmospheric plasma in skin biology. Free Radic. Biol. Med. **161**, 290–304 (2020)
45. S. Kumar, J. Koh, Physiochemical, optical and biological activity of chitosan-chromone derivative for biomedical applications. Int. J. Mol. Sci. **13**(5) (2012)
46. A.R. Costa-Pinto et al., Osteogenic differentiation of human bone marrow mesenchymal stem cells seeded on melt based chitosan scaffolds for bone tissue engineering applications. Biomacromolecules **10**(8), 2067–2073 (2009)
47. F. Wang, Y.-C. Zhang, H. Zhou, Y.-C. Guo, X.-X. Su, Evaluation of in vitro and in vivo osteogenic differentiation of nano-hydroxyapatite/chitosan/poly(lactide-co-glycolide) scaffolds with human umbilical cord mesenchymal stem cells. J. Biomed. Mater. Res. Part A **102**(3), 760–768 (2014)
48. M.-H. Ho, C.-J. Yao, M.-H. Liao, P.-I. Lin, S.-H. Liu, R.-M. Chen, Chitosan nanofiber scaffold improves bone healing via stimulating trabecular bone production due to upregulation of the Runx2/osteocalcin/alkaline phosphatase signaling pathway. Int. J. Nanomed. **10**, 5941–5954 (2015)
49. I. Han, E.H. Choi, The role of non-thermal atmospheric pressure biocompatible plasma in the differentiation of osteoblastic precursor cells, MC3T3-E1. Oncotarget **8**(22), 36399–36409 (2017)
50. Y. Li, M. Ho Kang, H. Sup Uhm, G. Joon Lee, E. Ha Choi, I. Han, Effects of atmospheric-pressure non-thermal bio-compatible plasma and plasma activated nitric oxide water on cervical cancer cells. Sci. Rep. **7**(1), 45781 (2017)
51. R. Foest, E. Kindel, A. Ohl, M. Stieber, K.-D. Weltmann, Non-thermal atmospheric pressure discharges for surface modification. Plasma Phys. Control. Fusion **47**(12B), B525–B536 (2005)
52. Y. Li, J.H. Kim, E.H. Choi, I. Han, Promotion of osteogenic differentiation by non-thermal biocompatible plasma treated chitosan scaffold. Sci. Rep. **9**(1), 3712 (2019a)
53. Y. Li, E.H. Choi, I. Han, Regulation of redox homeostasis by nonthermal biocompatible plasma discharge in stem cell differentiation. Oxid. Med. Cell. Longev. **2019**, 2318680 (2019b)
54. S. Kubinova et al., Non-thermal air plasma promotes the healing of acute skin wounds in rats. Sci. Rep. **7**, 45183 (2017)
55. X.-F. Wang et al., Potential effect of non-thermal plasma for the inhibition of scar formation: a preliminary report. Sci. Rep. **10**(1), 1064 (2020)
56. B.B.R. Choi, J.H. Choi, J. Ji, K.W. Song, H.J. Lee, G.C. Kim, Increment of growth factors in mouse skin treated with non-thermal plasma. Int. J. Med. Sci. **15**, 1203–1209 (2018)
57. J. Park et al., Non-thermal atmospheric pressure plasma is an excellent tool to activate proliferation in various mesoderm-derived human adult stem cells. Free Radic. Biol. Med. **134**, 374–384 (2019)
58. F. Tan, Y. Fang, L. Zhu, M. Al-Rubeai, Controlling stem cell fate using cold atmospheric plasma. Stem Cell Res. Ther. **11**(1), 368 (2020)
59. H.-J. Kim et al., Non-thermal plasma promotes hair growth by improving the inter-follicular macroenvironment. RSC Adv. **11**(45), 27880–27896 (2021)
60. Y.-W. Hung, L.-T. Lee, Y.-C. Peng, C.-T. Chang, Y.-K. Wong, K.-C. Tung, Effect of a nonthermal-atmospheric pressure plasma jet on wound healing: an animal study. J. Chin. Med. Assoc. **79**(6), 320–328 (2016)
61. A. L. Garner, T.A. Mehlhorn, A review of cold atmospheric pressure plasmas for trauma and acute care. Front. Phys. **9** (2021)
62. G.S. Dijksteel, M.M.W. Ulrich, M. Vlig, A. Sobota, E. Middelkoop, B.K.H.L. Boekema, Safety and bactericidal efficacy of cold atmospheric plasma generated by a flexible surface dielectric barrier discharge device against pseudomonas aeruginosa in vitro and in vivo. Ann. Clin. Microbiol. Antimicrob. **19**(1), 37 (2020)

63. G.K. Menon, A.M. Kligman, Barrier functions of human skin: a holistic view. Skin Pharmacol. Physiol. **22**(4), 178–189 (2009)
64. W.Z. Mostafa, R.A. Hegazy, Vitamin D and the skin: Focus on a complex relationship: a review. J. Adv. Res. **6**(6), 793–804 (2015)
65. J.P. Kuhtz-Buschbeck, W. Andresen, S. Göbel, R. Gilster, C. Stick, Thermoreception and nociception of the skin: a classic paper of Bessou and Perl and analyses of thermal sensitivity during a student laboratory exercise. Adv. Physiol. Educ. **34**(2), 25–34 (2010)
66. D. Filingeri, Neurophysiology of skin thermal sensations. Compr. Physiol. **6**(3), 1429 (2016)
67. D. Filingeri, Humidity sensation, cockroaches, worms, and humans: are common sensory mechanisms for hygrosensation shared across species? J. Neurophysiol. **114**(2), 763–767 (2015)
68. A.A. Romanovsky, Skin temperature: its role in thermoregulation. Acta Physiol. (Oxf) **210**(3), 498–507 (2014)
69. T.S. Poet, J.N. McDougal, Skin absorption and human risk assessment. Chem. Biol. Interact. **140**(1), 19–34 (2002)
70. M. Gallagher, C.J. Wysocki, J.J. Leyden, A.I. Spielman, X. Sun, G. Preti, Analyses of volatile organic compounds from human skin. Br. J. Dermatol. **159**(4), 780–791 (2008)
71. H.J. Hurley, J. Witkowski, Dye clearance and eccrine sweat secretion in human skin. J. Invest. Dermatol. **36**(4), 259–272 (1961)
72. Y. Peng, X. Cui, Y. Liu, Y. Li, J. Liu, B. Cheng, Systematic review focusing on the excretion and protection roles of sweat in the skin. Dermatology **228**(2), 115–120 (2014)
73. E. Guttman-Yassky, L. Zhou, J.G. Krueger, The skin as an immune organ: tolerance versus effector responses and applications to food allergy and hypersensitivity reactions. J. Allergy Clin. Immunol. **144**(2), 362–374 (2019)
74. E. Papadavid, A. Katsambas, Lasers for facial rejuvenation: a review. Int. J. Dermatol. **42**(6), 480–487 (2003)
75. S. Shuster, M.M. Black, E. McVitie, The influence of age and sex on skin thickness, skin collagen and density. Br. J. Dermatol. **93** (1975)
76. J. Sandby-Møller, T. Poulsen, H.C. Wulf, Epidermal thickness at different body sites: relationship to age, gender, pigmentation, blood content, skin type and smoking habits. Acta Derm. Venereol. **83**(6), 410–413 (2003)
77. P.R. Bergstresser, J.R. Taylor, Epidermal 'turnover time'—A new examination. Br. J. Dermatol. **96**(5), 503–509 (1977)
78. C.S. Potten, C. Booth, Keratinocyte stem cells: a commentary. J. Invest. Dermatol. **119**(4), 888–899 (2002)
79. C. Pincelli, A. Marconi, Keratinocyte stem cells: friends and foes. J. Cell. Physiol. **225**(2), 310–315 (2010)
80. I.M. Braverman, The cutaneous microcirculation. J. Investig. Dermatol. Symp. Proc. **5**(1), 3–9 (2000)
81. N.T. Evans, P.F. Naylor, The systemic oxygen supply to the surface of human skin. Respir. Physiol. **3**(1), 21–37 (1967)
82. G. Pizzino et al., Oxidative stress: harms and benefits for human health. Oxid. Med. Cell. Longev. **2017**, 8416763 (2017)
83. P.M. Krien, M. Kermici, Evidence for the existence of a self-regulated enzymatic process within the human stratum corneum-an unexpected role for urocanic acid. J. Invest. Dermatol. **115**(3), 414–420 (2000)
84. J.-P. Hachem, D. Crumrine, J. Fluhr, B.E. Brown, K.R. Feingold, P.M. Elias, pH directly regulates epidermal permeability barrier homeostasis, and stratum corneum integrity/cohesion. J. Invest. Dermatol. **121**(2), 345–353 (2003)
85. C.-H. Song, P. Attri, S.-K. Ku, I. Han, A. Bogaerts, E.H. Choi, Cocktail of reactive species generated by cold atmospheric plasma: oral administration induces non-small cell lung cancer cell death. J. Phys. D. Appl. Phys. **54**(18), 185202 (2021)
86. E.H. Choi, H.S. Uhm, N.K. Kaushik, Plasma bioscience and its application to medicine. AAPPS Bull. **31**(1), 10 (2021)

87. L. Gao, X. Shi, X. Wu, Applications and challenges of low temperature plasma in pharmaceutical field. J. Pharm. Anal. **11**(1), 28–36 (2021)
88. S. Herianto, C.-Y. Hou, C.-M. Lin, H.-L. Chen, Nonthermal plasma-activated water: a comprehensive review of this new tool for enhanced food safety and quality. Compr. Rev. Food Sci. Food Saf. **20**(1), 583–626 (2021)
89. H. Tanaka et al., Non-thermal atmospheric pressure plasma activates lactate in Ringer's solution for anti-tumor effects. Sci. Rep. **6**(1), 36282 (2016)
90. R.P. Guragain et al., Impact of plasma-activated water (PAW) on seed germination of soybean. J. Chem. **2021**, 7517052 (2021)
91. P. Galář et al., Non-thermal pulsed plasma activated water: environmentally friendly way for efficient surface modification of semiconductor nanoparticles. Green Chem. **23**(2), 898–911 (2021)
92. H. Tanaka, M. Hori, Medical applications of non-thermal atmospheric pressure plasma. J. Clin. Biochem. Nutr. **60**(1), 29–32 (2017)
93. S.N. Kutlu, F. Canatan, A. Güle, Plasma activated water for plasma medicine, in *2018 Medical Technologies National Congress (TIPTEKNO)* (2018), pp. 1–4
94. T. Kwon et al., Potential applications of non-thermal plasma in animal husbandry to improve infrastructure. In Vivo (Brooklyn) **33**(4), 999–1010 (2019)
95. G. Fridman, M. Peddinghaus, H. Ayan, A. Fridman, M. Balasubramanian, A. Gutsol, A. Brooks, G. Friedman, Plasma Chem. Plasma P. **26**, 425 (2006)
96. G. Daeschlein, T. von Woedtke, E. Kindel, R. Brandenburg, K.D. Weltmann, M. Juenger, Plasma Process. Polym. **7**, 224–230 (2010)
97. G. Daeschlein, S. Scholz, T. von Woedtke, M. Niggemeier, E. Kindel, R. Brandenburg, K.D. Weltmann
98. G. Daeschlein, S. Scholz, A. Arnold, T. von Woedtke, E. Kindel, M. Niggemeier, K.D. Weltmann, M. Juenger, IEEE Trans. Plasma Sci. **38**, 2969–2973 (2010); M. Juenger, IEEE Trans. Plasma Sci. **39**, 815–821 (2011)
99. C.N. Etufugh, T.J. Phillips, Clin. Dermatol. **25**, 121–130 (2007)
100. S. Emmert, F. Brehmer, H. Haenßle, A. Helmke, N. Mertens, R. Ahmed, D. Simon, D. Wandke, W. Maus-Friedrichs, G. Daeschlein, M.P. Schoen, W. Vioel, Clinic. Plasma Med. **1**, 24–26 (2013)
101. R. Edwards, K.G. Harding, Curr. Opin. Infect. Dis. **17**, 91–96 (2004)
102. D. Dobrynin, G. Fridman, G. Friedman, A. Fridman, New J. Phys. **11**, 115020 (2009)
103. T. Shimizu, B. Steffes, R. Pompl, F. Jamitzky, W. Bunk, K. Ramrath, M. Georgi, W. Stolz, H.U. Schmidt, T. Urayama, S. Fujii, G.E. Morfill, Plasma Process. Polym. **5**, 577–582 (2008)
104. G. Isbary, G. Morfill, H.U. Schmidt, M. Georgi, K. Ramrath, J. Heinlin, S. Karrer, M. Landthaler, T. Shimizu, B. Steffes, W. Bunk, R. Monetti, J.L. Zimmermann, R. Pompl, W. Stolz, Brit. J. Dermatol. **163**, 78–82 (2010)
105. K.M. Thoms, J. Baesecke, B. Emmert, J. Hermann, T. Roedling, P. Laspe, D. Leibeling, L. Truemper, S. Emmert, Scand. J. Clin Lab. Invest. **67**, 580–588 (2007)
106. B. Emmert, J. Buenger, K. Keuch, M. Muller, S. Emmert, E. Hallier, G.A. Westphal, Toxicology **228**, 66–76 (2006)
107. K.M. Thoms, C. Kuschal, E. Oetjen, T. Mori, N. Kobayashi, P. Laspe, L. Boekmann, M.P. Schoen, S. Emmert, Exp. Dermatol. **20**, 232–236 (2010)
108. S.I. Moriwaki, S. Ray, R.E. Tarone, K.H. Kraemer, L. Grossman, Mutat. Res. **364**, 117–123 (1996)
109. C.N. Parris, M.M. Seidman, Gene **117**, 1–5 (1992)
110. G. Busco, A. Valinataj Omran, L. Ridou, J.M. Pouvesle, E. Robert, C. Grillon, Cold atmospheric plasma induced acidification of tissue surface: visualization and quantification using agarose gel models. J. Phys. Appl. Phys. (2019). https://doi.org/10.1088/1361-6463/ab1119
111. J.-L. Brisset, B. Benstaali, D. Moussa, J. Fanmoe, E. Njoyim-Tamungang, Acidity control of plasma-chemical oxidation: applications to dye removal, urban waste abatement and microbial inactivation. Plasma Sour. Sci. Technol. **20**(3), 034021 (2011)

112. B. Benstaali, D. Moussa, A. Addou, J.-L. Brisset, Plasma treatment of aqueous solutes: some chemical properties of a gliding arc in humid air. Eur. Phys. J. AP **4**(2), 171–179 (1998). https://doi.org/10.1051/epjap:1998258
113. T. Borchardt, J. Ernst, A. Helmke, M. Tanyeli, A.F. Schilling, G. Felmerer, W. Viol, Effect of direct cold atmospheric plasma (diCAP) on microcirculation of intact skin in a controlled mechanical environment. Microcirculation **24** (8) (2017). https://doi.org/10.1111/micc.12399
114. K. Heuer, M.A. Hoffmanns, E. Demir, S. Baldus, C.M. Volkmar, M. Rohle, P.C. Fuchs, P. Awakowicz, C.V. Suschek, C. Oplander, The topical use of nonthermal dielectric barrier discharge (DBD): nitric oxide related effects on human skin. Nitric Oxide **44**, 52–60 (2015). https://doi.org/10.1016/j.niox.2014.11.015
115. L.A. Schneider, A. Korber, S. Grabbe, J. Dissemond, Influence of pH on woundhealing: a new perspective for wound-therapy? Arch. Dermatol. Res. **298**(9), 413–420 (2007). https://doi.org/10.1007/s00403-006-0713-x
116. T. Soleymani, J. Lanoue, Z. Rahman, A practical approach to chemical peels: a review of fundamentals and step-by-step algorithmic protocol for treatment. J. Clin. Aesthet. Dermatol. **11**(8), 21–28 (2018)
117. V. Cau, E. Pendaries, P.R. Lhuillier, G. Thompson, H. Serre, M.-C. Takahara, H. Méchin, M. Simon, Lowering relative humidity level increases epidermal protein deimination and drives human filaggrin breakdown. J. Dermatol. Sci. **86**(2), 106–113 (2017). https://doi.org/10.1016/j.jdermsci.2017.02.280
118. A.V. Rawlings, C.R. Harding, Moisturization and skin barrier function. Dermatol. Ther. **17**(s1), 43–48 (2004). https://doi.org/10.1111/j.1396-0296.2004.04S1005.x
119. E.H. Mojumdar, Q.D. Pham, D. Topgaard, E. Sparr, Skin hydration: interplay between molecular dynamics, structure and water uptake in the stratum corneum. Sci. Rep. **7**(1), 15712–15712 (2017). https://doi.org/10.1038/s41598-017-15921-5
120. M.J. Choi, H.I. Maibach, Role of ceramides in barrier function of healthy and diseased skin. Am. J. Clin. Dermatol. **6**(4), 215–223. https://doi.org/10.2165/00128071-200506040-00002
121. J.W. Fluhr, S. Sassning, O. Lademann, M.E. Darvin, S. Schanzer, A. Kramer, H. Richter, W. Sterry, J. Lademann, In vivo skin treatment with tissue-tolerable plasma influences skin physiology and antioxidant profile in human stratum corneum. Exp. Dermatol. **21**(2) (2012)
122. K.W. Foster, R.L. Moy, E.F. Fincher, Advances in plasma skin regeneration. J. Cosmet. Dermatol. **7**(3), 169–179 (2008). https://doi.org/10.1111/j.1473-2165.2008.00385.x
123. D.K. Athanasopoulos, P. Svarnas, A. Gerakis, Cold plasma bullet influence on the water contact angle of human skin surface. J. Electrost. **102**, 103378 (2019). https://doi.org/10.1016/j.elstat.2019.103378
124. C. Kaemling, A. Kaemling, S. Tümmel, W. Viol, Plasma treatment on finger nails prior to coating with a varnish. Surf. Coating. Technol. **200**(1), 668–671 (2005). https://doi.org/10.1016/j.surfcoat.2005.01.065
125. A.V. Nastuta, I. Topala, C. Grigoras, V. Pohoata & G. Popa, Stimulation of wound healing by helium atmospheric pressure plasma treatment. J. Phys. D: Appl. Phys. **44**(10), 105204 (2011). https://doi.org/10.1088/0022-3727/44/10/105204
126. G.M. Xu, X.M. Shi, J.F. Cai, S.L. Chen, P. Li, C.W. Yao, Z.S. Chang, G.J. Zhang, (2015) Dual effects of atmospheric pressure plasma jet on skin wound healing of mice. Wound. Repair. Regeneration. **23**(6), 878–884 (2015). https://doi.org/10.1111/wrr.12364
127. R. Bussiahn, R. Brandenburg, T. Gerling, E. Kindel, H. Lange, N. Lembke, K.D. Weltmann, Th. von Woedtke, T. Kocher, The hairline plasma: an intermittent negative dc-corona discharge at atmospheric pressure for plasma medical applications. Appl. Phys. Lett. **96**(14), 143701 (2010). https://doi.org/10.1063/1.3380811
128. J. Heinlin, J.L. Zimmermann, F. Zeman, W. Bunk, G. Isbary, M. Landthaler, T. Maisch, R. Monetti, G. Morfill, T. Shimizu, J. Steinbauer, W. Stolz, S. Karrer, Randomized placebo-controlled human pilot study of cold atmospheric argon plasma on skin graft donor sites. Wound. Repair. Regeneration. **21**(6), 800–807 (2013). https://doi.org/10.1111/wrr.12078

129. S. Salehi, A. Shokri, M.R. Khani, M. Bigdeli, B. Shokri, Investigating effects of atmospheric-pressure plasma on the process of wound healing. Biointerphases. **10**(2), 029504 (2015). https://doi.org/10.1116/1.4914377
130. E. García-Alcantara, R. López-Callejas, P.R. Morales-Ramírez, R. Peña-Eguiluz, R. Fajardo-Muñoz, A. Mercado-Cabrera, S.R. Barocio, R. Valencia-Alvarado, B.G. Rodríguez-Méndez, A.E. Muñoz-Castro, A. de. la. Piedad-Beneitez, I.A. Rojas-Olmedo, Accelerated mice skin acute wound healing In vivo by combined treatment of argon and helium plasma needle. Arch. Med. Res. **44**(3), 169–177 (2013). S0188440913000465. https://doi.org/10.1016/j.arcmed.2013.02.001
131. G. Fridman, M. Peddinghaus, A. Fridman, M. Balasubramanian, A. Gutsol, G. Friedman, Use of non-thermal atmospheric pressure plasma discharge for coagulation and sterilization of surface wounds. In 17th international Symposium on plasma chemistry. Toronto. 1–2 (2005)
132. J. Pan, et al., A novel method of tooth whitening using cold plasma microjet driven by direct current in atmospheric-pressure air. IEEE Trans. Plasma. Sci. **38**(11), 3143–3151 (2010)
133. A. Schmidt, S. Bekeschus, K. Wende, B. Vollmar, T. von Woedtke, A cold plasma jet accelerates wound healing in a murine model of full-thickness skin wounds. Exp. Dermatol. **26**(2), 156–162 (2017). https://doi.org/10.1111/exd.13156
134. S.A. Ermolaeva, et al., Bactericidal effects of non-thermal argon plasma in vitro, in biofilms and in the animal model of infected wounds. J. Med. Microbiol. **60**(1), 75–83 (2011)
135. S. Fathollah, et al., Investigation on the effects of the atmospheric pressure plasma on wound healing in diabetic rats. Sci. R. **6**(1), 1–9 (2016)
136. C. Chutsirimongkol, D. Boonyawan, N. Polnikorn, W. Techawatthanawisan, T. Kundilokchai, Non-Thermal plasma for acne and aesthetic skin improvement. Plasma. Med. **4**(1-4), 79–88 (2014). https://doi.org/10.1615/PlasmaMed.2014011952
137. M. Wirtz, et al., Actinic keratoses treated with cold atmospheric plasma. J. Eur. Acad. Dermatol. Venereol. **32**(1), e37–e39 (2018)
138. S. Kalghatgi, et al., Transdermal drug delivery using cold plasmas. 22nd International Symposium on Plasma Chemistry. **7** (2015)
139. X. Liu, L. Gan, M. Ma, S. Zhang, J. Liu, H. Chen, D. Liu, X. Lu, A comparative study on the transdermal penetration effect of gaseous and aqueous plasma reactive species. J. Phy. D: Appl. Phy. **51**(7), 075401 (2018). https://doi.org/10.1088/1361-6463/aaa419

Chapter 8
Clinical Studies on Cold Gas Plasma Applications: The Autonomous Patient and Getting Informed Consent for Treatment and Clinical Studies

Hans-Robert Metelmann, Philine Henriette Doberschütz, and Christian Seebauer

8.1 Background

This chapter is spreading the official Clinical Practice Guidelines (Leitlinien) of the Association of the Scientific Medical Societies in Germany (Arbeitsgemeinschaft der Wissenschaftlichen Medizinischen Fachgesellschaften e.V., AWMF) concerning Rational Therapeutic Use of Cold Physical Plasma (Rationaler therapeutischer Einsatz von kaltem physikalischem Plasma), AWMF 007-107, 23/Feb/2022. The intention of the chapter is to utilize for study purposes, especially for patient recruitment, the official template for medical briefing of patients as an obligatory part of an informed consent document.

The patient targeted briefing part is a complete citation of the official guidelines.[1] The footnotes are the new content, targeted scientific information for the doctor to be prepared for the patient consultation. This combination of official guidelines at the hands of a patient and scientific comments at the hands of a doctor is needed to support the recruitment of study patients for clinical research in plasma medicine.

[1] Deutsche Gesellschaft für Mund-, Kiefer- und Gesichtschirurgie (DGMKG) Rationaler therapeutischer Einsatz von kaltem physikalischem Plasma Version 1.0 vom 23. Februar 2022: https://www.awmf.org/leitlinien/detail/ll/053-054.html. Editor Hans-Robert Metelmann. Access 22/08/2022.

H.-R. Metelmann (✉) · C. Seebauer
Department of Oro-Maxillo-Facial and Plastic Surgery, Greifswald University Medicine, Greifswald, Germany
e-mail: metelmann@uni-greifswald.de

P. H. Doberschütz
Department of Orthodontics, Greifswald University Medicine, Greifswald, Germany

E. H. Choi (ed.), *Plasma Biosciences and Medicine*, Topics in Applied Physics 148,
https://doi.org/10.1007/978-981-19-7935-4_8

8.2 Template

Dear Patient,

You are consulting your doctor because of a medical or aesthetical problem, and you will probably participate in a clinical study. The key term plasma medicine has been mentioned. It became obvious to you, that plasma medicine has nothing to do with blood plasma. Now, you are interested to learn about plasma medicine, and why it makes sense to consider it for treating your problem.

The purpose of this document is to make you familiar with the basic clinical principles of plasma medicine. Please read this information carefully. Your doctor will inform you about treatment options with plasma medicine, typical risks and possible consequences, and the details of the medical intervention regarding your case. When you feel adequately informed and expressly wish to undergo plasma medicine treatment, please confirm your consent with your signature.

8.2.1 General Aspects of Plasma Medicine

Colloquially, the term "plasma medicine" often refers to tools that generate physical plasma or to products activated by physical plasma, mainly used for cosmetic purposes and by laypersons. Your doctor, on the other hand, is talking about cold physical atmospheric pressure plasma, abbreviated to cold plasma or CAP, generated by officially approved medical devices, and indicated with particular relevance for the medical therapy of chronic wounds and infected skin.

If you suffer from a severe skin infection or wound that is not healing, you have experienced the heavy burden on your health and well-being. These problems can sometimes be difficult to handle by established therapeutic procedures, calling for innovative treatment like CAP medicine.

A wound by itself is not a disease, and wound healing is just a natural process that does not require a targeted treatment. However, problems may arise

- when open wounds become severely infected by pathogens,
- when wound healing is retarded and the risk of infection is rapidly increasing,
- when wounds cannot heal because of consuming illness and show massive infection,
- when pain requires rapid healing of open wounds or infected skin,
- when general risk prevention requires rapid healing of open wounds or infected skin,
- when wounds and skin infections are health-threatening suppurative focuses, or
- when infected wounds contaminated with certain bacteria are causing smell and odor.

CAP medicine is covering all of these indications.

You might be interested to learn that CAP is ionized gas, generated by physical energy. CAP induces biochemical reactions and releases molecules that interact with human wound cells and with microbial cells, such as infectious bacteria and viruses. CAP therefore accelerates wound healing in two ways: by killing harmful germs at the wound surface (antisepsis) and by promoting the growth of healing cells (tissue regeneration and microcirculation). This double effect is a unique advantage of CAP treatment compared to conventional and established wound care measures.

CAP may look like bluish little flames, but with a temperature not higher than 40 °C, it works on the cells without causing thermal damage, ensuring a painless treatment.

Moreover, CAP application is a touch-free treatment that avoids unpleasant contact of the device with your wound or irritated skin and prevents the risk of unintentionally injuring numb wounds.

8.2.2 Selection of Patients

You have learned that CAP treatment is useful for you, in case you are suffering from problematic wounds or infected skin and mucosa. This includes patients with

- chronic and infected wounds,
- wounds with standstill of healing but without infection,
- skin and mucosa lesions at risk of serious progression,
- non-healing wounds by other reasons,
- skin and mucosa with certain local infections and purulent focuses.

Patients suffering from infective and inflammatory skin and mucosa diseases like herpes zoster, atopic eczema, (oral) lichen planus or acne also benefit from CAP application.

You may also belong to a group of patients considered at risk of poor wound healing, who benefit from CAP treatment as preventive measure. This includes patients

- with wounds that are not closing within 28 days,
- aged 60 years or older,
- after the menopause,
- under systemic steroid medication
- taking medications that inhibit wound healing (e.g. glucocorticoids, immunosuppressants, NSAID) or
- with cancer or a history of previous impaired wound healing.

You see that cold plasma application can be used to support the healing of lesions and acute surgical wounds in cases, where the patient's difficult health-status, biographic condition or medication push the risk of problematic wounds. Accelerating the wound healing can also help to reduce scar formation. Together with the potential to prevent wound infection, CAP treatment is a promising option to control the risk of surgical site infections in the field of plastic surgery and aesthetic medicine.

8.2.3 Choice of Plasma Devices

Your individual medical problem calls for individual treatment, and your doctor will propose and choose the most appropriate cold plasma device for your treatment task. You might be interested to learn that there are two types of medical devices in use, approved by the competent authorities since 2013.

One type is called jet plasma device: CAP is generated by electrical tension within a slim tubular handpiece. The resulting ionized gas is driven out by a propellant gas and looks like a jet flame. This "plasma cocktail" consists of atmospheric air, noble gases (argon, helium) and gas mixtures of the working gases.

The other type of medical device is based upon dielectric barrier discharges (DBD): CAP is generated within an electric field forming between the large surface of a flat handpiece and the surface of the skin. This "plasma cocktail" looks like a carpet and consists of atmospheric air.

Jet plasma devices with plasma flames shaped like the tip of a lancet are very suitable for precise interventional procedures under visual inspection. They are used on wound craters and rugged tissue, on regions with undercut, and for intraoral application. DBD plasma devices with plasma carpets are very convenient for the quick treatment of large and flat wounds and infected skin areas.

Rest assured that your doctor is only using CAP devices with CE certification as medical devices class IIa according to the European Council Directive 93/42/EEC. These devices work with plasma sources that have been extensively examined for their biological and physical properties and have been tested in detailed preclinical and clinical investigations.

8.2.4 Handling of Complications

You might have experienced that standard treatment of wounds and skin infections does not succeed in some cases. This is also true for cold plasma therapy. Even with well proven healing effectiveness of CAP medicine, there are some patients with insufficient treatment results. Especially in chronic wounds, plasma medicine plays an important role—but it is not the only player. Continuous debridement, proper wound dressings, and keeping relevant co-morbidities and current medication under control are important as well.

First CAP medical devices have been approved in 2013 and still there are no known serious side effects or complications of therapy. Any enhanced risk of genotoxic and mutagenic effects of CAP treatment has been excluded by well-established in vitro tests as well as by a long-term animal trial and long-term clinical observations.

In principle, complications in medical procedures are due to the general health and medical condition of the patient. Please help your doctor to identify any risk of complications by carefully reporting your health status and medical history.

8.2.5 Frequently Asked Questions

Dear Patient,

To sum up this information supported by scientific data, we would like to answer some of the frequently asked questions. (The footnotes might provide your doctor with scientific additional background information in case you will ask for more detailed medical consultation.)

1. Might cold plasma application be effective in my case?

Yes, we recommend the application of cold atmospheric pressure plasma for the curative treatment of chronic and infected wounds or prevention of surgical site infections.Randomized clinical studies and reviews have confirmed the effectiveness in decontamination and tissue regeneration even for prevention and in skin diseases caused by multidrug-resistant organisms.[2]

We suggest the palliative treatment of ulcerated, open, anaerobically contaminated tumor metastases with cold atmospheric pressure plasma as a measure of germ reduction to mitigate odor development and pain.[3]

If necessary, the treatment should be supplemented by appropriate wound debridement and by specialist care for relevant comorbidities.

2. How is plasma medicine working?

Medical cold plasma devices generate an ionized gas, visible as a tiny blue light with body temperature. The main active components of this plasma are reactive nitrogen and oxygen species (RNS, ROS), UV radiation and electric fields.[4] The ionized gas directed towards the medical target area will induce proliferation of relevant wound cells, stimulate blood perfusion of the compromised tissue and reduce significantly contamination and infection with pathogens.[5]

---

[2] This recommendation is based upon randomized clinical studies of cold atmospheric pressure plasma for the curative treatment of chronic and infected wounds [12, 68, 69, 89] and current expert consensus of 14 scientific medical societies in Germany actively involved in cold plasma medicine.

[3] This suggestion is based upon several pilot studies, case reports and clinical experience [67, 84].

[4] Certified plasma sources either generate a fine beam plasma (jet concept), or emit a flat, carpet-like plasma (Dielectric Barrier Discharge, DBD) [7, 13, 26–28, 42, 53, 73, 92, 93, 95, 100, 101, 105, 106]. Plasma jets are particularly suitable for precise application of plasma directed under visual control and without touch of the wound or tumor, and for treating deep wound craters, fistulas, and undercuts. DBD-devices are well suited for use on large, flat treatment areas. The composition of cold atmospheric pressure plasma depends on the source design and variables such as room air, humidity, and skin surface.

[5] Plasma devices are approved for treating delayed wound healing and microbially contaminated wound and tumor surfaces, skin, and mucous membranes [12, 15–17, 19, 20, 22, 29, 30, 33, 34, 36,

3. Is plasma medicine safe?

Yes, there are no scientific reports of carcinogenic, genotoxic, or mutagenic effects linked to the application of cold atmospheric pressure plasma.[6] Since plasma treatment is local and limited in time, the risk of side effects associated with the entry of ROS and RNS into the tissue is assumed to be extremely low under normal conditions.

4. Are plasma medical devices approved?

Your doctor is using an approved medical device, belonging to a number of plasma sources with comprehensive physical and biological characterization and detailed preclinical and clinical investigations to prove efficacy.[7]

This statement does not include several other plasma tools on the market that claim to be suitable for "plasma medicine" but have no or very inadequate physical, technical, biological, or clinical references to prove this.[8]

5. How is the risk of local or systemic side effects and complications?

Approved plasma devices are in clinical use since 2013. There are no case observations or clinical studies in the literature that report severe side effects of any kind, including carcinogenesis or genetic damage. Cold atmospheric pressure plasma has no clinically discernible thermal effect because, when applied correctly, it barely exceeds the skin temperature of the target area. Slight local effects have to be considered, such as minor pinprick or irritation related to the tip of the plasma plume when using plasma jets. In very rare cases and unclear connection, a brief and mild redness of the skin following unintended touch might occur.

6. Can cold plasma cause cancer?

In many laboratory and animal experiments, physical plasma was examined for a possible induction of cancer. Although natural damage to the DNA could be shown

38–40, 43, 44, 51, 52, 60, 65, 67–69, 74, 77, 81, 84, 89, 97]. Randomized clinical studies[12, 68, 69, 89] and reviews [6, 55, 88] have confirmed the effectiveness, even for skin diseases caused by multidrug-resistant organisms.

[6] The absence of mutagenic effects on mammalian cells has been demonstrated by means of established standard test methods [5, 11, 61, 107], in a long-term animal study [83], and in long-term clinical observations[66, 82]. The UV exposure associated with the use of cold atmospheric pressure plasma is well below the general limit values for personal and occupational safety [4, 14, 59, 76].

[7] The application for treatment purposes is authorized by CE certification as medical devices class-IIa according to the European Council Directive 93/42/EEC. These devices are approved for the treatment of chronic wounds and pathogen associated skin diseases. The approval is based on a comprehensive physical and biological characterization as well as detailed preclinical and clinical examinations [4, 35, 56, 62, 79, 85, 96, 103].

[8] Advances in clinical plasma medicine and its increasing visibility in the media gave rise to dubious providers who advertise devices and corresponding therapies under the name of plasma medicine. Only certified plasma devices whose effectiveness has been confirmed by scientific studies and expert consensus should be used in clinical plasma medicine.

in some cell experiments, cancer induction could not be demonstrated neither in animal experiments nor in long-term clinical studies.[9]

7. How is the plasma medicine procedure going on?

The treatment is following a basic standardization with some individual adaptations, and many application parameters are specifically dependent on the respective type of plasma source. We suggest delegating the application of cold atmospheric pressure plasma to a qualified nurse if circumstances permit.[10]

The effectiveness of cold plasma in healing of chronic wounds and treatment of infected skin is well documented. However, there are always a couple of patients without positive treatment results for unknown reasons. Plasma medicine plays an important role in wound healing—but it is not the only player. Steady debridement, proper wound dressings, restoration and perfusion of vessels, lymphatic drainage, and keeping relevant co-morbidities under control are important as well. This is especially true for chronic wounds.

8. Is the medical effect well controllable?

In wound healing the medical effect can easily be controlled by measuring the regain of skin cover and the shrinking of the wound surface. On-going photo documentation is important. Documents will include scale and date and follow the very basic requirements of scientific medical photography.

9. Does it hurt?

Some patients experience mild pain and an increased production of wound drainage.

The ozone odor linked to plasma treatment can be unpleasant for some patients, especially when used intraorally. Depending on the treatment region and duration of the individual application, it can be helpful to use a dental suction device and to ventilate the treatment room well.

When applying cold atmospheric pressure plasma to intraoral lesions, sensitive tooth areas can be covered with a cotton swab to alleviate stinging sensations. When used in the periocular region, the eye should be protected by a cover.

---

[9] No serious adverse effects (carcinogenesis or genotoxic and mutagenic effects) associated with the application of cold atmospheric pressure plasma have been reported [3, 5, 9, 11, 18, 31, 32, 41, 47, 54, 57, 58, 61, 66, 82, 83, 98, 107–109].

[10] Prior to application, it can be useful to remove any biofilm from the treatment area. No drying is required since plasma treatment is more effective when moisture-mediated [84, 102]. Due to the largely painless application, local anesthesia or cooling are not necessary during treatment.

Most clinicians have had good experience with an exposure time of 1 min/cm^2. According to the concept of hormesis, shorter applications tend to have a stimulating effect, longer applications tend to inhibit. The therapy plan for wound treatment should include a few applications per week (2–3 x) with a longer break in between (2–3 weeks). A pure antisepsis and decontamination treatment should include several applications in a row (daily for 1 week). The stimulation of tissue regeneration is independent of the antisepsis [89]. Plasma treatment should be supplemented by appropriate wound debridement. and by specialist care for relevant comorbidities. Once the epithelial cover of a wound is closed, the treatment can be completed. No maintenance therapy is necessary. In palliative medicine, the degree of olfactory relief serves as indicator of treatment progress.

10. Will I see a quick medical effect?

Wound healing is never quick. You have to know that it takes stamina by all persons involved and sometimes many weeks of repeated treatment to reach a reasonable result.

11. Can bacteria become resistant when treated by plasma?

One of the significant advantages of plasma medicine compared to other anti-microbial therapies is its effectiveness against multi-resistant skin and wound germs. From the opposite point of view, the development of new resistances when treating germs with plasma has never been described—neither in clinical cases and studies, nor in pre-clinical and basic research.

12. Is there an inhibitory effect on my normal flora?

Jet plasma devices are able to precisely direct the flame to the surface and extension of wounds without significantly touching unaffected skin or normal flora. DBD medical devices with a plasma carpet may have an overlapping field of action affecting skin with normal flora. However, in principle, there are no case reports or pre-clinical and basic research studies mentioning problematic effects on the normal flora in clinical plasma medicine.

13. Could it be done easier? Are there no alternative solutions?

Patients suffering from problematic wounds usually have experience with many alternative but fruitless solutions. The crucial point should therefore not be whether there is a simpler option, but which option is the most effective.

References

1. M. Ashrafi, T. Alonso-Rasgado, M. Baguneid, A. Bayat, The efficacy of electrical stimulation in lower extremity cutaneous wound healing: a systematic review. Experim. Dermatol. **26**, 171–178 (2017)
2. O. Assadian, K.J. Ousey, G. Daeschlein, A. Kramer, C. Parker, J. Tanner, D.J. Leaper, Effects and safety of atmospheric low-temperature plasma on bacterial reduction in chronic wounds and wound size reduction: a systematic review and meta-analysis. Int. Wound J. **16**, 103–111 (2019)
3. S. Becker, J.L. Zimmermann, P. Baumeister, T.F. Brunner, T. Shimizu, Y.F. Li, G.E. Morfill, U. Harréus, C. Welz, Effects of cold atmospheric plasma (CAP) on bacteria and mucosa of the upper aerodigestive tract. Auris Nasus Larynx **46**, 294–301 (2019)
4. S. Bekeschus, A. Schmidt, K.D. Weltmann, T. von Woedtke, The plasma jet kINPen—A powerful tool for wound healing. Clin. Plasma Med. **4**, 19–28 (2016)
5. S. Bekeschus, A. Schmidt, A. Kramer, H.R. Metelmann, F. Adler, T. von Woedtke, F. Niessner, K.D. Weltmann, K. Wende, High throughput image cytometry micronucleus assay to investigate the presence or absence of mutagenic effects of cold physical plasma. Environ. Mol. Mutagen. **59**, 268–277 (2018)
6. A.M. Bernhardt, T. Schlöglhofer, V. Lauenroth, F. Mueller, M. Mueller, A. Schoede, C. Klopsch et al., Prevention and early treatment of driveline infections in ventricular assist device patients—The DESTINE staging proposal and the first standard of care protocol. Crit. Care **56**, 106–112 (2020)

7. T. Bernhardt, M.L. Semmler, M. Schäfer, S. Bekeschus, S. Emmert, L. Boeckmann, Plasma medicine: applications of cold atmospheric pressure plasma in dermatology. Oxid. Med. Cell. Longev. **3873928**, 1–10 (2019)
8. L. Boeckmann, T. Bernhardt, M. Schäfer, M.L. Semmler, M. Kordt, A.C. Waldner, F. Wendt, S. Sagwal, S. Bekeschus, J. Berner, E. Kwiatek, A. Frey, T. Fischer, S. Emmert, Aktuelle Indikationen der Plasmatherapie in der Dermatologie. Hautarzt **71**, 109–113 (2020)
9. D. Boehm, P. Bourke, Safety implications of plasma-induced effects in living cells—A review of in vitro and in vivo findings. Biol Chem **400**, 3–17 (2019)
10. T. Borchardt, J. Ernst, A. Helmke, M. Tanyeli, A.F. Schilling, G. Felmerer, W. Viöl, Effect of direct cold atmospheric plasma (diCAP) on microcirculation of intact skin in a controlled mechanical environment. Microcirculation **24**, e12399 (2017)
11. V. Boxhammer, Y.F. Li, J. Köritzer, T. Shimizu, T. Maisch, H.M. Thomas, J. Schlegel, G.E. Morfill, J.L. Zimmermann, Investigation of the mutagenic potential of cold atmospheric plasma at bactericidal dosages. Mutat. Res. **753**, 23–28 (2013)
12. F. Brehmer, H.A. Haenssle, G. Daeschlein, R. Ahmed, S. Pfeiffer, A. Görlitz, D. Simon, M.P. Schön, D. Wandke, S. Emmert, Alleviation of chronic venous leg ulcers with a hand-held dielectric barrier discharge plasma generator (PlasmaDerm® VU-2010): results of a monocentric, two-armed, open, prospective, randomized and controlled trial (NCT01415622). Eur. Acad. Dermatol. Venereol. **29**, 148–155 (2015)
13. P.J. Bruggeman, F. Iza, R. Brandenburg, Foundations of atmospheric pressure non-equilibrium plasmas. Plasma Sour. Sci. Technol. **26**, 123002 (2017)
14. R. Bussiahn, N. Lembke, R. Gesche, T. von Woedtke, K.D. Weltmann, Plasmaquellen für biomedizinische Applikationen. Hyg. Med. **38**, 212–216 (2013)
15. A. Chuangsuwanich, T. Assadamongkol, D. Boonyawan, The healing effect of low-temperature atmospheric-pressure plasma in pressure ulcer: a randomized controlled trial. Int. J. Low Extrem. Wounds **15**, 313–319 (2016)
16. C. Chutsirimongkol, D. Boonyawan, N. Polnikorn, W. Techawatthanawisan, T. Kundilokchaie, Non-thermal plasma for acne treatment and aesthetic skin improvement. Plasma Med. **4**, 79–88 (2014)
17. C. Chutsirimongkol, D. Boonyawan, N. Polnikorn, W. Techawatthanawisan, T. Kundilokchai, C. Bunsaisup, P. Rummaneethorn, W. Kirdwichai, A. Chuangsuwanich, P. Powthong, Non-thermal atmospheric dielectric barrier discharge plasma, medical application studies in Thailand. Plasma Med. **6**, 429–446 (2016)
18. G. Daeschlein, S. Scholz, R. Ahmed, A. Majumdar, T. von Woedtke, H. Haase, M. Niggemeier, E. Kindel, R. Brandenburg, K.D. Weltmann, M. Jünger, Cold plasma is well-tolerated and does not disturb skin barrier or reduce skin moisture. J. Dtsch. Dermatol. Ges. **10**, 509–515 (2012)
19. G. Daeschlein, S. Scholz, R. Ahmed, T. von Woedtke, H. Haase, M. Niggemeier, E. Kindel, R. Brandenburg, K.D. Weltmann, M. Jünger, Skin decontamination by low-temperature atmospheric pressure plasma jet and dielectric barrier discharge plasma. Hosp. Infect **81**, 177–183 (2012)
20. G. Daeschlein, M. Napp, S. Lutze, A. Arnold, S. von Podewils, D. Guembel, M. Jünger, Haut- und Wunddekontamination bei multiresistenten bakteriellen Erregern durch Koagulation mit kaltem Atmosphärendruck-Plasma. J. German Soc. Dermatol. **13**, 143–149 (2015)
21. X. Dai, K. Bazaka, D.J. Richard, E.W. Thompson, K. Ostrikov, The emerging role of gas plasma in oncotherapy. Trends Biotechnol. **26**, 1183–1198 (2018)
22. C.N. Dang, R. Anwar, G. Thomas, Y.D.M. Prasad, A.J.M. Boulton, R.A. Malik, The Biogun. A novel way of eradicating methicillin-resistant Staphylococcus aureus colonization in diabetic foot ulcers. Diabet. Care **29**, 1176 (2006)
23. A. Dubuc, P. Monsarrat, F. Virard, N. Merbahi, J.P. Sarrette, S. Laurencin-Dalicieux, S. Cousty, Use of cold-atmospheric plasma in oncology: a concise systematic review. Ther. Adv. Med. Oncol. **10**, 1–12 (2018)
24. S. Emmert, F. Brehmer, H. Hänßle, A. Helmke, N. Mertens, R. Ahmed, D. Simon, D. Wandke, W. MausFriedrichs, G. Däschlein, M.P. Schön, W. Viöl, Atmospheric pressure plasma in dermatology: Ulcus treatment and much more. Clin. Plasma Med. **1**, 24–29 (2013)

25. S.K. Emmert, Wundversorgung mit kaltem atmosphärischen Plasma Beispiele und Handlungsanweisungen aus der klinischen Praxis (Springer, Berlin/Heidelberg, Germany, 2019), pp. 1–93
26. S. Emmert, L. Boeckmann, T. Fischer, T. Bernhardt, T. Borchardt, W. Viöl W, P. Wahl P, D. Wandke, H.R. Metelmann, K. Masur, S. Bekeschus, T. von Woedtke, K.D. Weltmann, Plasmamedizin für Hauterkrankungen: Wunden und Tumoren, in *Jubiläumsausgabe AKADP 2019, MEOX Projektmanagement GbR*, ed. by K. Horn (Jena, 2019a), pp. 213–225
27. S. Emmert, T. Fischer, T. Bernhardt, T. Borchardt, W. Viöl, P. Wahl, D. Wandke, H.R. Metelmann, K. Masur, S. Bekeschus, T. von Woedtke, K.D. Weltmann, L. Boeckmann, Plasmamedizin für chronische und akute Wunden: sicher und effektiv. Chir. Allgem. Z **20**, 521–526 (2019b)
28. A. Fridman, A. Chirokov, A. Gutsol, Non-thermal atmospheric pressure discharges. J. Phys. D: Appl. Phys. **38**, R1–R24 (2005)
78. T.A. Fuchsluger, Argon cold plasma—A novel tool to treat therapy-resistant corneal infections. Am. J. Ophthalmol. **190**, 150–163 (2018)
29. B. González-Mendoza, R. López-Callejas, B.G. Rodríguez-Méndez, R. Peña Eguiluz, A. Mercado-Cabrera, R. Valencia-Alvarado, M. Betancourt-Ángeles, R.-F. de Lourdes, D. Reboyo-Barrios, E. Chávez-Aguilar, Healing of wounds in lower extremities employing a non-thermal plasma. Clin. Plasma Med. **16**, 100094 (2019)
30. S. Hartwig, C. Doll, J.O. Voss, M. Hertel, S. Preissner, J.D. Raguse, Treatment of wound healing disorders of radial forearm free flap donor sites using cold atmospheric plasma: a proof of concept. J. Oral Maxillofac. Surg. **75**, 429–435 (2017)
31. S. Hasse, O. Hahn, S. Kindler, T. von Woedtke, H.R. Metelmann, K. Masur, Atmospheric pressure plasma jet application on human oral mucosa modulates tissue regeneration. Plasma Med. **4**, 117–129
32. S. Hasse, T. Tran, O. Hahn, S. Kindler, H.R. Metelmann, T. von Woedtke, K. Masur, Induction of proliferation of basal epidermal keratinocytes by cold atmospheric pressure plasma. Clin. Exp. Dermatol. **41**, 202–209 (2016)
33. J. Heinlin, G. Isbary, W. Stolz, F. Zeman, M. Landthaler, G. Morfill, T. Shimizu, J.L. Zimmermann, S. Karrer, A randomized two-sided placebo-controlled study on the efficacy and safety of atmospheric non-thermal argon plasma for pruritus. JEADV **27**, 324–331 (2013a)
34. J. Heinlin, J.L. Zimmermann, F. Zeman, W. Bunk, G. Isbary, M. Landthaler, T. Maisch, R. Monetti, G. Morfill, T. Shimizu, J. Steinbauer, W. Stolz, S. Karrer, Randomized placebo-controlled human pilot study of cold atmospheric argon plasma on skin graft donor sites. Wound Rep. Regen. **21**, 800807 (2013b)
35. F. Herbst, J. van Schalkwyk, M. Mc Govern, MicroPlaSter and SteriPlas, in *Comprehensive Clinical Plasma Medicine—Cold Physical Plasma for Medical Application*, 1st edn., ed. by H.R. Metelmann, T. von Woedtke, K.D. Weltmann (Springer, Berlin/Heidelberg, Germany, 2018), pp.495–502
36. L. Hilker, T. von Woedtke, K.D. Weltmann, H.G. Wollert, Cold atmospheric plasma: a new tool for the treatment of superficial driveline infections. Eur. J. Cardiothorac. Surg. **51**, 186–187 (2017)
37. L. Hilker, T. von Woedtke, K. Masur, K.D. Weltmann, H.G. Wollert, Kaltplasma-Anwendungen bei Wundinfektionen mit Fremdkörperbeteiligung in der Herzchirurgie. Wundmanagement **12**, 260–267 (2018)
38. G. Isbary, G. Morfill, H.U. Schmidt, M. Georgi, K. Ramrath, J. Heinlin, S. Karrer, M. Landthaler, T. Shimizu, B. Steffes, W. Bunk, R. Monetti, J.L. Zimmermann, R. Pompl, W. Stolz, A first prospective randomized controlled trial to decrease bacterial load using cold atmospheric argon plasma on chronic wounds in patients. Br. J. Dermatol. **163**, 78–82 (2010)
39. G. Isbary, G. Morfill, J. Zimmermann, T. Shimizu, W. Stolz, Cold Atmospheric plasma. A successful treatment of Lesions in Hailey-Hailey disease. Arch. Dermatol. **147**, 388–390 (2011)
40. G. Isbary, J. Heinlin, T. Shimizu, J.L. Zimmermann, G. Morfill, H.U. Schmidt, R. Monetti, B. Steffes, W. Bunk, Y. Li, T. Klaempfl, S. Karrer, M. Landthaler, W. Stolz, Successful and

safe use of 2 min cold atmospheric argon plasma in chronic wounds: results of a randomized controlled trial. Br. J. Dermatol. **167**, 404–410 (2012)
41. G. Isbary, J. Köritzer, A. Mitra, Y.F. Li, T. Shimizu, J. Schroeder, J. Schlegel, G.E. Morfill, W. Stolz, J.L. Zimmermann, Ex vivo human skin experiments for the evaluation of safety of new cold atmospheric plasma devices. Clin. Plasma Med. **1**(2), 36–44 (2013e)
42. G. Isbary, T. Shimizu, Y.F. Li, W. Stolz, H.M. Thomas, G.E. Morfill, J.L. Zimmermann, Cold atmospheric plasma devices for medical issues. Expert Rev. Med. Devices **10**, 367–377 (2013b)
43. G. Isbary, T. Shimizu, J.L. Zimmermann, H.M. Thomas, G.E. Morfill, W. Stolz, Cold atmospheric plasma for local infection control and subsequent pain reduction in a patient with chronic postoperative ear infection. New Microbe New Infect **1**, 41–43 (2013d)
44. G. Isbary, W. Stolz, T. Shimizu, R. Monetti, W. Bunk, H.U. Schmidt, G.E. Morfill, T.G. Klämpfl, B. Steffes, H.M. Thomas, J. Heinlin, S. Karrer, M. Landthaler, J.L. Zimmermann, Cold atmospheric argon plasma treatment may accelerate wound healing in chronic wounds: results of an open retrospective randomized controlled study in vivo. Clin. Plasma Med. **1**, 25–30 (2013c)
45. G. Isbary, J.L. Zimmermann, T. Shimizu, Y.F. Li, G.E. Morfill, H.M. Thomas, B. Steffes, J. Heinlin, S. Karrer, W. Stolz, Non-thermal plasma—More than five years of clinical experience. Clin. Plasma Med. **1**, 19–23 (2013a)
46. M. Izadjoo, Z. Sullivan, H. Kim, J. Skiba, Medical applications of cold atmospheric plasma: state of the science. Wound Care **27**, S4–S10 (2018)
47. L. Jablonowski, T. Kocher, A. Schindler, K. Müller, F. Dombrowski, T. von Woedtke, T. Arnold, A. Lehmann, S. Rupf, M. Evert, K. Evert, Side effects by oral application of atmospheric pressure plasma on the mucosa in mice. PLoS ONE **14**, e0215099 (2019)
48. S. Karrer, S. Arndt, Plasmamedizin in der Dermatologie Wirkmechanismen und Anwendungsmöglichkeiten. Hautarzt **66**, 819–828 (2015)
49. T. Kisch, S. Schleusser, A. Helmke, K.L. Mauss, E.T. Wenzel, B. Hasemann, P. Mailaender, R. Kraemer, The repetitive use of non-thermal dielectric barrier discharge plasma boosts cutaneous microcirculatory effects. Microvasc. Res. **106**, 8–13 (2016a)
50. T. Kisch, A. Helmke, S. Schleusser, J. Song, E. Liodaki, F.H. Stang, P. Mailaender, R. Kraemer, Improvement of cutaneous microcirculation by cold atmospheric plasma (CAP): results of a controlled, prospective cohort study. Microvasc. Res. **104**, 55–62 (2016b)
51. M. Klebes, J. Lademann, S. Philipp, C. Ulrich, A. Patzelt, M. Ulmer, F. Kluschke, A. Kramer, K.D. Weltmann, W. Sterry, B. Lange-Asschenfeldt, Effects of tissue-tolerable plasma on psoriasis vulgaris treatment compared to conventional local treatment: a pilot study. Clin. Plasma Med. **2**, 22–27 (2014)
52. M. Klebes, C. Ulrich, F. Kluschke, A. Patzelt, S. Vandersee, H. Richter, A. Bob, J. von Hutten, J.T. Krediet, A. Kramer, J. Lademann, B. Lange-Asschenfeld, Combined antibacterial effects of tissue-tolerable plasma and a modern conventional liquid antiseptic on chronic wound treatment. Biophotonics **8**, 382–391 (2015)
53. U. Kogelschatz, Atmospheric-pressure plasma technology. Plasma Phys. Control. Fusion **46**, B63–B75 (2004)
54. S. Kos, T. Blagus, M. Cemazar, G. Filipic, G. Sersa, U. Cvelbar, Safety aspects of atmospheric pressure helium plasma jet operation on skin: in vivo study on mouse skin. PLoS ONE **12**, e0174966 (2017)
55. A. Kramer, J. Dissemond, C. Willy, S. Kim, D. Mayer, R. Papke, R. Tuchmann, G. Daeschlein, O. Assadian, Auswahl von Wundantiseptika—Aktualisierung des Expertenkonsensus 2018. Wundmanagement **13**, 5–22 (2019)
56. M. Kuchenbecker, N. Bibinov, A. Kaemling, D. Wandke, P. Awakowicz, W. Viöl, Characterization of DBD plasma source for biomedical applications. J. Phys. D Appl. Phys. **42**, 045212 (2009)
57. J. Lademann, H. Richter, A. Alborova, D. Humme, A. Patzelt, A. Kramer, K.D. Weltmann, B. Hartmann, C. Ottomann, J.W. Fluhr, P. Hinz, G. Hübner, O. Lademann, Risk assessment of the application of a plasma jet in dermatology. Biomed. Opt. **14**, 054025 (2009)

58. J. Lademann, C. Ulrich, A. Patzelt, H. Richter, F. Kluschke, M. Klebes, O. Lademann, A. Kramer, K.D. Weltmann, B. Lange-Asschenfeldt, Risk assessment of the application of tissue-tolerable plasma on human skin. Clin. Plasma Med. **1**, 5–10 (2013)
59. A. Lehmann, F. Pietag, T. Arnold, Human health risk evaluation of a microwave-driven atmospheric plasma jet as medical device. Clin. Plasma Med. **7–8**, 16–23 (2017)
60. Y.F. Li, D. Taylor, J.L. Zimmermann, W. Bunk, R. Monetti, G. Isbary, V. Boxhammer, H.U. Schmidt, T. Shimizu, H.M. Thomas, G.E. Morfill, In vivo skin treatment using two portable plasma devices: comparison of adirect and an indirect cold atmospheric plasma treatment. Clin. Plasma Med. **1**, 35–39 (2013)
61. T. Maisch, A.K. Bosserhoff, P. Unger, J. Heider, T. Shimizu, J.L. Zimmermann, G.E. Morfill, M. Landthaler, S. Karrer, Investigation of toxicity and mutagenicity of cold atmospheric argon plasma. Environ. Mol. Mutagen. **58**, 172–177 (2017)
62. M.S. Mann, R. Tiede, K. Gavenis, G. Daeschlein, R. Bussiahn, K.D. Weltmann, S. Emmert, T. von Woedtke, R. Ahmed, Introduction to DIN-specification 91315 based on the characterization of the plasma jet kINPen® MED. Clin. Plasma Med. **4**, 35–45 (2016)
63. K. Masur, J. Schmidt, E. Stürmer, T. von Woedtke, Kaltes Plasma zur Heilung chronischer Wunden. Wundmanagement **12**, 253–259 (2018)
64. H.R. Metelmann, T. von Woedtke, R. Bussiahn, K.D. Weltmann, M. Rieck, R. Khalili, F. Podmelle, P.D. Waite, Experimental recovery of CO_2-laser skin lesions by plasma stimulation. Am. J. Cosmetic Surg. **29**, 52–56 (2012)
65. H.R. Metelmann, D.S. Nedrelow, C. Seebauer, M. Schuster, T. von Woedtke, K.D. Weltmann, S. Kindler, P.H. Metelmann, S.E. Finkelstein, D.D. Von Hoff, F. Podmelle, Head and neck cancer treatment and physical plasma. Clin. Plasma Med. **3**, 17–23 (2015)
66. H.R. Metelmann, T.T. Vu, H.T. Do, T.N.B. Le, T.H.A. Hoang, T.T.T. Phi, T.M.L. Luong, V.T. Doan, T.T.H. Nguyen, T.H.M. Nguyen, D.Q. Le, T.K.X. Le, T. von Woedtke, R. Bussiahn, K.D. Weltmann, R. Khalili, F. Podmelle, Scar formation of laser skin lesions after cold atmospheric pressure plasma (CAP) treatment: a clinical long term observation. Clin. Plasma Med. **1**, 30–35 (2013)
67. H.R. Metelmann, C. Seebauer, V. Miller, A. Fridman, G. Bauer, D.B. Graves, J.M. Pouvesle, R. Rutkowski, M. Schuster, S. Bekeschus, K. Wende, K. Masur, S. Hasse, T. Gerling, M. Hori, H. Tanaka, E.H. Choi, K.D. Weltmann, P.H. Metelmann, D.D. Von Hoff, T. von Woedtke, Clinical experience with cold plasma in the treatment of locally advanced head and neck cancer. Clin. Plasma Med. **9**, 6–13 (2018)
68. S. Mirpour, S. Fathollah, P. Mansouri, B. Larijani, M. Ghoranneviss, M.M. Therani, M.R. Amini, Cold atmospheric plasma as an effective method to treat diabetic foot ulcers: a randomized clinical trial. Sci. Rep. **10**(1), 10440 (2020)
69. M. Moelleken, F. Jockenhöfer, C. Wiegand, J. Buer, S. Benson, J. Dissemond, Pilot study on the influence of cold atmospheric plasma on bacterial contamination and healing tendency of chronic wounds. Dt. Dermatol. Gesell. **18**(10), 1094–1101 (2020)
70. A. Nishijima, T. Fujimoto, T. Hirata, J. Nishijima, A new energy device for skin activation to acute wound using cold atmospheric pressure plasma: a randomized controlled clinical trial. Biomed. Sci. Tech. Res. **21**, 15494–15501 (2019a)
71. A. Nishijima, T. Fujimoto, T. Hirata, J. Nishijima, Effects of cold atmospheric pressure plasma on accelerating acute wound healing: a comparative study among 4 different treatment groups. Modern Plastic Surg. **9**, 18–31 (2019b)
72. C. Opländer, Physkalisches plasma: science Fiction oder eine neue option in der Wundbehandlung? Wundmanagement **12**, 247–252 (2018)
73. G.Y. Park, S.J. Park, M.Y. Choi, I.G. Koo, J.H. Byun, J.W. Hong, J.Y. Sim, G.J. Collins, J.K. Lee, Atmosphericpressure plasma sources for biomedical applications. Plasma Sour. Sci. Technol. **21**, 043001 (2012)
74. S. Preissner, I. Kastner, E. Schütte, S. Hartwig, A.M. Schmidt-Westhausen, S. Paris, R. Preissner, M. Hertel, Adjuvant antifungal therapy using tissue tolerable plasma on oral mucosa and removable dentures in oral candidiasis patients: a randomised double-blinded split-mouth pilot study. Mycoses **59**, 467–475 (2016)

75. A. Privat-Maldonado, A. Schmidt, A. Lin, K.D. Weltmann, K. Wende, A. Bogaerts, S. Bekeschus, ROS from physical plasmas: redox chemistry for biomedical therapy. Ox Med. Cell Longev. **9062098**, 2019 (2019)
76. P. Rajasekaran, C. Opländer, D. Hoffmeister, N. Bibinov, C.V. Suschek, D. Wandke, P. Awakowicz, Characterization of dielectric barrier discharge (DBD) on mouse and histological evaluation of the plasma-treated tissue. Plasma Process Polym. **8**, 246–255 (2011)
77. Reitberger HH, Czugala M, Chow C, Mohr A, Burkovski A, Gruenert AK, Schoenebeck R
79. S. Reuter, T. von Woedtke, K.D. Weltmann, The kINPen—A review on physics and chemistry of the atmospheric pressure plasma jet and its applications. J. Phys. D Appl. Phys. **1**, 233001 (2018)
80. H. Rotering, M. Al Shakaki, H. Welp, A.M. Dell'Aquila, Preliminary results of a new treatment strategy for relapsed left ventricular assist device-specific infections. Ann. Thorac. Surg. **4**, 1302–1307 (2020a)
81. H. Rotering, U. Hansen, H. Welp, A.M. Dell'Aquila, Kaltes atmosphärisches Plasma und „advanced negative pressure wound therapy. Behandlungskonzept für komplexe Wunden in der Herzchirurgie. Herz-Thorax-Gefäßchir **34**, 52–61 (2020b)
82. R. Rutkowski, G. Daeschlein, T. von Woedtke, R. Smeets, M. Gosau, H.R. Metelmann, Long-term risk assessment for medical application of cold atmospheric pressure plasma. Diagnostics **10**, 210 (2020)
83. A. Schmidt, T. von Woedtke, J. Stenzel, T. Lindner, S. Polei, B. Vollmar, S. Bekeschus, One year follow up risk assessment in SKH-1 mice and wounds treated with an argon plasma jet. Int. J. Mol. Sci. **18**, 868 (2017)
84. M. Schuster, C. Seebauer, R. Rutkowski, A. Hauschild, F. Podmelle, C. Metelmann, B. Metelmann, T. von Woedtke, S. Hasse, K.D. Weltmann, H.R. Metelmann, Visible tumor surface response to physical plasma and apoptotic cell kill in head and neck cancer. J. Cranio-Maxillofac. Surg. **44**, 14451452 (2016)
85. R. Schönebeck, kINPen MED®, in *Comprehensive Clinical Plasma Medicine—Cold Physical Plasma for Medical Application*, 1st edn., ed. by H.R. Metelmann, T. von Woedtke, K.D. Weltmann (Springer, Berlin/Heidelberg, 2018), pp.485–494
86. B. Schwertlick, Kaltplasmatherapie – ein vielversprechender Therapieansatz für die Behandlung peripherer Ulcerationen und multiresistenter Erreger. Spitzenforschung in der Dermatologie—Innovationen und Auszeichnungen 2017/2018. Lebendige Wissenschaft, APHA InformationsgGmbH, Lampertsheim 52–53 (2018)
87. M.L. Semmler, S. Bekeschus, M. Schäfer, T. Bernhardt, T. Fischer, K. Witzke, C. Seebauer, H. Rebl, E. Grambow, B. Vollmar, J.B. Nebe, H.R. Metelmann, T. von Woedtke, S. Emmert, L. Boeckmann, Molecular mechanisms of the efficacy of cold atmospheric pressure plasma (CAP) in cancer treatment. Cancers **12**, 269 (2020)
88. H. Sorg, D.J. Tilkorn, S. Hager, J. Hauser, U. Mirastschijski, Skin wound healing: an update on the current knowledge and concepts. Eur. Surg. Res. **58**, 81–94 (2017)
89. B. Stratmann, T.C. Costea, C. Nolte, J. Hiller, J. Schmidt, J. Reindel, K. Masur, W. Motz, J. Timm, W. Kerner, D. Tschoepe, Effect of cold atmospheric plasma therapy vs standard therapy placebo on wound healing in patients with diabetic foot ulcers. A randomized clinical trial. JAMA Netw. Open **3**, e2010411 (2020)
90. R. Strohal, G. Hämmerle, Kaltplasma als neue Behandlungsoption bei häufig auftretenden
91. Wundsituationen im klinischen Alltag: eine Pilot-Fallserie. Wundmanagement **12**, 275282 (2018)
92. H. Tanaka, K. Ishikawa, M. Mizuno, S. Toyokuni, H. Kajiyama, F. Kikkawa, H.R. Metelmann, M. Hori, State of the art in medical applications using non-thermal atmospheric pressure plasma. Rev. Mod. Plasma Phys. **1**, 3 (2017)
93. C. Tendero, C. Tixier, P. Tristant, J. Desmaison, P. Leprince, Atmospheric pressure plasmas: a review. Spectrochimica Acta Part B **61**, 2–30 (2006)
94. R. Tiede, J. Hirschberg, G. Daeschlein, T. von Woedtke, W. Vioel, S. Emmert, Plasma applications: a dermatological view. Contrib. Plasma Phys. **54**, 118–130 (2014a)

95. R. Tiede, M.S. Mann, W. Viöl, G. Däschlein, C. Welz, H. Wolff, T. von Woedtke, J. Lademann, S. Emmert, Plasmamedizin in der Dermatologie. HAUT **6**, 283–289 (2014b)
96. R. Tiede, J. Hirschberg, W. Viöl, S. Emmert, A μs-pulsed dielectric barrier discharge source: physical characterization and biological effects on human skin fibroblasts. Plasma Process. Polym. **13**, 775–787 (2016)
97. C. Ulrich, F. Kluschke, A. Patzelt, S. Vandersee, V.A. Czaika, H. Richter, A. Bob, J. von Hutten, C. Painsi, R. Hügel, A. Kramer, O. Assadian, J. Lademann, B. Lange-Asschenfeldt, Clinical use of cold atmospheric pressure argon plasma in chronic leg ulcers: a pilot study. Wound Care **24**, 196–203 (2015)
98. J. van der Linde, K.R. Liedtke, R. Matthes, A. Kramer, C.D. Heidecke, L.I. Partecke, Repeated cold atmospheric plasma application to intact skin does not cause sensitization in a standardized murine model. Plasma Med. **7**, 383–393 (2017)
99. S. Vandersee, H. Richter, J. Lademann, M. Beyer, A. Kramer, F. Knorr, B. Lange-Asschenfeldt, Laser scanning microscopy as a means to assess the augmentation of tissue repair by exposition of wounds to tissue tolerable plasma. Laser Phys. Lett. **11**, 115701 (2014)
100. T. von Woedtke, S. Reuter, K. Masur, K.D. Weltmann, Plasmas for medicine. Phys. Rep. **530**, 291–320 (2013)
101. T. von Woedtke, H.R. Metelmann, K.D. Weltmann, Clinical plasma medicine: state and perspectives of in vivo application of cold atmospheric plasma. Contrib. Plasma Phys. **54**, 104–117 (2014)
102. T. von Woedtke, A. Schmidt, S. Bekeschus, K. Wende, Wissenschaftliche Grundlagen, Stand und Perspektiven der Plasmamedizin, in *Plasmamedizin – Kaltplasma in der medizinischen Anwendung*, 1st edn., ed. by H.R. Metelmann, T. von Woedtke, K.D. Weltmann (Springer, Berlin/Heidelberg, Germany, 2016), pp.17–32
103. D. Wandke, PlasmaDerm®—Based on di_CAP technology, in *Comprehensive Clinical Plasma Medicine—Cold Physical Plasma for Medical Application*, 1st edn, eds. by H.R. Metelmann, T. von Woedtke, K.D. Weltmann. (Springer, Berlin/Heidelberg, Germany, 2018), pp. 495–502
104. C. Weishaupt, S. Emmert, Connecting basic cold plasma technology to dermato-oncology. Clin. Plasma Med. **10**, 16–19 (2018)
105. K.D. Weltmann, E. Kindel, T. von Woedtke, M. Hähnel, M. Stieber, R. Brandenburg, Atmosphericpressure plasma sources: prospective tools for plasma medicine. Pure Appl. Chem. **82**, 1223–1237 (2020)
106. K.D. Weltmann, T. von Woedtke, Plasma medicine—Current state of research and medical application. Plasma Phys. Control. Fusion **59**, 014031 (2017)
107. K. Wende, S. Bekeschus, A. Schmidt, L. Jatsch, S. Hasse, K. Masur, T. von Woedtke, Risk assessment of a cold argon plasma jet in respect to its mutagenicity. Mutat. Res. Genet. Toxicol. Environ. Mutagen. **798**, 48–54 (2016)
108. K. Wende, A. Schmidt, S. Bekeschus, Safety aspects of non-thermal plasmas, in *Comprehensive Clinical Plasma Medicine*, 1st edn, eds. by H.R. Metelmann, T. von Woedtke, K.D. Weltmann (Springer, Berlin/Heidelberg, Germany, 2018), pp. 83–109
109. J.J. Zhang, J.O. Jo, D.L. Huynh, M. Ghosh, N. Kim, S.B. Lee, H.K. Lee, Y.S. Mok, T. Kwon, D.K. Jeong, Lethality of inappropriate plasma exposure on chicken embryonic development. Oncotarget **8**, 85642–85654 (2017)

Chapter 9
Safety Aspects and Standardization

Jinsung Choi, Young June Hong, Junsup Lim, Kai Masur, and Eun Ha Choi

Abstract Recently, medical devices using atmospheric pressure plasma have been introduced to people. Medical devices using atmospheric pressure plasma have different safety points from traditional medical devices, so a new standard is needed. In this chapter, we will introduce the measurement factors for the safety of atmospheric pressure plasma medical devices.

9.1 Background

Atmospheric pressure plasma is being used in various fields such as agriculture, medicine, and semiconductor industry [1–3]. Among them, it is being applied as a treatment device of a new concept in the medical field. Conventional medical devices diagnose or treat diseases using heat, light (laser, LED, lamp), radiation, and electrical stimulation [4–9]. On the other hand, plasma medical devices treat a patient's condition using electrons, active species, UV, etc. generated from electric discharge (Fig. 9.1).

It is difficult to apply the 60,601-1 based standard to plasma medical devices. Accurate safety standards for plasma active species, plasma current, plasma temperature, etc. and standards for measuring these factors are required. Therefore, plasma-based medical devices are suitable for plasma medical devices that use a mechanism different from that of existing medical devices. A new standard is needed [8, 9].

Various plasma medical devices are being launched worldwide. ADTEC Healthcare's adtec steriplas, Germany's neoplas tools, Wacker, Cinogy, Terraplasma, etc. are presenting medical device products for the purpose of wound treatment.

According to a report by Mordor Intelligence, the size of the plasma medical device market, where various plasma medical devices are being released, is expected

J. Choi · Y. J. Hong · J. Lim · E. H. Choi (✉)
Plasma Bioscience Research Center, Department of Electrical and Biological Physics, Kwangwoon University, Seoul, Korea
e-mail: ehchoi@kw.ac.kr

K. Masur
Leibniz-Institute for Plasma Science and Technology, Greifswald, Germany

E. H. Choi (ed.), *Plasma Biosciences and Medicine*, Topics in Applied Physics 148,
https://doi.org/10.1007/978-981-19-7935-4_9

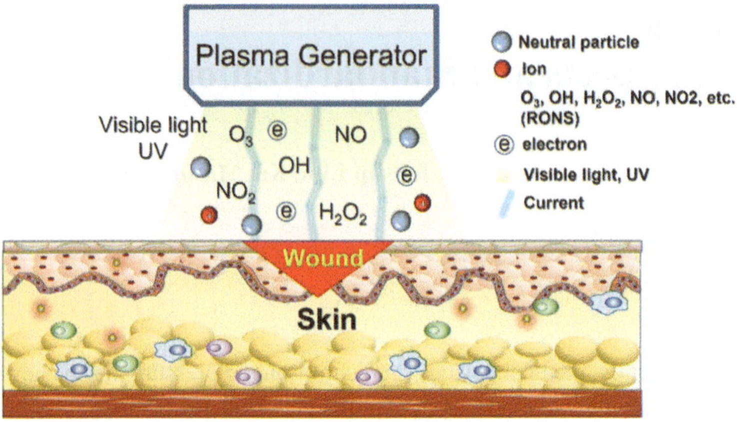

Fig. 9.1 Schematic diagram of the interaction between atmospheric pressure plasma and wound treatment

to grow from 2.17 billion USD in 2019 to 4.59 billion USD in 2024. This shows a growth potential of 16.2% annually. The market share according to the scope of application is 37% for wound treatment equipment, 25% for hemostatic equipment, and 20% for dental equipment [10].

What is difficult to apply to atmospheric plasma as an existing international standard is that there are no measurement methods and standards for various active species generated from atmospheric pressure plasma generators. The measurement methods and standards in the existing ISO 6768:1998, ISO 7996:1985, ISO 10313:1993, and ISO13964: 1998 standards are for atmospheric ozone, NO, and NO_2, and are applicable to the plasma medical devices discussed here. It is difficult to do. IEC 60,601-2-76 is a standard for hemostasis and does not contain regulations on active species.

Standards for plasma medical devices can be found in the German national standard DIN spec 91,315 and the guidelines published by the Korean Ministry of Food and Drug Safety. DIN spec 91,315 results in physical properties such as temperature, thermal output, optical emission spectroscopy, gas emission and leakage current. In addition, the results of biological efficacy such as Microbial inhibition zone assays, Treatment of microbial suspensions, and Treatment of human cells are presented. Two guidelines in Korea suggest methods and standards for measuring the density, ozone, nitrogen species generation, temperature, and leakage current of plasma medical devices. Currently, in order to sell medical and cosmetic devices using plasma in Korea, you must satisfy the criteria of these two guidelines to obtain a license.

9.2 Confirmation of Plasma ME for Wound Treatment

As various medical devices using atmospheric pressure plasma are released, the need for a definition of a device using plasma is emerging. Atmospheric pressure plasma medical devices are different from traditional medical devices, and they have the characteristics of plasma and are being medically applied. It has a stark difference from devices using negative ions or photocatalysts. Based on the unique characteristics of atmospheric pressure plasma, we are going to propose a method that can define atmospheric plasma medical devices.

Atmospheric pressure plasma tries to use the spectrum seen during discharge. Among traditional medical devices, the spectrum of a medical device using an LED that generates light and a medical device using a LASER have different characteristics from those generated by plasma. In addition, the spectrum of plasma shows characteristics different from the spectrum of light generated by an anion generator or a photocatalyst.

Figure 9.2 shows the spectrum of plasma, LED and laser. As shown in the figure, the spectrum of LED has a bandwidth of about tens of nm with respect to the center wavelength and shows low light output. LED basically emits light in RGB color, but it emits light in various colors depending on the purpose. The spectrum of LASER has a linewidth of less than 1 nm and shows high light output. On the other hand, the spectrum of atmospheric pressure plasma shows the spectrum of N_2 SPS (300–400 nm) and N_2 FPS (500–700 nm) of nitrogen molecules when discharging using the atmosphere, and when discharging using argon, Ar * atoms (700–850 nm).

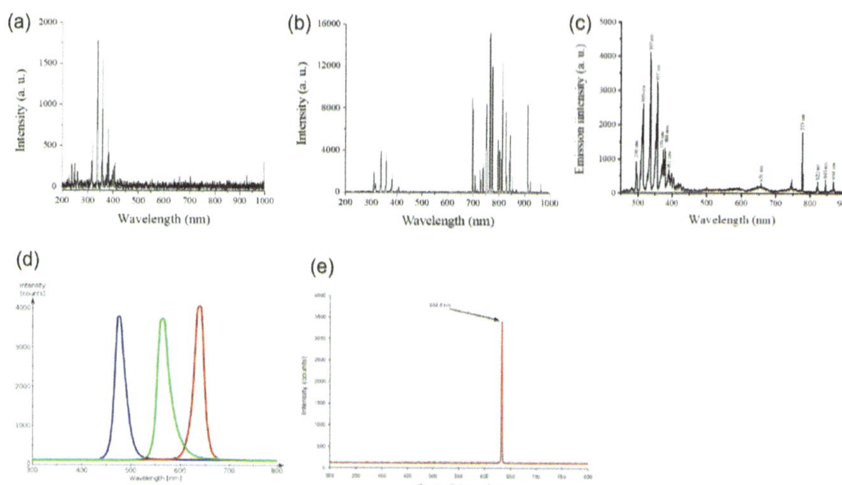

Fig. 9.2 Spectrum of **a** N_2 DBD plasma, **b** Ar DBD plasma, **c** Air soft jet plasma, **d** LED and **e** laser

In this way, by using the spectral characteristics of atmospheric pressure plasma, it is intended to present a standard that can be named as atmospheric pressure plasma medical device only when certain spectral conditions are satisfied.

9.3 The Role of RONS in Cancer Therapy Protection Against Excessive Reactive Species

Plasma discharge at atmospheric pressure generates reactive oxygen species (ROS) such as ozone, OH, and H_2O_2 and reactive nitrogen species (RNS) such as NO and NO_2 [11]. Plasma-generated reactive nitrogen-oxygen species (RONS) play an important role in biological interactions [12, 13]. Some RONS are toxic and require caution in biological applications. For biological applications, control of RONS generated by atmospheric plasma is essential. In order to secure the safety of medical staff and patients using atmospheric plasma medical devices, it is necessary to present safety standards and accurate measurement methods for active species.

In general, ozone and nitrogen dioxide are known as air pollutants that have a detrimental effect on the respiratory system of humans and animals [14, 15]. Long-term exposure to ozone is reported to be related to the occurrence of asthma [16]. If the ozone concentration in the atmosphere is 0.02 ppm or more, you can smell it. The safety standard recommended by the Korean Ministry of Environment and the American Conference of Governmental Industrial Hygienists (ACGIH) in the United States is 0.05 ppm. Living and working below this standard are not dangerous. If more than 0.44 ppm NO_2 is produced, people can smell it. NO_2 gas can cause cardiovascular disease with prolonged exposure [17, 18]. The safety standard recommended by the Korean Ministry of Environment and the US ACGIH is 3 ppm, which should not be exceeded for human and animal health.

On the other hand, NO gas is known as an essential substance for maintaining homeostasis of the human body [19, 20]. In particular, NO is reported as an antibacterial substance that plays a particularly important role in the immune system to protect the human body from microorganisms [21].

The active species measurement method of atmospheric pressure plasma medical devices can be divided into two. The first is the definition of the site for measuring the active species. The second is the definition of the measurement distance. Each active species has an individual lifetime. Therefore, measurement results according to the measurement location or distance are required.

First, the measurement location will be described. The measurement method of RONS suggested in the plasma medical device standard is measured in three places as shown in Fig. 9.3 [22]. Figure 9.3a shows the method for measuring RONS at the front of the plasma generator. Figure 9.3b, c are methods for confirming the effect on the medical staff including the patient when the RONS generated from the plasma generator is diffused.

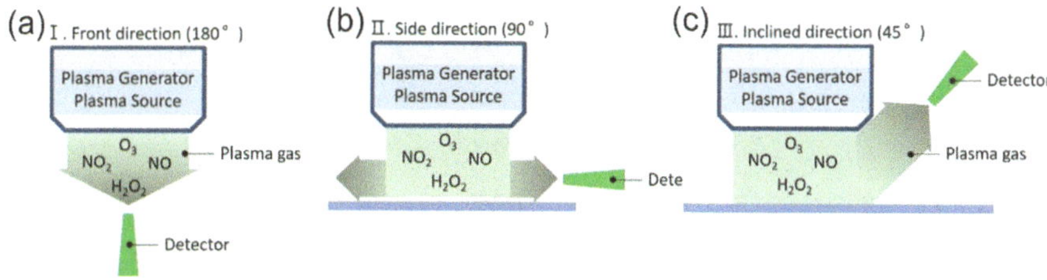

Fig. 9.3 Measurement method of gas emission at different angle, **a** 180°, **b** 90°, **c** 45° [23]

Let's discuss the measurement distance. The distance between the plasma medical device (or quartz glass) and the active species detector is divided into use on the face and on the body other than the face. If the medical device is used on the face, the measurement position is measured at a distance of 5, 40, 50, and 45 mm from the edge of the device. These distances are assumed to be used under the nose, cheeks, forehead, and chin. When using a medical device on the body except for the face, measure at the distance set by the manufacturer according to the product characteristics and purpose of use. The measurement time is during the use time of individual medical devices to measure the amount of RONS. When used on the face, the measurement time for each location is determined by dividing the intended use time by the number of parts to be measured [24] (Fig. 9.4).

Figure 9.5a is a schematic diagram of ozone measurement of soft plasma jet using detector (200 series, aeroqual) [23]. Figure 9.5a is the result of ozone measurement according to distance. In Fig. 9.5b, ozone generated from the soft plasma jet was measured to increase with distance, and slightly increased with increasing off time.

Fig. 9.4 Example of the measurement position when using it on the face [24]

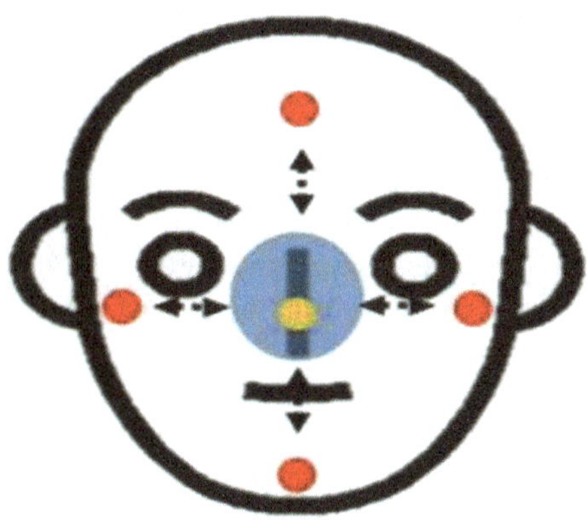

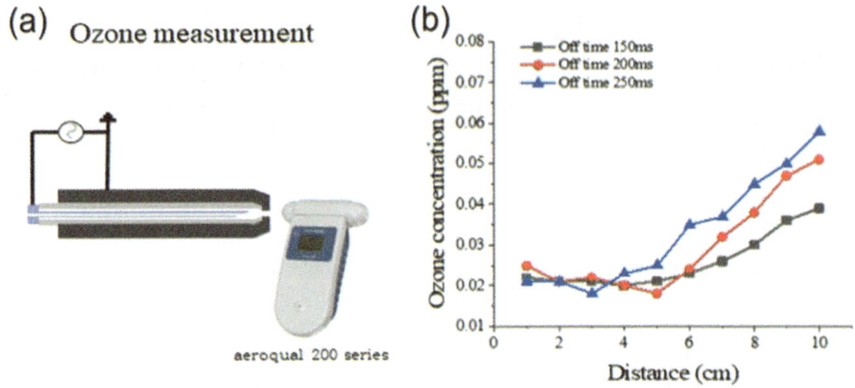

Fig. 9.5 **a** Ozone measurement, **b** ozone concentration versus distance under several off-time durations in discharge for soft plasma jet [23]

9.4 Plasma Current

The measurement of plasma current is an electrical safety device. It is a safety standard to protect against electric shock that can be applied to a patient from a plasma medical device.

The electric current between the plasma and the skin is also one of the biological effects of plasma. Among the existing medical devices, there is a device that expects a therapeutic effect by flowing an electric current to the patient. Safety limits and measurement methods for the current flowing through the patient are specified in IEC 60,601-1-1. The safe value of patient current given in IEC 60,601-1-1 is 100 μA [25]. Sensitive people may feel uncomfortable by sensing current even if the tolerance is met. Detection thresholds vary from person to person, and vary by age and gender. From a physiological point of view, the tolerances given in IEC60601-1 do not pose a health problem [25].

The measurement was proposed based on the standards related to medical devices, IEC 60,601-1 and IEC 60,601-2-76. The figure shows a schematic diagram of a typical leakage current measurement. Plasma current values are defined by placing a copper plate opposite the plasma device and measuring the current in the copper plate.

The Fig. 9.6 shows the plasma current measurement result for the soft jet. UNIMET® 800ST, BENDER was used as a plasma current measurement device, and the current flowing through the copper plate was defined as the soft jet plasma current [22]. Plasma current was measured when the distance between the soft jet and the copper plate was increased. As in Fig. 9.6b, the current was measured only up to 2 mm from the soft jet nozzle, and the current beyond 3 mm from the nozzle was not measured anymore. The lower measurement limit of the instrument is 1uA [23].

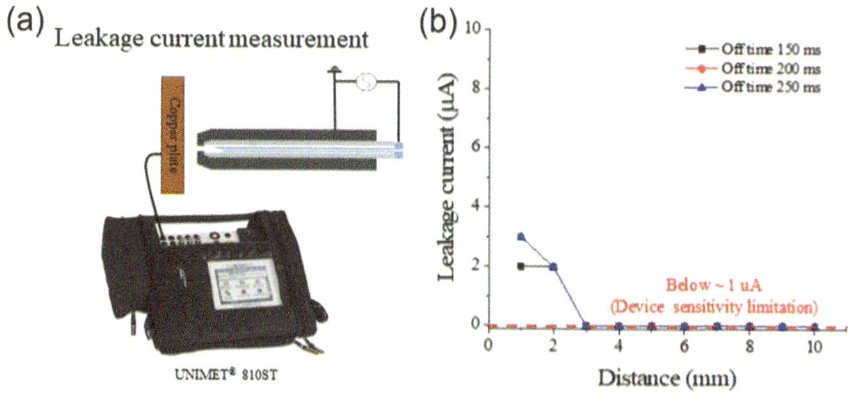

Fig. 9.6 **a** Schematic diagram of plasma current, **b** plasma current versus distance under off-time [23]

9.5 Plasma Temperature

Heat is also generated during atmospheric pressure plasma discharge. There should be no burns due to heat generated by atmospheric pressure plasma during wound healing. Therefore, it is necessary to measure the thermal energy generated from atmospheric pressure plasma or the temperature transferred to the skin. According to IEC 60,601-1-1, the temperature of the medical device must not exceed 40 °C. However, a slight increase in temperature can cause proliferation of live keratinocytes [21]. Therefore, atmospheric plasma treatment can actively induce wound healing and tissue regeneration if the temperature does not exceed 40 °C. It is also well known that above this temperature can cause protein denaturation and membrane destruction.

The plasma temperature of soft jet is shown in Fig. 9.7 [23]. The plasma temperature is measured according to the distance from plasma devices. Figure 9.7b shows the temperature according to the distance of the soft jet. The soft jet can be safely used regardless of the distance when the off time is 200 and 250 ms. When the off time was 150 ms, the temperature was measured to be less than 40 °C at 4 mm. This distance will be the recommended use distance for the soft jet. It should be used on patients at this distance to keep the plasma temperature below 40 °C [26, 27]. These are the biologically safe temperatures allowed by IEC60601-1-1.

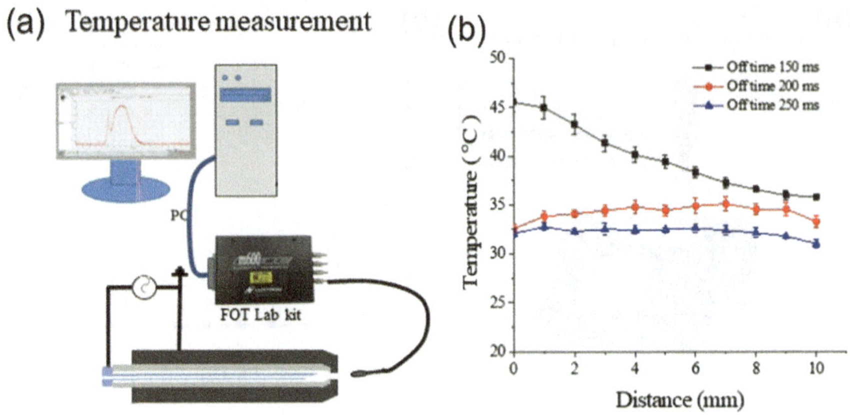

Fig. 9.7 **a** Schematic diagram of plasma temperature, **b** temperature versus distance under off-time [23]

Acknowledgements This work was funded by the National Research Foundation (NRF) of Korea, funded by the Korean government (2021R1A6A1A03038785).

References

1. N. Puač, M. Gherard, M. Shiratani, Plasma agriculture: a rapidly emerging field. Plasma Process Polym. **15**, e1700174 (2018)
2. K.D. Weltmann, E. Kindel, T. von Woedtke, M. Hähnel, M. Stieber, R. Brandenburg, Atmospheric-pressure plasma sources: prospective tools for plasma medicine. Pure Appl. Chem. **82**(6), 1223 (2010)
3. P. Dimitrakellis, E. Gogolides, Hydrophobic and superhydrophobic surfaces fabricated using atmospheric pressure cold plasma technology: a review. Adv. Coll. Interface Sci. **254**, 1 (2018)
4. S. Taavoni, S. Abdolahian, H. Haghani, Effect of sacrum-perineum heat therapy on active phase labor pain and client satisfaction: a randomized. Control. Trial Stud. Pain Med. **14**, 1301 (2013)
5. A. Hofer, L. Cerroni, H. Kerl, P. Wolf, Narrowband (311 nm) UV-B therapy for small plaque parapsoriasis and early-stage mycosis fungoides. ARCH Dermatol. **135**, 1377 (1999)
6. H. Kim, M.S. Huq, R. Lalonde, C.J. Houser, S. Beriwal, D.E. Heron, Early clinical experience with varian halcyon V2 linear accelerator: dual-isocenter IMRT planning and delivery with portal dosimetry for gynecological cancer treatments. J. Appl. Clin. Med. Phys. **20**(11), 111 (2019)
7. N.D. Sisterson, T.A. Wozny, V. Kokkinos, A. Constantino, R.M. Richardson, Closed-loop brain stimulation for drug-resistant epilepsy: towards an evidence-based approach to personalized medicine. Neurotherapeutics **16**, 119 (2019)
8. T. von Woedtke, S. Reuter, K. Masur, K.D. Weltmann, Plasmas for medicine. Phys. Rep. **530**, 291 (2013)
9. K.D. Weltmann, T. von Woedtke, Plasma medicine—Current state of research and medical application. Plasma Phys. Control. Fusion **59**, 014031 (2017)
10. Mordor Intelligence, Cold plasma in healthcare market-growth, trends, and forecast (2019–2024) (2018)

11. S. Emmert, F. Brehmer, H. Hänßle, A. Helmke, N. Mertens, R. Ahmed, D. Simon, D. Wandke, W. Maus-Friedrichs, G. Däschlein, M.P. Schön, W. Viöl, Atmospheric pressure plasma in dermatology: Ulcus treatment and much more. Clin. Plasma Med. **1**, 1 (2013)
12. A. Moldgy, G. Nayak, H.A. Aboubakr, S.M. Goyal, P.J. Bruggeman, Inactivation of virus and bacteria using cold atmospheric pressure air plasmas and the role of reactive nitrogen species. J. Phys. D: Appl. Phys. **53**, 434004 (2020)
13. M.J. Nicol, T.R. Brubaker, B.J. Honish II, A.N. Simmons, A. Kazemi, M.A. Geissel, C.T. Whalen, C.A. Siedlecki, S.G. Bilén, S.D. Knecht, G.S. Kirimanjeswara, Antibacterial effects of low temperature plasma generated by atmospheric-pressure plasma jet are mediated by reactive oxygen species. Sci. Rep. **10**, 3066 (2020)
14. E.C. Filippidou, A. Koukouliata, Ozone effects on the respiratory system. Prog. Health Sci. **1**, 2 (2011)
15. J. Gamble, W. Jones, S. Minshall, Epidemiological-environmental study of diesel bus garage workers: acute effects of NO_2 and respirable particulate on the respiratory system. Environ. Res. **42**, 1 (1987)
16. W.F. McDonnell, D.E. Abbey, N. Nishino, M.D. Lebowitz, Long-term ambient ozone concentration and the incidence of asthma in nonsmoking adults: the AHSMOG study. Environ. Res. **80** (1999)
17. O. Raaschou-Nielsen, Z.J. Andersen, S.S. Jensen, M. Ketzel, M. Sørensen, J. Hansen, S. Loft, A. Tjønneland, K. Overvad, Traffic air pollution and mortality from cardiovascular disease and all causes: a Danish cohort study. Environ. Health **11**, 60 (2012)
18. G. Walford, J. Loscalzo, Nitric oxide in vascular biology. J. Thromb. Haemost. 1 (2003)
19. J.D. Luo, A.F. Chen, Nitric oxide: a newlydiscovered function on wound healing. Acta Pharmacol. Sin. 26 (2005)
20. Y. Yang, P.K. Qi, Z.L. Yang, N. Huang, Nitric oxide based strategies for applications of biomedical devices. Biosurface Biotribol. **1**, 3 (2015)
21. P. Boukamp, R.T. Petrussevska, D. Breitkreutz, J. Hornung, A. Markham, N.E. Fusenig, Normal keratinization in a spontaneously immortalized aneuploid human keratinocyte cell line. J. Cell Biol. **106**, 3 (1988)
22. M.S. Mann, R. Tiede, K. Gavenis, G. Daeschein, R. Bussiahn, K.-D. Weltmann, S. Emmert, T. von Woedtke, R. Ahmed, Introduction to DIN-specification 91315 based on the characterization of the plasma jet kINPen® MED. Clin. Plasma Med. **4** (2016)
23. E.H. Choi, N.K. Kaushik, Y.J. Hong, J.S. Lim, J.S. Choi, I. Han, Plasma bioscience for medicine, agriculture and hygiene applications. J. Korean Phys. Soc. **80**(8), 817–851 (2022)
24. Guidelines for plasma medical devices for use on skin, (2021)
25. IEC 60601-1-1, Medical electrical equipment—Part 1-1: general requirements for safety—Collateral standard: safety requirements for medical electrical systems
26. R. Bussiahn, N. Lembke, R. Gesche, T. von Woedtke, K.D. Weltmann, Plasmaquellen für biomedizinische Applikationen—plasma sources for biomedical applications. Hyg. Med. **38**, 5 (2013)
27. H.R. v Metelmann, T. Woedtke, R. Bussiahn, K.D. Weltmann, M. Rieck, R. Khalili, F. Podmelle, P.D. Waite, Experimental recovery of CO_2-laser skin lesions by plasma stimulation. Am. J. Cosmet. Surg. **29** (2012)

Chapter 10
Biological Effects of Pulsed High-Power Microwaves

Sohail Mumtaz, Junsup Lim, Nagendra Kumar Kaushik, and Eun Ha Choi

Abstract Microwaves have been incorporated into nearly every part of our life because to the tremendous innovations in science, technology, and inventive high-power microwave (HPMW)-based systems. Due to improvements in electronics and novel microwave-based systems, microwave radiation has become a necessary part of modern life and is difficult to avoid their exposure. Humans are swimming like fish in a vast ocean of different radiations in this environment, resulting in frequent exposure. As a result, studying the biological impacts of these radiations has become an important subject of study. Microwave radiations have positive, negative, and neutral effects, which are highly dependent on electromagnetic field strengths, operational frequencies, and exposure times. With advancements in medical technologies, microwaves have played a major role in the treatment and detection of early-stage tumors; however, they can also have adverse effects on the human central nervous system, including neurotransmitters, which play a key role in passing singles inside the human body. The primary objective of this chapter is to outline how microwave radiation affects living things and the processes by which they do so. By contrasting microwave frequencies and power densities, this chapter also highlighted new methods for assessing how microwave radiation affects biological systems. Today, advancements in pulsed HPMW technology are being made specifically for military applications. This chapter also provided an overview of recent approaches for studying the effects of HPMW. In order to establish correlated safety standards by maximizing beneficial and minimizing detrimental effects of microwaves, it is essential to consider the health effects of particular frequencies when developing microwave-based applications. New strategies and a number of other factors also need to be subjected to further experimental studies.

S. Mumtaz · J. Lim · N. K. Kaushik · E. H. Choi (✉)
Department of Electrical and Biological Physics, Plasma Bioscience Research Center (PBRC), Kwangwoon University, Seoul, Korea
e-mail: ehchoi@kw.ac.kr

E. H. Choi (ed.), *Plasma Biosciences and Medicine*, Topics in Applied Physics 148,
https://doi.org/10.1007/978-981-19-7935-4_10

10.1 Introduction

Microwaves have been incorporated into nearly every part of our life because to the tremendous innovations in science, technology, and inventive high-power microwave (HPMW)-based systems. In this atmosphere, people move around like fish every day in the vast electromagnetic (EM) wave ocean. A wide range of applications of microwaves has led us to investigate their biological effects. The EM field interaction with biological systems and its consequences for human health were described in this chapter.

Non-ionizing radiation refers to microwaves (frequency range: 300 MHz–300 GHz). In its broadest sense, the term "*microwaves*" refers to a number of bands that are separated by various frequency ranges. Additionally, microwaves operating in the 1–300 GHz frequency band with peak power exceeding 100 MW are often classified as HPMW. The HPMW is used in different fields of modern technologies and become an essential part of our commercial, military, and medical life [1, 2] and modern science [3–10]. It is worth noting that both low power microwaves (LPMW) and high-power microwaves (HPMW) have been considered for the development of new medical devices and the modification of existing systems. In the medical domains microwaves with frequency ranges between 400 kHz and 10 GHz are now intensively being investigated. They are also being researched for diagnostic uses, including early cancer diagnosis, organ imaging, tumor detection, and more.

10.1.1 Origin of Pulsed HPMW

The HPMW has arisen in recent decades as a revolutionary technology that allows new applications and gives cutting-edge approaches for improving those that already exist. Intense relativistic electron beam (IREB) technology is currently used to produce HPMW, which taps into immense stores of power and energy. In the 1880s, Hertz invented artificial microwaves. The radio was developed in the early twentieth century working at low frequencies with grid tubes. Most scientists realized in the 1930s that using various resonant cavities, greater frequencies may be achievable. By following this idea, a klystron device is developed as the first cavity device in 1939 [11]. In the 1960s, electrical technology has been developed with *pulsed power*, which leads to beam production with a high voltage pulse exceeding 1 MV and a flowing current of 10 kA. This relativistic electron beams (REB) were helpful for high-energy density physics research projects including simulating the effects of nuclear weapons and inertial confinement fusion. The obtainability of REB and understanding of wave-particle interaction from plasma physics helps to generate HPMW.

10.2 Applications of HPMW

The HPMW was influential in a multitude of sectors from which major applications are industrial, military, communications, medical, radar technologies, ultra-wideband, fusion heating, linear collider, accelerators, and astronomy. A brief explanation of major applications of HPMW is described below.

10.2.1 Military Based Applications

The military uses HPMW devices that cover radar technology, communication systems, electronic countermeasures, electronic warfare. Also, HPMW can be used as a "nonlethal weapon". The radar technology expanded with the availability of HPMW with high power and a shorter wavelength. The detection range of the radar is extended which helps to detect a target of a small cross-section. Also, the HPMW is useful as a direct energy weapon, which damages their selected target with the highly focused energy of HPMW. These weapons are useful to target, missiles, optical and electrical devices, personnel, and vehicles.

10.2.2 Industry Based Applications

Penetrating and delivering energy through specific materials is a key characteristics of microwave. To understand this a simple example is a microwave oven. Due to this amazing property, the microwave has a significant impact and becomes a part of industrial needs. HPMW is useful in many industries for different reasons. The microwave for industrial applications including agriculture controls the insects by heating for certain periods with high temperature, chemicals, foods, papa rand textiles, automotive industry, and power transmissions of satellites, etc. [12]. Microwaves can be used in the separation of isotopes which resulted in many practical. This necessitates extremely stringent criteria for fixed phase and frequency stability. The most common application of microwaves in the industry at all generated frequencies is diagnostic for the process control.

10.2.3 Medical Applications

Microwaves have several applications in everyday life [13], most commonly in the medical industry to induce localized dielectric heating to desiccate human tissues, which is known as microwave ablation. Microwaves have played a vital role against cancer in recent decades. The microwave-based technologies provide the facility to monitor and treat cancer diseases at early stages. However, microwave ablation is mostly used to remove the unwanted tissue masses, for example, liver tumor, lung tumor, and prostrate ablation, and in the treatment of large tumors. Microwave ablation can benefit cancer patients who are experiencing a critical condition and unable to undergo surgery. The microwaves can also be used for imaging, sensing, bone imaging, early diagnostics, detection of the tumor, blood clot and blood stroke detection, heart imaging, and early detection of breast cancer [14–17]. The most modern and common areas of microwave medical applications include cardiology, oncology, gynecology, Rhizotomy, otolaryngology, ophthalmology, cosmetic treatment, and dental treatment [12, 13].

10.2.4 Communication Satellite and Astronomy-Based Applications

Applications of communication satellites for transmission at frequencies between 35 and 94 GHz, when the transmission window occurs in the natural environment. There are two factors to study this development. First, the saturation of the present band at a lower frequency, and second, rise in the bandwidth at high frequencies and advancement in the directionality. These applications demand low power, light weight, great reliability, and consistent gain characteristics [18]. Moreover, the existence of these sources will help the astronomers to achieve millimeter and submillimeter investigation of the space [18, 19].

10.2.5 Spectroscopy

Microwaves are useful in electron paramagnetic resonance spectroscopy. In chemical schemes, it delivers the information on the unpaired electrons like transition metal ions of free radical. The microwave was also used in rotational spectroscopy and together with electrochemistry, which is known as microwave enhanced electrochemistry.

10.3 Important High Power Microwave Sources

Many HPMW sources exist which are fundamentally operating with different working principles, oscillating frequencies, and power ranges [20, 21]. Generally, HPMW devices produced microwave radiations by the emission of IREB [22–24]. The IREB radiates because of their oscillations which are transverse to the motion of beam direction due to the external force. In addition to having the capacity to operate at high oscillation frequencies, HPMW devices have significant peak power output capabilities. A *virtual cathode oscillator*, or *vircator*, is regarded as the finest class of oscillators to make HPMW at low voltages out of various existing HPMW sources [25].

10.3.1 Backward Wave Oscillator (BWO)

The BWO, which is a vacuum tube that is part of a traveling wave tube that produces HPMW, was first shown in 1955. It is a wide tuning range oscillator. In concept, an electron gun could be employed to generate an electron beam capable of interacting with the slow-wave structure. By reversing the propagation of a moving wave against the beam direction, this slow-wave structure kept the oscillations going. The resultant EM wave power has a group velocity that is perpendicular to electron mobility. The output power is related outward near the electron gun.

10.3.2 Gyrotrons

A kind of free-electron maser is the gyrotron, that produces high-frequency electromagnetic waves (20–527 GHz) by activating electron resonance owing to the existing of a powerful magnetic field. Because the gyrotron dimensions are significantly larger than the wavelength, it can create high microwave power in the millimeter wavelength area ranging from kilowatts to megawatts. A high anode voltage has accelerated this beam, which is going through a large resonant cavity with a strong axial magnetic field. Because of the magnetic field, electrons pass helically through the tube. The electrons emit the EM wave in a transverse direction when the magnetic field in the tube reaches its maximum magnitude. These millimeter EM waves generate a standing wave in the drift tube, and a collector at the end of the guiding tube collects the waste electrons.

10.3.3 Magnetrons

The HPMW is produced by a magnetron, which is a high-power vacuum tube. In 1910, an early magnetron device was constructed, and in 1935, an advance magnetron with many cavities was invented. The electron stream interacts with the magnetic field as it moves to a sequence of open metal cavities known as cavity resonators in the magnetron concept. The electrons in the beam go through the aperture to these resonant cavities, causing the radio wave inside the cavity to oscillate and produce the microwave frequency. Unlike other vacuum tubes such as the TWT or klystron, the magnetron cannot be employed as an amplifier to boost the power of an applied microwave signal. From the direct current provided to the vacuum tube, the magnetron acts as an oscillator to produce HPMW.

10.3.4 Vircator

Figure 10.1 shows the basic vircator used to generate HPMW. The vircator is a special class of oscillator as compared with other HPMW devices because it can produce HPMW with the gigawatt level power ranges in the centimeter wavelength regimes [18, 26, 27]. To fulfill the need for HPMW in the future, vircator is considered as a hopeful oscillator because of some advantageous factors: it can produce high power of microwaves, potential to work at high frequencies, being capable of functioning in the absence of an external magnetic field, operate at low impedance and produce high power at relatively low given voltages, simple in construction, easy to build, and easy to understand the mechanism [28–30]. Despite the oscillator's advantageous characteristics, the low conversion efficiency of IREB energy to HPMW radiation energy (traditionally <10%) prevents its use. Because of this, vircator efficiency enhancement is a significant and active area of study [31–33]. The suitable approaches are being used by researchers to boost the efficiency of this oscillator [3, 25, 29, 30, 34–42]. Recently, the vircator was developed with record efficiency by forming multi virtual cathodes [28, 43].

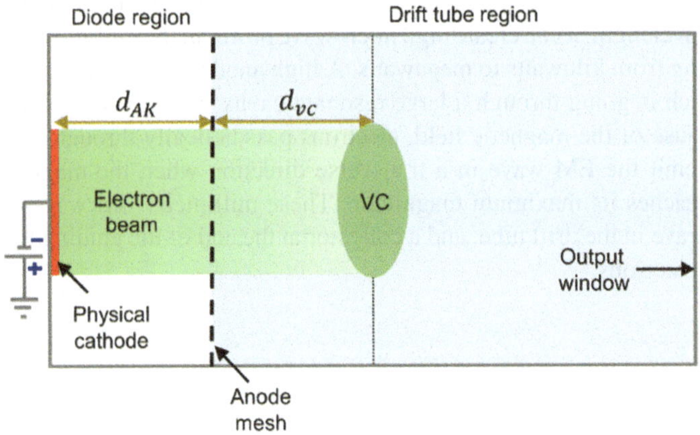

Fig. 10.1 The basic schematic of vircator. The real cathode, anode, and virtual cathode make up the vircator. The letters d_{AK} and d_{vc} stand for the distance between the anodic foil and cathode and the anode and virtual cathode, respectively

10.3.5 Basic Concept of Vircator

The conventional vircator is composed of a cathode for IREB injection, an anode grid (mesh), and a virtual cathode (VC). This VC is known as the source of the HPMW [27, 30, 44]. The requirement that the injected beam current surpasses the limiting current must be met in order to produce HPMW from vircator [24]. The basic schematic structure of the vircator was shown in Fig. 10.1, where d_{AK} indicating the distance kept between cathode and anode which is commonly named as A-K gap distance, and d_{vc} is the spacing between anodic foil and VC position.

In the working principle of the vircator, an IREB was emitted explosively from the cathode whose potential is $-V_0$, and these injected beam electrons are accelerated toward the anode position. Since the anode is transparent due to its mesh construction, the majority of accelerated electrons that reach its location pass through it and enter the drift tube area. If the IREB current I_b is higher than limiting current, IREB was forced to be pinched by a tremendous self-rotating magnetic field [24]. A cloud of electrons with a potential $-V_0$, similar to the actual cathode has therefore emerged behind the anode [45], which is defined as VC [21, 24, 46–48]. Immediately following the production of VC, the electrons start to move back to the anode location and start to oscillate. The main electron reflection, VC-wide oscillation, or both considered to characterize the mechanism of HPMW production from the vircator [46, 47].

10.4 Introduction of Vircator Based Pulsed Power Generator, "Chundoong"

A relativistic pulsed power generator *"Chundoong,"* was used in our laboratory to produce HPMW which is shown in Fig. 10.2. The *"Chundoong"* is a Korean term that means "Thunder" in English. This device uses an IREB with maximum values of 600 kV, 88 kA, and 60 ns for the voltage, current, and pulse length, respectively. The water-filled pulse forming a line and the vacuum diode have a characteristic matching impedance of 6.8 Ω and a maximum charging DC voltage of ±50kV. The chundoong has the ability to generate HPMW at power outputs between several hundred megawatts and gigawatts. The chundoong contains three major parts; Marx generator, pulse forming line, and vacuum diode region. To produce a high voltage pulse, a Marx generator with 12 stages of capacitors linked in series is employed. The high voltage long pulse is reduced to a nanosecond pulse and sent to the vacuum diode area via the pulse forming line. In the vacuum diode chamber, an IREB was produced by the field emission from the surface of cathode and created a VC behind the anode area where HPMW is produced.

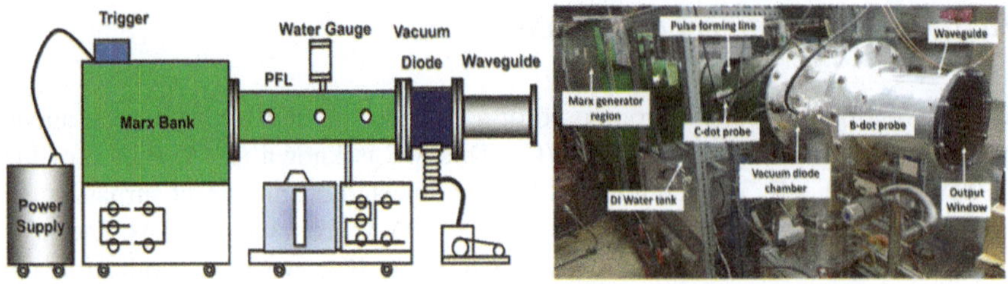

Fig. 10.2 Schematic and real image of pulse power generator, Chundoong [38]

10.5 Formation of Virtual Cathode and HPMW Generation

A dense VC was produced when the injected IREB current I_{beam} significantly surpassed the critical value of the space-charge limiting current I_{SCL} [24]. The condition $I_{\text{beam}} \gg I_{\text{SCL}}$ is essential for the development of a VC in vircator systems [24, 27]. When the cathode emits an IREB in an explosive manner, the beam is propelled toward the anode location, which is largely transparent for IREB. Owing to the anode transparency, most of the electrons pass through it and enter the drift tube region. In vircators, when the I_{beam} into the drift, tube exceeds the I_{SCL}, two forces, radial electric force, and a self-generated magnetic pinching force, are present inside the waveguide area and affect beam propagation which is as shown in Fig. 10.3. Due to the predominant self-pinching force, the beam is compressed [49], caused by the strong self-rotating magnetic field [48, 50]. Hence, the kinetic energy of the beam electrons becomes minimum to propagate along with the axial position z. In a certain area of the drift tube, these electrons gather as a cloud, and a deep potential VC is formed with a potential that is practically identical to the real cathode. Where the location of VC oscillates back and forth to its mean position, also a portion of beam electrons is reflected towards the anode mesh position. Many electrons are reflected by the VC in the direction of the diode, where they are finally reflected once more by cathode potential. This oscillating behavior of the electrons is referred to as reflexing [21, 51–55]. The microwave generation mechanism is considered by these two factors, electron reflection and oscillation of whole VC itself [21, 24, 27, 51, 56, 57].

10.6 Electromagnetic (EM) Field Interaction with Biological Systems

Due to the presence of EM and other types of radiation in our environment, biological systems are frequently exposed to such radiations. In past decades, hyperthermia and radiometry are the major subjects to determine the effects of microwaves in biology [58]. The pieces of evidences given by researchers through in vitro and in vivo studies

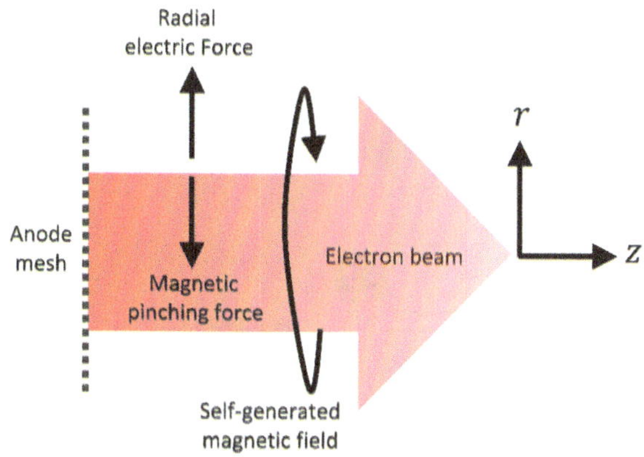

Fig. 10.3 Formation of VC inside the drift space

indicate that the EM field directly affects the biological systems. EM waves penetrate through the human body and cause various changes as shown in Fig. 10.4. The biological changes caused by the EM field might be harmful, beneficial, or none of both. Specific EM energy absorption in biological systems, particularly in the head and neck of humans, has drawn significant attention [59]. Many publications are available on this topic that described the interaction of EM waves with biological systems [58, 60, 61]. The possible mechanisms and interactions of EM waves with biological systems have recently been provided [62]. EM wave impacts have been seen at all biological levels, including microbial cells, animals, and the human [63]. Microwaves bring different physiological changes due to different frequencies and power. The in-depth mechanism of microwave bio-interaction exhibited that EM waves act as promoting agents to induce genetic changes in the biosystem [63]. The microwave is highly interactive with the human nervous system [64]. There are still unresolved problems addressing the proportional contributions of indirect to physiological modifications heat effects and potentially direct non-thermal interactions when examining EM wave (microwave) absorption on animals or people. Despite these uncertainties, in vivo and in vitro research shows that microwaves have direct impacts [58]. Studies that were carried out in temperature-controlled environments indicate that a number of cellular endpoints are directly affected at different frequencies and intensities [58]. The bulky molecular structures, such as cell-membrane receptors and enzyme complexes, and the dielectric characteristics of biomacromolecules are related to the biological interactions that take place at the microscopic level. The double layer of fat molecules that makes up the plasma membrane, has an electrical gradient (the membrane potential) of roughly 0.1 V throughout width in the majority of cells [65]. It has been hypothesized that this electrical gradient (10^5 V/cm) acts as a reliable barrier against cell activation by weak EM fields in the surrounding fluid [65].

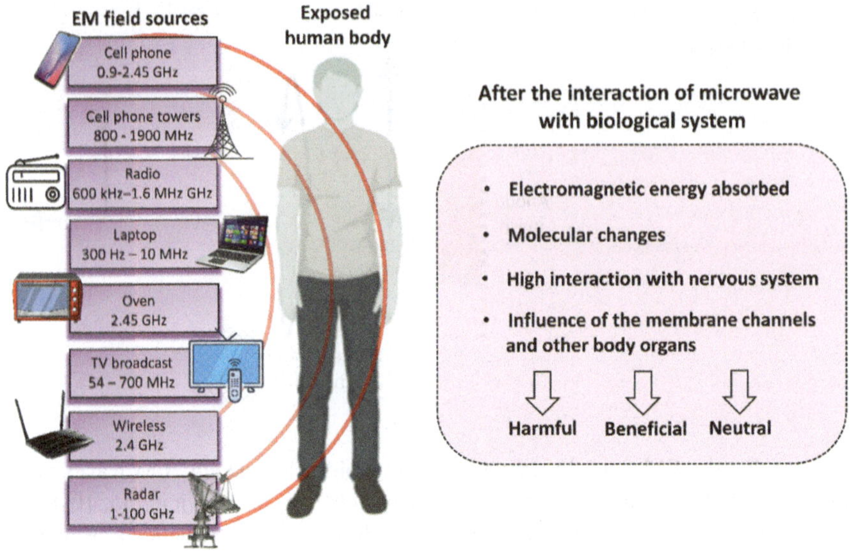

Fig. 10.4 Illustrations of typical home and commercial radiation sources, together with their operational frequency bands and potential health consequences

10.6.1 Mechanism for Action of EM Fields in Biology

When an electric field is present, opposing charges polarize and swing in the opposite directions from one another. However, this polarization also applies to particles with no net electric charge, as well as to free charges that erratically exist in biological tissue. In response to this polarization, electric dipoles were formed. The constant repolarization of dipoles in an alternating electric field consumes a lot of resources and absorbs the energy of the electric field. The water, which is present in large amounts in living tissue and whose particles behave like permanent electric dipoles. Dipoles constantly rotate about their axes in an alternating electric field, which makes them absorb electrical energy [66]. Several hypotheses may describe the effect of EM radiations in biology. The generation of powerful EM waves may cause the temperature of biological tissues to increase. Biochemical alterations brought on by EM waves that have less energy than necessary to directly ionize atoms can have a variety of impacts. It is acceptable to assume that all imaginable mechanisms depend on resonance, coherence, signal averaging, magnetic field heterogeneity, non-linear effects, and magnetic fields with lower powers than those required [67].

The majority of ions are bound to water; as a result, the energy dispersion when impacting water particles increases the system's loss at radio frequencies, limiting the degree of strengthening that may be produced in the resonance. The production of additional potentials on the cell membranes, which obstruct ion transport, is one of the theories frequently used to support the impact of radio waves on biological cells [66]. Only when external fields are sufficiently strong and produce voltages of several hundred microvolts, significantly greater than those generated by the membranes of organelles like mitochondria, is it feasible to change how ions are transported across cellular membranes. Exposure to the non-physiological voltage of the whole-cell organelles revealed that more energy is passed through the organelle membrane when it is thicker than the cellular membrane and the organelles contain greater ion concentrations [66]. The basis of this idea, which helps to explain how radio EM waves influence cells, is that changes in molecular bonds alter the action of protein enzymes [68]. Because biological proteins have a broad variety of structural differences, it is reasonable to predict that exposure to EM fields will have an effect depending on the protein structure [68]. Furthermore, several proteins are electrostatically bounded, therefore, the EM fields may influence the protein structures within the cells. It was already verified in investigations that EM fields can affect the stability, denaturation, and aggregation of proteins [69, 70]. The structure of a protein influences how efficiently it acts as an enzyme. In proteins, some side chains of the amino acids are polar and will react differently when exposed to varied EM fields.

10.7 The Biological Effects of EM Field of HPMW

The medical industry uses microwaves in a broad variety of ways [10, 71–76]. Induction of apoptosis in cancer cells, direct elimination of tumor cells, or reduction in nodule volume have all been achieved by the use of microwave-based hyperthermia and its combination with chemoradiotherapy as a noninvasive cancer treatment [77–83]. Neurotransmitter disruption, hippocampal damage, and cognitive decline were all observed in animals exposed to microwave radiation in the 860–2450 MHz frequency range [80–89]. By causing morphological and functional damage to the natural killer (NK-92) cells, pulsed HPMW impacts the autoimmunity [90]. The biochemical and morphological levels are harmed by prolonged EM field exposure. [91]. At the oscillation frequency of 2.856 GHz, the impact of pulsed HPMW on bone marrow cells was also investigated recently [92]. As pulse HPMW is used more often over time, concerns regarding the effects it may have on human health rapidly arise and call for in-depth research.

10.7.1 Effect of EM Field on Skin

Because skin is the outermost organ exposed to radiation on a regular basis, the effects of EM field on skin are particularly significant. It has been established that microwaves contribute to the development of skin and brain malignancies. According to one in vivo investigation, persistent microwave radiation in mouse skin at a frequency of 10 GHz caused substantial alterations in the molecular markers of the adaptive stress response [93]. In other studies, it was demonstrated that 25 GHz

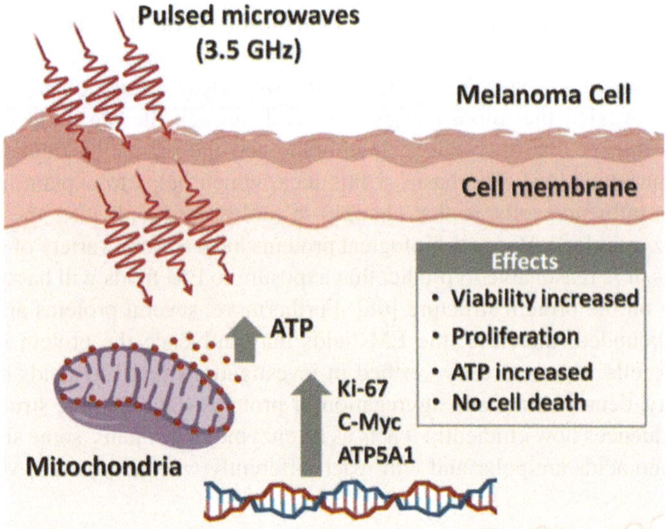

Fig. 10.5 Effect of HPMW on skin cancer cells. Reused with permission from [96]. Copyright 2020, Elsevier

microwave irradiation did not cause apoptosis or changes in pro-survival signaling pathways [94]. According to a research, microwave exposure causes skin cancer in mice and damages brain cells [95]. However, it is not well known how pulsed HPMW radiation affects the biology of the skin. Because the skin is the body's outermost layer and is continually exposed to various radiations, it's vital to look at how HPMW radiations affect skin. In a recent research, the operating frequency of the pulsed HPMW was 3.5 GHz, produced by employing the "*Chundoong*" device, to evaluate the effects on skin normal fibroblast and melanoma cells [96]. Regarding cellular development and energetics, potential impacts at the cellular and molecular levels were assessed [96]. Interestingly, HPMW does not show any effect on skin normal fibroblast cells, however, cell proliferation and increase in ATP levels were observed in melanoma as shown in Fig. 10.5. These results demonstrate that HPMW exposure can act as a stimulant for skin malignancies for up to 24 h. The exposure of HPMW to skin cancer patients should be limited.

10.7.2 Effects of EM Field on the Reproductive system

At some specific frequencies, HPMW has been suspected of having negative impacts on several human and animal body components. It was discovered that certain forms of EM fields had negative impacts on the reproductive system and, in other circumstances, neutral effects [97, 98]. To examine the biological impacts of 1.5 GHz HPMWs on the mouse reproductive system, an in vivo research was recently conducted [98]. The study team examined the effects of 1.5 GHz HPMW exposure on testicles and spermatozoa in C57BL/6 mice. For that, two 15 min exposures of 1.5 GHz HPMW with average absorption rates of 3, 6, and 12 W/kg were given to the mouse groups [98]. The findings of this reported investigation show no appreciable pathogenic or ultrastructural alterations in mouse blood testosterone levels, testicles, or spermatozoa.

The reproductive system of mice exposed to 1.5 GHz HPMW over their entire body did not undergo any discernible harm. The fertility potential of the male rat was also adversely impacted by the 10 GHz EM field [99]. There are several studies indicating the dependence of neutral or adverse effects of EM fields due to different frequencies, on reproductive systems. EM field exposure with a frequency of 6 GHz to chickens in vivo has been researched and published before [100], and no obvious changes observed in terms of growth rate, feed efficiency, egg quality and production, hatchability, or mortality. According to one study, the male reproductive system is more prone to inflammation and testicular failure when exposed to 2.45 GHz EM field [101]. In vitro study on human spermatozoa revealed enhanced cluster in genetic level and protein expression as well as DNA breakage after exposure to an 850 MHz EM field [102]. Similarly, in 900–1800 MHz, DNA fragmentation was observed to be increased after EM field exposure [103].

When the level of mitochondrial ROS in human spermatozoa grows, viability and motility drop, and DNA fragmentation occurs. At this stage, it is challenging to formulate meaningful predictions concerning the damaging, positive, or neutral effects of EM fields on human reproductive capacity. It is possible to predict if the EM energy doses will have a negative, neutral, or favourable effect. Only when doses are increased beyond a certain threshold can deadly consequences occur. The research currently available is insufficient for drawing conclusive findings regarding the amount and kind of EM radiation that causes harm to individuals. The value of the EM field intensity must be quantified in order to do numerical assessments of energy absorption; hence, research in this area is essential. Because the cellular membranes of the reproductive organs differ between species, animal research cannot be directly paralleled to human studies. Exposure to EM fields at higher frequencies, according to studies done on both people and animals, increases the chance of a variety of health issues.

10.7.3 Effect of EM Field on Brain

Microwave radiation has several beneficial consequences on modern civilization [104]. In many nations, cerebrovascular injuries are the leading cause of physical abnormalities and fatality. For example, by limiting chance, the frequency of brain strokes can be reduced, and the discovery of relevant solutions should be prioritized. Stroke increases the dynamic electric permittivity of brain tissues, which may be identified by microwave tomography [105]. Microwaves were utilized to efficiently treat a cold injury, avoiding amputation. Microwaves have been demonstrated to be advantageous [106]. Indeed, microwave imaging is a fresh and developing technique for the early detection of many disorders, among other benefits [107].

On the other hand, microwave radiation dose was discovered to be most detrimental to the brain [108], with the hippocampal being especially susceptible [109–112]. Previous research has demonstrated that hippocampal neurons are arranged in orderly rows, with borders that are distinct, nuclei that are clear, nucleoli that can be seen, and pyramidal cells that do not clearly display necrosis in unexposed control rats. On the other hand, neurons of radiation-treated rats show edema and are organized erratically. Additionally seen are nuclear pyknosis and capillary congestion [113, 114]. In particular, microwaves can harm the brain (one of the two major parts of the human CNS), especially the neurotransmitters that are crucial for signal transmission inside the body [115]. Studies that have particularly examined whether the CNS of children are more sensitive to EM field have been undertaken. This is because a child's CNS is at growing stage and more vulnerable to EM energy. According to certain scientific findings in this regard, children's CNSs are shown to be more vulnerable to EM radiation than adults [116, 117]. In light of this, microwave radiations have the potential to slow down the signaling process, which might cause the body further impairment. On the other hand, microwaves have a lot of applications in the medical arena, including the early detection and identification of tumors. Microwaves have been demonstrated to affect exposed biological systems in a variety of ways, including positively, negatively, and neutrally. Numerous other reports have demonstrated the detrimental effects of microwave radiation on the human brain, and Fig. 10.6 depicts a schematic of the most prevalent negative effects of microwave radiation on the brain and neurons.

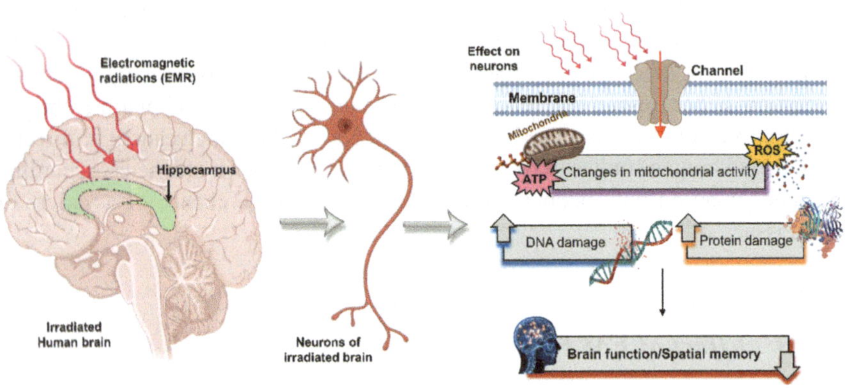

Fig. 10.6 The representation of the responses of brain to microwave exposure [62]

10.7.4 Biological Effect of High-Power Short Pulses of EM Field

The research has been undertaken actively from few decades on the interaction of biological systems with millimeter-wave radiation as a promising anticancer therapy. Millimeter radiations have several therapeutic uses, mainly in Eastern Europe, for the treatment of over fifty diseases, particularly cancer. It was stated that more than three million patients had favorable outcomes [118]. Eyes and skin are often the main targets of 60 GHz EM radiation [119]. The cornea, which has a 75% free water content and a thickness of 0.5 mm, absorbs the EM energy when exposed to short-wavelength EM waves. To identify the effects of millimeter EM waves (60 GHz) on the eyes, a recent study was carried out [120]. The obtained results show that physiological changes are not evidently induced by millimeter EM waves with a frequency of 60 GHz [120].

Exposure of cancer cells by millimeter wavelength range of EM waves causes increased cell mortality. To determine if high-power EM field short pulses have deleterious effects on healthy mice, a recent investigation was carried out on mice [118]. To achieve this, a free-electron laser device was used to subject the skin of healthy mice to dose-dependent exposure to 101 GHz millimeter EM waves (20–50 pulses). The results demonstrated that the biological parameters of mice were within normal limits after exposure. The physical, physiological, or behavioral state of the mice did not significantly alter following exposure. Additionally, following the exposure, there were no discernible changes in locomotor, exploratory, or anxious behavior, and no pathological modifications were found after hematological and biochemical blood examination [118]. According to these findings, millimeter 101 GHz EM waves have no substantial physiologically damaging impacts [118].

10.7.5 Effect of Long-Time Exposure of EM Field

The influence of EM field exposure on the human neurological system is a developing public issue. The effects of prolonged exposure to EM fields on brain function and associated pathways have recently been revealed [121]. After exposure, the morphology of the brain was studied. The reported research determines that mice's hippocampus and cerebral cortex may be damaged by EM field exposure at 1.5 W/m^2, along with cholinergic dysfunction, cell death, and oxidative damage. Additionally, the power density of EM field and the duration of radiation exposure were positively correlated with the deadly consequences. These findings indicate that extended exposure to the EM fields of HPMW may be damaging to the neurological systems of humans [121].

Long-term exposure to a 50 Hz EM field reduced the diameter and increased the number of seminiferous tubules per unit area of the testes, but had no discernible effect on sperm concentration, testes, or viability [122]. In order to create HPMW

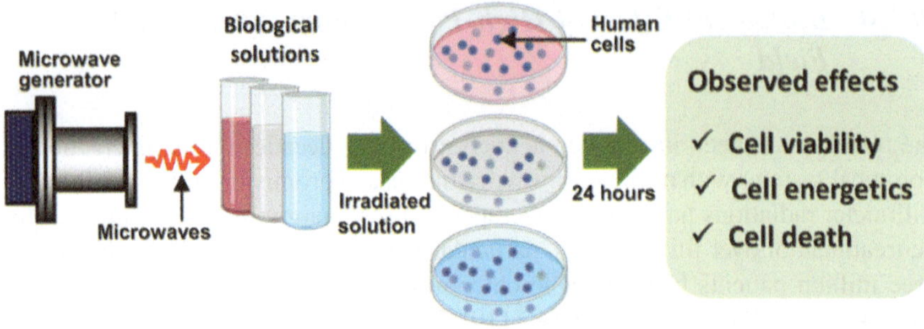

Fig. 10.7 The treatment of physiological solutions, and their application on human cells

from our "*Chundoong*" pulsed power generator, an axial vircator was built. According to Fig. 10.7, the physiological solutions (cell culture medium, distilled water (DW), and PBS) were exposed to HPMW radiation. As a result, 1 mJ of EM energy was applied to physiological solutions at the time of each pulse [123].

10.7.6 The Electric Field of HPMW Generated by Chundoong

In the vacuum, in the air, and in the DI water, the HPMW electric field was examined. A simulation of a three-dimensional particle in cell was used to examine the electric field within the vacuum [123]. The electric field has a 70 kV/cm^2 magnitude inside vacuum region. From the HPMW energy flowing Poynting vector, the maximum electric field (E_{max}) has been calculated to be around ~11 kV/cm [123]. At each HPMW pulse, the air–liquid solution interfacial area is affected by the electric field of around 11 kV/cm produced by HPMW, which was shown to be similar to the electric field of typical nonthermal atmospheric pressure plasma jet [124].

The electric field in the air and water areas of a test tube is determined to be comparable ($E = 11$ kV/cm) from the electric field distributions for vacuum, air, and physiological solution derived using HFSS (High-frequency structure simulator) code, as shown in Fig. 10.8a. Additionally, it was discovered that the electric field distribution in the water was in resonant nodal patterns along the test tube's vertical axis, with regular intervals of 2 cm, which is precisely equal to $\lambda/4$. Under the microwave frequency $f = 3.5$ GHz, both of these field distributions demonstrate that the water dielectric constant is almost identical to that of air. The observed electric field of HPMW is 11 kV/cm in air, which is comparable to the 10 kV/cm reading from HFSS and the water's interior at the same location. Therefore, the HPMW electric field of 11 kV/cm in both the test tube's air and water interiors would be the primary cause of the excitation of water and nitrogen molecules. Additionally, following exposure to HPMW, the temperature of the physiological fluids remained unaffected, suggesting that the HPMW had no thermal effects.

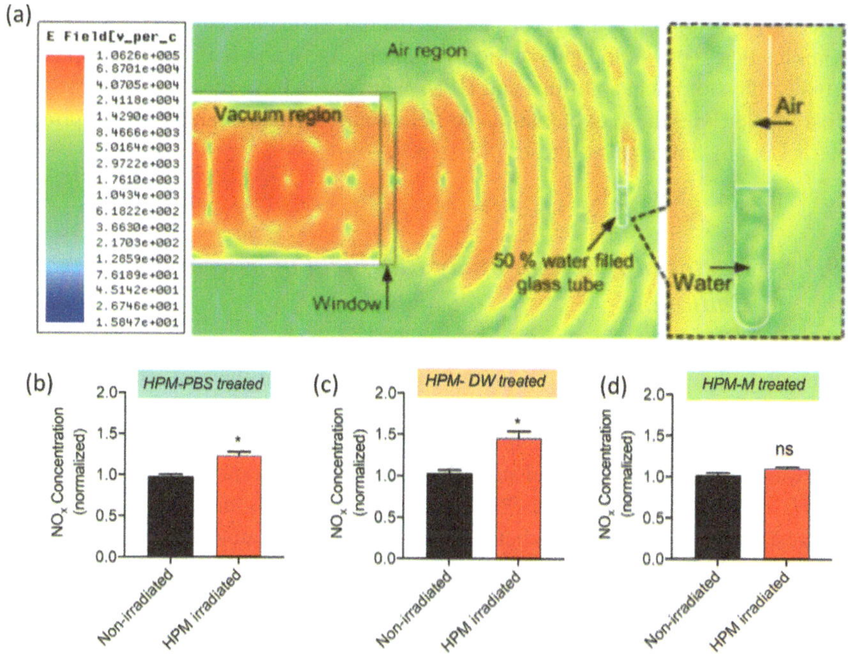

Fig. 10.8 **a** The HPMW electric field in a vacuum, in air, and in DI water, and **b–d** the concentration of NOx in a biological solution following HPMW exposure [123]

10.7.7 Generation of Reactive Species by HPMW Exposure

It's interesting to note that following exposure to HPMW, the NOx levels in a recent research were seen in the DW, slightly rose in PBS, but remained the same in cell culture medium [123]. These results are extremely useful and interesting for the production of NO_x by using HPMW. The interaction of the HPMW-induced electric field with the gases in the surrounding air and the water itself may be used to explain why NO_x is produced in the physiological solution. The ambient atmosphere and the inside of water naturally include molecular oxygen (O_2), water vapor (H_2O), and molecular nitrogen (N_2), respectively. The molecules oxygen and nitrogen that are present in the air and the inside of water interact with the about ~11 kV/cm HPMW electric field. As shown in Fig. 10.8b–d, this electric field interaction transforms them into their atomic nitrogen N and atomic oxygen O species, which are then mixed to form NO_x and are absorbed into the liquid. The interaction of an electric field comparable to that found in nonthermal atmospheric pressure plasma jets, the HPMW electric field (11 kV/cm), was used to validate the generation of NO_x [124]. Cellular NO homeostasis can be impacted and might undergo considerable variations by exogenous NO_x species supplementation [125–129]. Similarly, following HPMW exposure, a modest rise in H_2O_2 levels was seen in DW, but not in PBS or cell culture medium. The make-up of PBS and its ability to act as a buffer can account for these variations in NO_x and H_2O_2 levels. Due to the questionable resonance excitation of biomolecules in physiological solutions, theoretical explanations for the interactions between physiological solutions and HPMW appear to have few possibilities [130]. Most of the time, these explanations are clear and include details on how proteins interact in physiological solutions [131–133].

10.7.8 Bacterial Inactivation by EM Field of Microwave Radiation

Research in this area is crucial because it examines the non-thermal effects of pulsed HPMW on bacterial systems. From simple to complicated biological systems, it is crucial to design HPMW applications. Using a chundoong machine, HPMW exposure was administered to two distinct bacterial strains: gram-negative Escherichia coli (*E. coli*) and gram-positive Staphylococcus aureus (*S. aureus*) in recent study. As a result of direct interactions between particular (polar) molecules at bacterial cell surface and the electric field of 8 kV/cm produced by HPMW within the PBS with a power density of 17 kV/cm^2 at the sample point, changes in cell shape occur, which play role for intracellular oxidative defense failure and the inactivation of bacterial cells caused by DNA damage [134].

Increased dosages led to almost a 6-log decrease in E. coli and a 4-log reduction in S. aureus, which were the largest ratios of viable count reductions ever recorded. Additionally, as shown in Fig. 10.9, scanning electron microscopy showed demonstrated surface damage in both bacterial strains following HPMW treatment. DNA damage and the deactivation of oxidation-regulating genes were implicated in the inactivation of the bacterium.

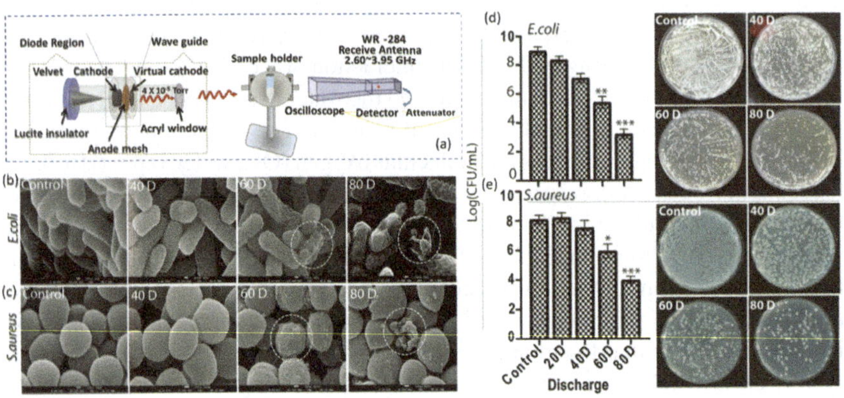

Fig. 10.9 **a** The experimental setup for HPMW exposure is shown in (**a**), along with (**b**) a SEM study of the morphology of E. coli and Sauers at various discharges of HPMW radiation and (**d**) and (**e**) the inactivation of E. coli and S. aureus bacterial strains after HPMW exposure [134]

Following HPMW treatment, levels of intracellular ROS grew, ultimately causing the bacteria to experience fatal damage [134]. The killing of all bacteria present on things is affected not only by the exposure period and item type, but also by the kind of bacteria and microwave power intensity. It is believed that *Bacillus stubilis var. nigar* is the best indicator bacteria for HPMW energy disinfection [135]. This finding provides substantial support for the choice of indicator bacteria for microwave-based disinfection as a new technological standard. According to estimates, HPMW sterilization has several benefits over traditional sterilizing and has the capability of being utilized in any industry.

Biofilms have the potential to pose serious problems in the food and medical industries [136]. Biofilms in processing equipment harm product safety and generate health concerns among customers. Biofilms, such as those seen in surgical implants, have been linked to approximately 80% of clinical infections [137]. Elevated temperatures reduced the elastic modulus and stiffness of staphylococcal biofilms, which may be advantageous for biofilm removal. A magnetic field has been shown in several studies to successfully eliminate biofilms; moreover, magnetic hyperthermia can alter biofilm damage [134, 138]. This impact has been seen in (gram—positive and gram—negative) bacterial biofilms, as well as including *methicillin-resistant Staphylococcus aureus* and *Pseudomonas aeruginosa* biofilms [139].

10.8 Summary

With the progress of science and advanced technologies, HPMWs have been integrated into approximately every aspect of human lives. In this environment, humans are swimming like fish in the ocean of EM waves every day. Due to its many applications, HPMW has become a necessary component of daily living, raising questions about its potential health impacts. Owing to the increasing number of HPMW based technologies, it raises the concern to investigate its biological effects. The generation and biological application of HPMW becomes an interesting and future important field of research. Our HPMW device "*Chundoong*" utilizes an IREB to generate HPMW and to study its biological effects. Pulsed HPMW showed nonthermal effects on biological samples [96, 123, 134]. HPMW does not show any effect on skin normal fibroblast cells, however, cell proliferation and increase in ATP levels in melanoma were measured 24 h after exposure which drops to non-significant at 48 h. These findings suggest that the HPMW exposure at high doses, it can act as a stimulant for skin malignancies for up to 24 h. The exposure of HPMW (3.5 GHz) to skin cancer patients should be limited [96]. In today's contemporary style of life, it is impossible to completely prevent EM field exposure during domestic and professional activities, but individuals should be aware of the biological risk posed by EM fields. The embryonic development, the function of the gonadal organs, pregnancy, and fetal growth have all been discovered to be altered by EM field exposure [140]. The whole body of mice exposed to 1.5 GHz HPMW did not cause noticeable injury or damage to the reproductive system [98]. The frequency, power, and duration of the EM field amount of exposure time may all have a significant impact on these EM field effects. To protect the human reproductive system, it is vital to minimize unwanted exposure to EM fields and to develop techniques for shielding against or relieving EM radiation.

In summary, the impacts of EM fields can be favourable, unfavourable, or neutral, and these effects are highly influenced by the field's intensity, frequency, and exposure duration. The biological consequences vary depending on the EM field intensity, frequency, and exposure length. In addition, it was crucial to set safety guidelines for EM field exposure through trials in order to produce beneficial effects and reduce potentially dangerous ones.

Acknowledgements This work was supported by the National Research Foundation of Korea (NRF) grant funded by the Korean government (MIST) [grant number NRF-2022R1A2C1004257, and partially by NRF-2021R1A6A1A03038785]; also, in part by Kwangwoon University, Seoul, Korea, 2022.

References

1. H. Gubler, M. Hiller, The use of microwave FMCW radar in snow and avalanche research. Cold Reg. Sci. Technol. **9**, 109–119 (1984). https://doi.org/10.1016/0165-232X(84)90003-X
2. J.M. Osepchuk, A history of microwave heating applications. IEEE Trans. Microw. Theory Tech. **32**, 1200–1224 (1984). https://doi.org/10.1109/TMTT.1984.1132831
3. A.M. Afzal, Y. Javed, N. Akhtar Shad et al., Tunneling-based rectification and photoresponsivity in black phosphorus/hexagonal boron nitride/rhenium diselenide van der Waals heterojunction diode. Nanoscale **12**, 3455–3468 (2020). https://doi.org/10.1039/C9NR07971H
4. A.M. Afzal, M.Z. Iqbal, S. Mumtaz, I. Akhtar, Multifunctional and high-performance GeSe/PdSe2 heterostructure device with a fast photoresponse. J. Mater. Chem. C (2020). https://doi.org/10.1039/D0TC00004C
5. A.M. Afzal, Y. Javed, S. Hussain et al., Enhancement in photovoltaic properties of bismuth ferrite/zinc oxide heterostructure solar cell device with graphene/indium tin oxide hybrid electrodes. Ceram Int. **46** (2020). https://doi.org/10.1016/j.ceramint.2019.12.166
6. A.M. Afzal, M.Z. Iqbal, G. Dastgeer et al., WS2/GeSe/WS2 bipolar transistor-based chemical sensor with fast response and recovery times. ACS Appl. Mater. Interfaces **12**, 39524–39532 (2020). https://doi.org/10.1021/acsami.0c05114
7. S. Yoon, K. Jeong, S. Mumtaz, H. Choi, Electromagnetic pulse shielding effectiveness of circular multi-waveguides for fluids. Results Phys. **16**, 102946 (2020). https://doi.org/10.1016/j.rinp.2020.102946
8. A.M. Afzal, S. Mumtaz, M.Z. Iqbal et al., Fast and high photoresponsivity gallium telluride/hafnium selenide van der Waals heterostructure photodiode. J. Mater. Chem. C **9**, 7110–7118 (2021). https://doi.org/10.1039/D1TC00867F
9. F. Sterzer, Microwave medical devices. IEEE Microw. Mag. **3**, 65–70 (2002). https://doi.org/10.1109/6668.990689
10. S. Semenov, J. Kellam, P. Althausen et al., Microwave tomography for functional imaging of extremity soft tissues: feasibility assessment. Phys. Med. Biol. **52**, 5705–5719 (2007). https://doi.org/10.1088/0031-9155/52/18/015
11. R.H. Varian, S.F. Varian, A high frequency oscillator and amplifier. J. Appl. Phys. **10**, 321–327 (1939). https://doi.org/10.1063/1.1707311

12. M.A. Stuchly, S. Stuchly, Industrial, scientific, medical and domestic applications of microwaves. IEE Proc. A (Phys. Sci. Meas. Instrum. Manag. Educ. Rev. **130**, 467–503 (1983)
13. J. Vrba, Medical applications of microwaves. Electromagn. Biol. Med. **24**, 441–448 (2005). https://doi.org/10.1080/15368370500382214
14. R. Chandra, H. Zhou, I. Balasingham, R.M. Narayanan, On the opportunities and challenges in microwave medical sensing and imaging. IEEE Trans. Biomed. Eng. **62**, 1667–1682 (2015). https://doi.org/10.1109/TBME.2015.2432137
15. E.J. Bond, X. Li, S.C. Hagness, V.B.D. Van, Microwave imaging via space-time beamforming for early detection of breast cancer. IEEE Trans. Antennas Propag. **51**, 1690–1705 (2003). https://doi.org/10.1109/TAP.2003.815446
16. E.C. Fear, X. Li, S.C. Hagness, M.A. Stuchly, Confocal microwave imaging for breast cancer detection: localization of tumors in three dimensions. IEEE Trans. Biomed. Eng. **49**, 812–822 (2002). https://doi.org/10.1109/TBME.2002.800759
17. Y. Nikawa, Application of microwaves in medical sensing and treatment, in *2013 Asia-Pacific Microwave Conference Proceedings (APMC)* (2013), pp. 62–64
18. V.L. Granastein, I. Alexeff, High power microwave sources, in *Second. Artech House*, Ch. 3, 4 and 13 (Boston, 1987)
19. J. Benford, Space applications of high-power microwaves. IEEE Trans. Plasma Sci. **36**, 569–581 (2008). https://doi.org/10.1109/TPS.2008.923760
20. E. Schamiloglu, High power microwave sources and applications, in *2004 IEEE MTT-S International Microwave Symposium Digest (IEEE Cat. No.04CH37535)* (vo. 2, 2004), pp. 1001–1004
21. J. Benford, J. Swegle, E. Schamiloglu, *High Power Microwaves*, 3rd edn. (Taylor & Francis, New York, 2016)
22. R.Q. Twiss, Radiation transfer and the possibility of negative absorption in radio astronomy. Aust. J. Phys. 564–579 (1958). https://doi.org/10.1071/PH580564
23. H. Motz, Applications of the radiation from fast electron beams. J. Appl. Phys. **22**, 527–535 (1951). https://doi.org/10.1063/1.1700002
24. L.S. Bogdankevich, A.A. Rukhadze, Stability of relativistic electron beams in a plasma and problem of critical currents. Sov. Phys. Uspekhi **14**, 163 (1971). https://doi.org/10.1070/PU1971v014n02ABEH004456
25. E.-H. Choi, M.-C. Choi, Y. Jung et al., High-power microwave generation from an axially extracted virtual cathode oscillator. IEEE Trans. Plasma Sci. **28**, 2128–2134 (2000). https://doi.org/10.1109/27.902240
26. S. Burkhart, Multigigawatt microwave generation by use of a virtual cathode oscillator driven by a 1–2 MV electron beam. J. Appl. Phys. **62**, 75–78 (1987). https://doi.org/10.1063/1.339163
27. W. Jiang, K. Masugata, K. Yatsui, Mechanism of microwave generation by virtual cathode oscillation. Phys. Plasmas **2**, 982–986 (1995). https://doi.org/10.1063/1.871377
28. S. Mumtaz, H. Uhm, J.S. Lim, E.H. Choi, Output-power enhancement of vircator based on second virtual cathode formed by wall charge on a dielectric reflector. IEEE Trans. Electron. Devices (2022). https://doi.org/10.1109/TED.2022.3149455
29. S. Mumtaz, P. Lamichhane, J.S. Lim et al., Enhancement in the power of microwaves by the interference with a cone-shaped reflector in an axial vircator. Results Phys. **15** (2019). https://doi.org/10.1016/j.rinp.2019.102611
30. S. Mumtaz, J.S. Lim, B. Ghimire et al., Enhancing the power of high power microwaves by using zone plate and investigations for the position of virtual cathode inside the drift tube. Phys. Plasmas **25**, 103113 (2018). https://doi.org/10.1063/1.5043595
31. S. Mumtaz, B.C. Adhikari, I.V. Minin et al., Particle in cell simulation for the power enhancement by forming the second virtual cathode in an axial vircator. Results Phys. **24**, 104126 (2021). https://doi.org/10.1016/j.rinp.2021.104126
32. W. Jiang, M. Kristiansen, Theory of the virtual cathode oscillator. Phys. Plasmas **8**, 3781–3787 (2001). https://doi.org/10.1063/1.1382643

33. W. Jiang, H. Kitano, L. Huang et al., Effect of longitudinal magnetic field on microwave efficiency of virtual cathode oscillator. IEEE Trans. Plasma Sci. **24**, 187–192 (1996). https://doi.org/10.1109/27.491758
34. K. Nagao, W. Takatsu, P. Van Thuan et al., High-power microwave generation by double-anode virtual cathode oscillator. Electr. Eng. Jpn. **210**, 11–18 (2020). https://doi.org/10.1002/eej.23253
35. W. Jeon, K.Y. Sung, J.E. Lim et al., A diode design study of the virtual cathode oscillator with a ring-type reflector. IEEE Trans. Plasma Sci. **33**, 2011–2016 (2005). https://doi.org/10.1109/TPS.2005.860901
36. K.B. Song, J.E. Lim, Y. Seo, E.H. Choi, Output characteristics of the axially extracted virtual cathode oscillator with a cathode-wing. IEEE Trans. Plasma Sci. **37**, 304–310 (2009). https://doi.org/10.1109/TPS.2008.2010547
37. S. Mumtaz, L.N. Nguyen, H. Uhm et al., A novel approach to form second virtual cathode by installing a floating zone plate inside the drift tube. Results Phys. **17**, 103052 (2020). https://doi.org/10.1016/j.rinp.2020.103052
38. J.H. Jang, S. Mumtaz, S.W. Lee et al., Focus of high-power microwaves with positive and negative zone plate to increase the receiving power in axial virtual cathode oscillator. Curr. Appl. Phys. **29**, 89–96 (2021). https://doi.org/10.1016/j.cap.2021.06.006
39. A.E. Dubinov, V.P. Tarakanov, PIC simulation of the dynamics of electrons in a conical vircator. IEEE Trans. Plasma Sci. **44**, 1391–1395 (2016). https://doi.org/10.1109/TPS.2016.2580608
40. A.E. Dubinov, V.P. Tarakanov, Simulated formation of a virtual cathode chain in a conical drift tube. Tech. Phys. Lett. **45**, 754–756 (2019). https://doi.org/10.1134/S106378501908008X
41. A.A. Badarin, A.V. Andreev, S.A. Kurkin, Photonic crystal as a section of modulation and interaction with a virtual cathode in two-section vircator. IEEE Trans. Electron. Devices **1–6** (2020). https://doi.org/10.1109/TED.2020.3037278
42. N.S. Frolov, S.A. Kurkin, A.A. Koronovskii et al., High-efficiency virtual cathode oscillator with photonic crystal. Appl. Phys. Lett. **113**, 23503 (2018). https://doi.org/10.1063/1.5038277
43. S. Mumtaz, E.H. Choi, An efficient vircator with high output power and less drifting electron loss by forming multi virtual cathodes. IEEE Electron. Device Lett. **1** (2022). https://doi.org/10.1109/LED.2022.3200395
44. D. Biswas, R. Kumar, Efficiency enhancement of the axial vircator. IEEE Trans. Plasma Sci. **35**, 369–378 (2007). https://doi.org/10.1109/TPS.2007.891623
45. S. Humphries, *Charged Particle Beams* (Wiley, New Mexico, 1990)
46. R.B. Miller, Ch. 2, 3 and 5, in *An Introduction to the Physics of Intense Charged Particle Beam.* (Plenum, New York, p Ch. 2, 3 and 5)
47. J. Benford, J. Swegle, E. Schamiloglu, in *High Power Microwaves*, 2nd edn. (Taylor & Francis, 2006)
48. H. Sze, J. Benford, W. Woo, B. Harteneck, Dynamics of a virtual cathode oscillator driven by a pinched diode. Phys. Fluids **29**, 3873–3880 (1986). https://doi.org/10.1063/1.865771
49. S. Yang, C. Tang, S. Chen, Y. Xia, Formation condition of virtual cathode in the relativistic electron beam-plasma system. IEEE Trans. Plasma Sci. **47**, 4984–4987 (2019). https://doi.org/10.1109/TPS.2019.2946937
50. J. Benford, H. Sze, W. Woo, B. Harteneck, Virtual-cathode oscillator emission by a pinched diode. Phys. Rev. Lett. **56**, 344–346 (1986). https://doi.org/10.1103/PhysRevLett.56.344
51. R.B. Miller, *An Introduction to the Physics of Intense Charged Particle Beams, First* (Plenum press, New York, 1982)
52. M. Mahto, P.K. Jain, Study of virtual cathodes formation during beam-wave interaction in the reltron oscillator. Phys. Plasmas **24**, 93107 (2017). https://doi.org/10.1063/1.4991793
53. M. Mahto, P.K. Jain, PIC simulation study of the formation mechanism of periodic virtual cathodes in the reltron. IEEE Trans. Plasma Sci. **46**, 518–523 (2018). https://doi.org/10.1109/TPS.2018.2800054
54. S.A. Kurkin, A.A. Badarin, A.A. Koronovskii, A.E. Hramov, The development and interaction of instabilities in intense relativistic electron beams. Phys. Plasmas **22** (2015). https://doi.org/10.1063/1.4938216

55. A.E. Dubinov, V.P. Tarakanov, Simulation of a magnetically isolated vircator with an under-limit electron beam. Plasma Phys. Rep. **46**, 570–573 (2020). https://doi.org/10.1134/S1063780X20040029
56. S.A. Kurkin, N.S. Frolov, A.O. Rak et al., High-power microwave amplifier based on overcritical relativistic electron beam without external magnetic field. Appl. Phys. Lett. **106**, 153503 (2015). https://doi.org/10.1063/1.4918713
57. A.E. Dubinov, S.K. Saikov, V.P. Tarakanov, Multivircator as a new highly effective microwave generator with multiple virtual cathodes: concept and PIC-simulation. IEEE Trans. Plasma Sci. **48**, 141–145 (2020). https://doi.org/10.1109/TPS.2019.2956833
58. A. Vander Vorst, A. Rosen, Y. Kotsuka, *RF/Microwave Interaction with Biological Tissues*. (John Wiley & Sons, Inc., 2006)
59. M. Okoniewski, M.A. Stuchly, A study of the handset antenna and human body interaction. IEEE Trans. Microw. Theory Tech. **44**, 1855–1864 (1996). https://doi.org/10.1109/22.539944
60. L. Dubois, J.-P. Sozanski, V. Tessier et al., Temperature control and thermal dosimetry by microwave radiometry in hyperthermia. IEEE Trans. Microw. Theory Tech. **44**, 1755–1761 (1996). https://doi.org/10.1109/22.539932
61. F. Gao, Q. Zheng, Y. Zheng, Electrical circuit modeling and analysis of microwave acoustic interaction with biological tissues. Med. Phys. **41**, 53302 (2014). https://doi.org/10.1118/1.4871783
62. S. Mumtaz, J.N. Rana, E.H. Choi, I. Han, Microwave radiation and the brain: mechanisms, current status, and future prospects. Int. J. Mol. Sci. **23**
63. S. Banik, S. Bandyopadhyay, S. Ganguly, Bioeffects of microwave—A brief review. Bioresour. Technol. **87**, 155–159 (2003)
64. V.A. Vander, F. Duhamel, 1990–1995 advances in investigating the interaction of microwave fields with the nervous system. IEEE Trans. Microw. Theory Tech. **44**, 1898–1909 (1996). https://doi.org/10.1109/22.539948
65. J. Lin, *Electromagnetic Interaction with Biological SYSTEMS* (PLenum Press, New York, Springer Science & Business Media, 2012)
66. A. Wdowiak, P.A. Mazurek, A. Wdowiak, I. Bojar, Effect of electromagnetic waves on human reproduction. Ann. Agric. Environ. Med. **24**
67. A.R. Sheppard, M.L. Swicord, Q. Balzano, Quantitative evaluations of mechanisms of radiofrequency interactions with biological molecules and processes. Health Phys. **95** (2008)
68. J. Friedman, S. Kraus, Y. Hauptman et al., Mechanism of short-term ERK activation by electromagnetic fields at mobile phone frequencies. Biochem. J. **405**, 559–568 (2007). https://doi.org/10.1042/BJ20061653
69. M. Porcelli, G. Cacciapuoti, S. Fusco et al., Non-thermal effects of microwaves on proteins: thermophilic enzymes as model system. FEBS Lett. **402**, 102–106 (1997). https://doi.org/10.1016/S0014-5793(96)01505-0
70. D.I. de Pomerai, B. Smith, A. Dawe et al., Microwave radiation can alter protein conformation without bulk heating. FEBS Lett. **543**, 93–97 (2003). https://doi.org/10.1016/S0014-5793(03)00413-7
71. A. Fhager, S. Candefjord, M. Elam, M. Persson, Microwave diagnostics ahead: saving time and the lives of trauma and stroke patients. IEEE Microw. Mag. **19**, 78–90 (2018). https://doi.org/10.1109/MMM.2018.2801646
72. L.E. Solberg, Ø. Aardal, T. Berger et al., Experimental investigation into radar-based central blood pressure estimation. IET Radar Sonar Navig. **9**, 145–153 (2015). https://doi.org/10.1049/iet-rsn.2014.0206
73. S.Y. Semenov, A.E. Bulyshev, V.G. Posukh et al., Microwave tomography for detection/imaging of myocardial infarction. I Excised Canine Hearts. Ann. Biomed. Eng. **31**, 262–270 (2003)
74. D. Obeid, S. Sadek, G. Zaharia, G. El Zein, Multitunable microwave system for touchless heartbeat detection and heart rate variability extraction. Microw. Opt. Technol. Lett. **52**, 192–198 (2010). https://doi.org/10.1002/mop.24877

75. M. Zakrzewski, H. Raittinen, J. Vanhala, Comparison of center estimation algorithms for heart and respiration monitoring with microwave doppler radar. IEEE Sens. J. **12**, 627–634 (2012). https://doi.org/10.1109/JSEN.2011.2119299
76. P.M. Meaney, T. Zhou, D. Goodwin et al., Bone dielectric property variation as a function of mineralization at microwave frequencies. Int. J. Biomed. Imaging **2012**, 649612 (2012). https://doi.org/10.1155/2012/649612
77. B. Hildebrandt, P. Wust, O. Ahlers et al., The cellular and molecular basis of hyperthermia. Crit. Rev. Oncol. Hematol. **43**, 33–56 (2002). https://doi.org/10.1016/S1040-8428(01)00179-2
78. S. Jha, P.K. Sharma, R. Malviya, Hyperthermia: role and risk factor for cancer treatment. Ach. Life Sci. **10**, 161–167 (2016). https://doi.org/10.1016/j.als.2016.11.004
79. Y.-J. Liu, L.-X. Qian, D. Liu, J.-F. Zhao, Ultrasound-guided microwave ablation in the treatment of benign thyroid nodules in 435 patients. Exp. Biol. Med. (Maywood) **242**, 1515–1523 (2017). https://doi.org/10.1177/1535370217727477
80. Y.-H. Hao, L. Zhao, R.-Y. Peng, Effects of microwave radiation on brain energy metabolism and related mechanisms. Mil. Med. Res. **2**, 4 (2015). https://doi.org/10.1186/s40779-015-0033-6
81. L. Xiong, C.F. Sun, J. Zhang et al., Microwave exposure impairs synaptic plasticity in the rat hippocampus and PC12 cells through over-activation of the NMDA receptor signaling pathway. Biomed. Environ. Sci. **28**, 13–24 (2015). https://doi.org/10.3967/bes2015.002
82. H. Wang, R. Peng, L. Zhao et al., The relationship between NMDA receptors and microwave-induced learning and memory impairment: a long-term observation on Wistar rats. Int. J. Radiat. Biol. **91**, 262–269 (2015). https://doi.org/10.3109/09553002.2014.988893
83. A.W. Guy, C. Chou, Effects of high-intensity microwave pulse exposure of rat brain. Radio. Sci. **17**, 169S-178S (1982). https://doi.org/10.1029/RS017i05Sp0169S
84. H. Hinrikus, M. Bachmann, R. Tomson, J. Lass, Non-thermal effect of microwave radiation on human brain. Environmentalist **25**, 187–194 (2005)
85. P.S. Deshmukh, K. Megha, N. Nasare et al., Effect of low level subchronic microwave radiation on rat brain. Biomed. Environ. Sci. **29**, 858–867 (2016). https://doi.org/10.3967/bes2016.115
86. E. Odaci, H. Hanci, A. Ikinci et al., Maternal exposure to a continuous 900 MHz electromagnetic field provokes neuronal loss and pathological changes in cerebellum of 32-day-old female rat offspring. J. Chem. Neuroanat. **75**, 105–110 (2016). https://doi.org/10.1016/j.jchemneu.2015.09.002
87. M.L. Pall, Microwave frequency electromagnetic fields (EMFs) produce widespread neuropsychiatric effects including depression. J. Chem. Neuroanat. **75**, 43–51 (2016). https://doi.org/10.1016/j.jchemneu.2015.08.001
88. M. Terzi, B. Ozberk, O.G. Deniz, S. Kaplan, The role of electromagnetic fields in neurological disorders. J. Chem. Neuroanat. **75**, 77–84 (2016). https://doi.org/10.1016/j.jchemneu.2016.04.003
89. E.G. Kivrak, B.Z. Altunkaynak, I. Alkan et al., Effects of 900 MHz radiation on the hippocampus and cerebellum of adult rats and attenuation of such effects by folic acid and Boswellia sacra. J. Microsc. Ultrastruct. **5**, 216–224 (2017). https://doi.org/10.1016/j.jmau.2017.09.003
90. L. Zhao, J. Li, Y.H. Hao et al., Microwave-induced apoptosis and cytotoxicity of NK cells through ERK1/2 signaling. Biomed. Environ. Sci. **30**, 323–332 (2017). https://doi.org/10.3967/bes2017.043
91. O.G. Deniz, E.G. Kivrak, A.A. Kaplan, B.Z. Altunkaynak, Effects of folic acid on rat kidney exposed to 900 MHz electromagnetic radiation. J. Microsc. Ultrastruct. **5**, 198–205 (2017). https://doi.org/10.1016/j.jmau.2017.06.001
92. C. Wang, X. Wang, H. Zhou et al., Effects of pulsed 2.856 GHz microwave exposure on BM-MSCs isolated from C57BL/6 mice. PLoS ONE **10**, e0117550–e0117550 (2015). https://doi.org/10.1371/journal.pone.0117550
93. S. Verma, G.K. Keshri, S. Karmakar et al., Effects of microwave 10 GHz radiation exposure in the skin of rats: an insight on molecular responses. Radiat. Res. **196**, 404–416 (2021). https://doi.org/10.1667/RADE-20-00155.1

94. V. Franchini, E. Regalbuto, A. De Amicis et al., Genotoxic effects in human fibroblasts exposed to microwave radiation. Health Phys. **115** (2018)
95. S. Szmigielski, A. Szudzinski, A. Pietraszek et al., Accelerated development of spontaneous and benzopyrene-induced skin cancer in mice exposed to 2450 MHz microwave radiation. Bioelectromagnetics **3**, 179–191 (1982). https://doi.org/10.1002/bem.2250030202
96. S. Mumtaz, P. Bhartiya, N. Kaushik et al., Pulsed high-power microwaves do not impair the functions of skin normal and cancer cells in vitro: A short-term biological evaluation. J. Adv. Res. **22**, 47–55 (2020). https://doi.org/10.1016/j.jare.2019.11.007
97. K.K. Kesari, A. Agarwal, R. Henkel, Radiations and male fertility. Reprod. Biol. Endocrinol. **16**, 118 (2018). https://doi.org/10.1186/s12958-018-0431-1
98. G. Dong, H. Zhou, Y. Gao et al., Effects of 1.5 GHz high-power microwave exposure on the reproductive systems of male mice. Electromagn. Biol. Med. **40**, 311–320 (2021). https://doi.org/10.1080/15368378.2021.1891091
99. S. Kumar, J. Behari, R. Sisodia, Influence of electromagnetic fields on reproductive system of male rats. Int. J. Radiat. Biol. **89**, 147–154 (2013). https://doi.org/10.3109/09553002.2013.741282
100. P.A. Kondra, G.C. Hodgson, M.A. Hamid, Effects of microwave radiation on growth and reproduction of two stocks of chickens. Can J. Anim. Sci. **52**, 317–320 (1972). https://doi.org/10.4141/cjas72-035
101. B. Bilgici, S. Gun, B. Avci et al., What is adverse effect of wireless local area network, using 2.45 GHz, on the reproductive system? Int. J. Radiat. Biol. **94**, 1054–1061 (2018). https://doi.org/10.1080/09553002.2018.1503430
102. A. Zalata, A.Z. El-Samanoudy, D. Shaalan et al., In vitro effect of cell phone radiation on motility, DNA fragmentation and clusterin gene expression in human sperm. Int. J. Fertil. Steril. **9**, 129–136 (2015). https://doi.org/10.22074/ijfs.2015.4217
103. I. Gorpinchenko, O. Nikitin, O. Banyra, A. Shulyak, The influence of direct mobile phone radiation on sperm quality. Cent. Eur. J. Urol. **67**, 65–71 (2014). https://doi.org/10.5173/ceju.2014.01.art14
104. M. Oghbaei, O. Mirzaee, Microwave versus conventional sintering: a review of fundamentals, advantages and applications. J. Alloys Compd. **494**, 175–189 (2010). https://doi.org/10.1016/j.jallcom.2010.01.068
105. R. Scapaticci, O.M. Bucci, I. Catapano, L. Crocco, Differential microwave imaging for brain stroke followup. Int. J. Antennas Propag. **2014**, 312528 (2014). https://doi.org/10.1155/2014/312528
106. M. Bayat, N. Karimi, M. Karami et al., Chronic exposure to 2.45 GHz microwave radiation improves cognition and synaptic plasticity impairment in vascular dementia model. Int. J. Neurosci. 1–12 (2021). https://doi.org/10.1080/00207454.2021.1896502
107. E. Porter, A. Raterink, A. Farshkaran, Microwave-based detection of the bladder state as a support tool for urinary incontinence [Bioelectromagnetics]. IEEE Antennas Propag. Mag. **64**, 112–122 (2022). https://doi.org/10.1109/MAP.2021.3129687
108. J.C. Lin, The microwave auditory effect BT, in *Auditory Effects of Microwave Radiation*, ed. by J. C. Lin. (Springer International Publishing, Cham, 2021), pp. 127–173
109. K.K. Kesari, J. Behari, Fifty-gigahertz microwave exposure effect of radiations on rat brain. Appl. Biochem. Biotechnol. **158**, 126 (2008). https://doi.org/10.1007/s12010-008-8469-8
110. D.A. Sylvestre, Y. Otoki, A.H. Metherel et al., Effects of hypercapnia/ischemia and dissection on the rat brain metabolome. Neurochem. Int. **156**, 105294 (2022). https://doi.org/10.1016/j.neuint.2022.105294
111. Y. Hao, W. Li, H. Wang et al., Microwave radiation induces neuronal autophagy through miR-30a-5p/AMPKα2 signal pathway. Biosci. Rep. BSR20212584 (2022). https://doi.org/10.1042/BSR20212584
112. J.A. Payne, R.A. Barnes, A.X. Downey et al., Temperature dynamics in rat brains exposed to near-field waveguide outputs at 2.8 GHz. Bioelectromagnetics **43**, 14–24 (2022). https://doi.org/10.1002/bem.22377

113. L. Zhao, R.Y. Peng, S.M. Wang et al., Relationship between cognition function and hippocampus structure after long-term microwave exposure. Biomed. Environ. Sci. **25**, 182–188 (2012). https://doi.org/10.3967/0895-3988.2012.02.009
114. W.-J. Zhi, L.-F. Wang, X.-J. Hu, Recent advances in the effects of microwave radiation on brains. Mil Med Res **4**, 29 (2017). https://doi.org/10.1186/s40779-017-0139-0
115. R. de Seze, C. Poutriquet, C. Gamez et al., Repeated exposure to nanosecond high power pulsed microwaves increases cancer incidence in rat (2019)
116. M. Otto, K.E. von Mühlendahl, Electromagnetic fields (EMF): Do they play a role in children's environmental health (CEH)? Int. J. Hyg. Environ. Health **210**, 635–644 (2007). https://doi.org/10.1016/j.ijheh.2007.07.007
117. D. Aydin, M. Feychting, J. Schüz et al., Predictors and overestimation of recalled mobile phone use among children and adolescents. Prog. Biophys. Mol. Biol. **107**, 356–361 (2011). https://doi.org/10.1016/j.pbiomolbio.2011.08.013
118. O. Furman, K. Komoshvili, J. Levitan et al., The lack of toxic effect of high-power short-pulse 101 GHz mm waves on healthy mice. Bioelectromagnetics **41**, 188–199 (2020). https://doi.org/10.1002/bem.22247
119. M. Zhadobov, N. Chahat N, R. Sauleau et al., Millimeter-wave interactions with the human body: state of knowledge and recent advances. Int. J. Microw. Wirel. Technol. **3**, 237–247 (2011). https://doi.org/10.1017/S1759078711000122
120. H.A. Kues, S.A. D'Anna, R. Osiander et al., Absence of ocular effects after either single or repeated exposure to 10 mW/cm^2 from a 60 GHz CW source. Bioelectromagnetics **20**, 463–473 (1999). https://doi.org/10.1002/(SICI)1521-186X(199912)20:8<463::AID-BEM1>3.0.CO;2-T
121. Y. Lin, P. Gao, Y. Guo et al., Effects of long-term exposure to l-band high-power microwave on the brain function of male mice. Biomed. Res. Int. **2021**, 2237370 (2021). https://doi.org/10.1155/2021/2237370
122. A. Bahaodini, M. Owjfard, A. Tamadon, S.M. Jafari, Low frequency electromagnetic fields long-term exposure effects on testicular histology, sperm quality and testosterone levels of male rats. Asian Pacific J. Reprod. **4**, 195–200 (2015). https://doi.org/10.1016/j.apjr.2015.06.001
123. P. Bhartiya, S. Mumtaz, J.S. Lim et al., Pulsed 3.5 GHz high power microwaves irradiation on physiological solution and their biological evaluation on human cell lines. Sci. Rep. **11**, 8475 (2021). https://doi.org/10.1038/s41598-021-88078-x
124. M. Mirzaee, M. Simeni Simeni, P.J. Bruggeman, Electric field dynamics in an atmospheric pressure helium plasma jet impinging on a substrate. Phys. Plasmas **27**, 123505 (2020)
125. T.T. Mensinga, G.J.A. Speijers, J. Meulenbelt, Health implications of exposure to environmental nitrogenous compounds. Toxicol. Rev. **22**, 41–51 (2003). https://doi.org/10.2165/00139709-200322010-00005
126. J.O. Lundberg, E. Weitzberg, M.T. Gladwin, The nitrate–nitrite–nitric oxide pathway in physiology and therapeutics. Nat. Rev. Drug Discov. **7**, 156–167 (2008). https://doi.org/10.1038/nrd2466
127. C.W. Chow, A. Kapus, R. Romanek, S. Grinstein, NO_3—Induced pH changes in mammalian cells: evidence for an NO_3–H+ cotransporter. J. Gen. Physiol. **110**, 185–200 (1997)
128. B. Ghimire, G.J. Lee, S. Mumtaz, E.H. Choi, Scavenging effects of ascorbic acid and mannitol on hydroxyl radicals generated inside water by an atmospheric pressure plasma jet. AIP Adv. **8** (2018). https://doi.org/10.1063/1.5037125
129. J.S. Lim, Y.J. Hong, B. Ghimire et al., Measurement of electron density in transient spark discharge by simple interferometry. Results Phys. **20**, 103693 (2021). https://doi.org/10.1016/j.rinp.2020.103693
130. R.K. Adair, Vibrational resonances in biological systems at microwave frequencies. Biophys. J. **82**, 1147–1152 (2002)
131. J. Kmetko, M. Warkentin, U. Englich, R.E. Thorne, Can radiation damage to protein crystals be reduced using small-molecule compounds? Acta Crystallogr. Sect. D Biol. Crystallogr. **67**, 881–893 (2011)

132. R.K. Adair, Biophysical limits on athermal effects of RF and microwave radiation. Bioelectromagn. J. Bioelectromagn. Soc. Soc. Phys. Regul. Biol. Med. Eur. Bioelectromagn. Assoc. **24**, 39–48 (2003)
133. R. Weissenborn, K. Diederichs, W. Welte et al., Non-thermal microwave effects on protein dynamics? An X-ray diffraction study on tetragonal lysozyme crystals. Acta Crystallogr. Sect. D Biol. Crystallogr. **61**, 163–172 (2005)
134. P. Shaw, N. Kumar, S. Mumtaz et al., Evaluation of non-thermal effect of microwave radiation and its mode of action in bacterial cell inactivation. Sci. Rep. **11**, 14003 (2021). https://doi.org/10.1038/s41598-021-93274-w
135. Q. Wu, Effect of high-power microwave on indicator bacteria for sterilization. IEEE Trans. Biomed. Eng. **43**, 752–754 (1996). https://doi.org/10.1109/10.503183
136. M. van Wolferen, A. Orell, S.-V. Albers, Archaeal biofilm formation. Nat. Rev. Microbiol. **16**, 699–713 (2018). https://doi.org/10.1038/s41579-018-0058-4
137. L. Caputo, L. Quintieri, M.M. Cavalluzzi et al., Antimicrobial and antibiofilm activities of citrus water-extracts obtained by microwave-assisted and conventional methods. Biomedicines **6** (2018)
138. J. Li, R. Nickel, J. Wu et al., A new tool to attack biofilms: driving magnetic iron-oxide nanoparticles to disrupt the matrix. Nanoscale **11**, 6905–6915 (2019). https://doi.org/10.1039/C8NR09802F
139. Q. Wang, J. Vachon, B. Prasad et al., Alternating magnetic fields and antibiotics eradicate biofilm on metal in a synergistic fashion. NPJ Biofilms Microbio. **7**, 68 (2021). https://doi.org/10.1038/s41522-021-00239-y
140. M.C. Gye, C.J. Park, Effect of electromagnetic field exposure on the reproductive system. Clin. Exp. Reprod. Med. **39**, 1–9 (2012). https://doi.org/10.5653/cerm.2012.39.1.1

Index

E. H. Choi (ed.), *Plasma Biosciences and Medicine*, Topics in Applied Physics 148, https://doi.org/10.1007/978-981-19-7935-4

J

K

L

M

N

The manufacturer's authorised representative in the EU is Springer Nature Customer Service Centre GmbH, Europaplatz 3, 69115 Heidelberg, Germany. If you have any concerns regarding our products, please contact ProductSafety@springernature.com

Printed and bound by CPI Group (UK) Ltd, Croydon, CR0 4YY
30/09/2026
02225242-0011